ILLUSTRATED ADVANCED BIOLOGY

Microbes in Action

CJ Clegg

JOHN MURRAY

Also in this series:

Genetics & Evolution 0 7195 7552 4
Mammals: Structure & Function 0 7195 7551 6

Data and artwork credits

The author would like to thank the following for giving permission to redraw or adapt artwork and data:

Figure 2.4 (p.4); **Figure 10.24 (p.77)**; **Figure 10.30 (p.79)** adapted from Madigan, M.T., Martinko, J.M. and Parker, J. (2000) *Brock's Biology of Microorganisms*, 9th edition, Prentice Hall International, New Jersey; **Figure 2.13 (p.10)**; **Figure 10.25 (p.77)** adapted from Clegg, C.J., Mackean, D.G., Openshaw, P.H. and Reynolds, R.C. (1996) *Advanced Biology Study Guide*, John Murray (Publishers) Ltd, London; **Figure 6.9 (p.33)** adapted from M.G.R. Cannell, R.C. Dewar and P.G. Pyatt (1993) *Conifer Plantations on Drained Peatlands in Britain: A Net Gain or Loss of Carbon?* Forestry, Centre for Ecology and Hydrology, Huntingdon, UK; **Figure 8.12 (p.60)** adapted from James, C. (2000) *Global Review of Commercialised Transgenic Crops*, International Service from the Acquisition of Agribiotech Applications (ISAAA), Ithaca, New York, USA; **Figure 10.4 (p.68)** Crown copyright material is reproduced under Class Licence number CO2P00060 with the permission of the Controller of HMSO; **Figures 10.5–10.8 (pp.68–69)** adapted from Garbutt, J. (1997) *Essentials of Food Microbiology*, Arnold, London; **Figure 10.18 (p.73)** adapted from Kavanagh, S.J.G. and Denning, D.W. (1995) Tuberculosis: the global challenge. *Biological Sciences Review*, **8**; **Figure 10.31 (p.80)** www.who.int; **Figure 10.35 (p.81)** adapted from Pantaleo, G., Graziosi, C. and Fauci, A.S. (1993) The immunopathogenesis of human immunodeficiency virus infection. *New England Journal of Medicine*, **325**(5); **Figure 10.36 (p.82)** adapted from World Malaria Situation in 1993, *Weekly Epidemiological Record*, **71**, (1996); **Figure 10.46 (p.86)** adapted from Mestel, R., Special News Report, Putting prions to the test. *Science*, **273**, (1996) 184–89, by permission of the author; **Figure A.1 (p.87)** adapted from Monger, G. (ed.) (1974) *Nuffield Biology* (revised edition) *Teacher's Guide 1 – Introducing Living Things*, Longman, London.

Photo credits

Thanks are due to the following copyright holders for permission to reproduce photographs:

Cover Microfield Scientific Ltd/Science Photo Library; **p.2** Science Photo Library/Kwangshim Kim; **p.4** *both* Gene Cox; **p.5** *top* Science Photo Library/Dr Linda Stannard/UCI, *bottom* Gene Cox; **p.6** Science Photo Library/Antonia Reeve; **p.7** Oxoid Ltd.; **p.9** Science Photo Library/Geoff Tompkinson; **p.11** Science Photo Library/Charles D. Winters; **p.16** Gene Cox; **p.17** Science Photo Library/Peter Ryan/Scripps; **p.18** Science Photo Library/PR Mainsonneuve/Publiphoto Diffusion; **p.19** *left & centre* Holt Studios, *right* from J.D. Dodge (1968), *An Atlas of Biological Ultrastructure*, Arnold, London; **p.20** *top* Science Photo Library/Eye of Science, *bottom* Science Photo Library/NIBSC; **p.21** Science Photo Library/Institute Pasteur/CNRI; **p.25** *top left* Scientific American, *top right & bottom* Dr C.J. Clegg; **p.26** *both* Gene Cox; **p.27** *top* Science Photo Library/Noble Proctor, *bottom* Science Photo Library/David Scharf; **p.28** *top* Science Photo Library/Eric Grave, *bottom* Science Photo Library; **p.29** *top* Biophoto Associates, *centre & bottom* Gene Cox; **p.31** Natural Visions; **p.32** Biophoto Associates; **p.33** Dr C.J. Clegg; **p.34** Thames Water; **p.35** *top* London Aerial Photo Library, *bottom* Natural Visions; **p.36** Gene Cox; **p.37** Dr C.J. Clegg; **p.38** Gene Cox; **p.40** *both* S.W. Watson, Woods Hole Oceanographic Institution; **p.42** Panos Pictures; **p.43** *top* Dr C.J. Clegg, *bottom* South American Picture Library; **p.44** *all* Anthony Blake Photo Library; **p.45** *all* Anthony Blake Photo Library; **p.46** Oxford Scientific Films; **p.47** *all* John Townson/Creation; **p.48** Roger Scruton; **p.49** Science Photo Library/Michael Abbey; **p.51** Powerstock; **p.53** Science Photo Library/James King-Holmes; **p.56** Science Photo Library/Dr Gopal Murti; **p.58** Science Photo Library/Will & Dent McIntyre; **p.60** Holt Studios; **p.62** *top left & bottom right* Dr C.J. Clegg, *top right* Science Photo Library/John Howard, *bottom left* Robert Harding Picture Library; **p.63** *top* Dr C.J. Clegg, *centre left* Science Photo Library/Dr Chris Somerville, *centre right* Science Photo Library/Mauro Femariello, *bottom* Science Photo Library; **p.65** *top* Holt Studios, *bottom* Science Photo Library/Marcus Lopez; **p.70** Science Photo Library/Moredun Animal Health Ltd.; **p.71** Dr C.J. Clegg; **p.72** *top right* Science Photo Library/Alex Rakosy/Custom Medical Stock, *bottom left* Science Photo Library/Matt Meadows/Peter Arnold Inc., *bottom right* Science Photo Library/Simon Fraser; **p.73** *all* Science Photo Library; **p.76** *top* Science Photo Library/Jean-Loup Charmet, *bottom* Wellcome Trust; **p.78** Science Photo Library/Dr Linda Stannard; **p.81** Science Photo Library/Institute Pasteur; **p.84** *top* Science Photo Library/Dr P Marazzi, *centre* Science Photo Library, *bottom left* Science Photo Library/E. Gueho/CNRI, *bottom right* Science Photo Library/Dr P. Marazzi; **p.85** *top* Holt Studios, *bottom* Microfield Scientific Ltd/Science Photo Library; **p.86** Farmers Weekly.

Every effort has been made to contact copyright holders but if any have been inadvertently overlooked the Publishers will be pleased to make the necessary arrangements at the earliest opportunity.

First published in 2002
by John Murray (Publishers) Ltd
50 Albemarle Street
London W1S 4BD

Illustrations by Art Construction
Layouts by Eric Drewery
Cover design by John Townson/Creation

Typeset in 10/12pt Galliard by Wearset Ltd, Boldon, Tyne and Wear
Printed and bound in Spain by Bookprint S.L., Barcelona

A catalogue entry for this title is available from the British Library

ISBN 0 7195 7554 0

Contents

	Preface	iv
Chapter 1:	Introducing microbes	1
Chapter 2:	Bacteria	2
Chapter 3:	Viruses	18
Chapter 4:	Fungi	24
Chapter 5:	Protoctista	28
Chapter 6:	Microorganisms and the environment	30
Chapter 7:	Microorganisms and biotechnology	44
Chapter 8:	Genetic engineering of microorganisms	54
Chapter 9:	Microorganisms, biodeterioration and preservation	62
Chapter 10:	Microorganisms and disease	66
	Answers	87
	Glossary	90
	Index	91

Front cover

Below the leaf surface a tiny fungal parasite feeds! Spore-producing hyphae have grown out between the guard cells of a stomate and will release spores that may be carried to new host plants. Parasites of crop plants attract special attention, but many wild plants (this is the common dock) are also parasitised (page 85).

Preface

Microbes in Action explores the range of microorganisms that feature in key topics of Advanced level biology and other post-GCSE programmes. The importance of microorganisms can easily be underestimated because of their extremely small size. Yet they occur in vast numbers, and their dynamic metabolism makes them major players in our environment. Microorganisms influence the way we live!

Today, there are new roles for microbes in agriculture, industry, biotechnology and the world of human health, and new developments arise all the time. Traditionally, microbes are blamed for the diseases that some cause, and this also makes us inclined to try to avoid them. But without microbes, life as we know it would not occur.

All these aspects are covered here in an up-to-date and novel way. By means of labelled photomicrographs and electron micrographs, annotated diagrams and flow charts, all linked by bridging text, the essentials of microbiology are reviewed. No significant aspect is overlooked, and the presentation aids understanding of specification essentials.

Related publications

In the same series:

Clegg, C.J. (1998) *Mammals: Structure & Function*. London: John Murray.
Clegg, C.J. (1999) *Genetics & Evolution*. London: John Murray.

In preparation:

Green Plants, The Inside Story

Taking your studies further

You can help yourself keep up-to-date with developments by reading articles in:

- *New Scientist*, a weekly review of science and technology, including their occasional 'Inside Science' pull-outs
- *Biological Science Review*, a journal designed and written for A level students
- *Scientific American*, a review journal, sometimes with articles that are useful at this level.

A good general read is:

- Postgate, John (1992) *Microbes and Man*, 3rd edn. Cambridge, UK: Cambridge University Press.

Checking details you cannot locate elsewhere is possible in either:

- Lansing, M., Prescott, J., Harley, P. and Klein, D.A. (1999) *Microbiology*, 4th edn. New York/London: McGraw-Hill.
- Madigan, M.T., Martinko, J.M. and Parker, J. (2000) *Brock's Biology of microorganisms*, 9th edn. New Jersey: Prentice Hall International.

Using the internet:

- National Centre for Biotechnology Education, University of Reading, at www.reading.ac.uk/NCBE

Acknowledgements

To all the known and unknown researchers, teachers, illustrators and writers who have influenced my own teaching, I gladly acknowledge my debt. Where copyright material has been used it is acknowledged on page ii. If the intellectual property of anyone has been used without prior agreement, I ask that John Murray (Publishers) are contacted so that correction can be made.

I have benefited from critical comments from Katie Mackenzie Stuart, Science Publisher at John Murray, and from Dean Madden at the University of Reading National Centre for Biotechnology Education. Nevertheless, the remaining errors are my responsibility.

At John Murray, the production team have brought together text, photomicrographs and drawings skilfully, and I am most grateful to them.

Dr Chris Clegg
Salisbury, Wiltshire, 2001

Introducing microbes

Microbes are **microorganisms** – all the organisms too small to be studied by the naked eye. They are an extraordinary group, many with a very long evolutionary history. Microbes show great power to adapt and change in a changing environment, and they can achieve some incredible biochemical feats, despite their size. Relatively few of the millions of different species of microbes are dangerous to us because they cause serious disease, but mostly microbes are more than just harmless – they carry out processes essential for the survival of all life.

The range of microorganisms

Microbes occur in huge numbers – mostly too large to contemplate. They occur on and in ourselves, and in the environment at large – more-or-less everywhere. There are many more different species of microorganisms than there are other forms of life. However, these fall into one of four groups (of which more later):

- **bacteria** – Chapter 2, page 2
- many of the **fungi** – Chapter 4, page 24
- **viruses** – Chapter 3, page 18
- unicellular **protoctista** – Chapter 5, page 28.

The one universal characteristic of these organisms is their tiny **size**. But there are large differences in size between them, from the viruses (in the range 10–400 nm), to the larger unicellular algae (10 to several 100 µm). The units by which microscopic structures are measured are defined in Table 1.1.

Table 1.1 Units of length used in microscopy

The metre (**m**) is the agreed standard unit of length (an **SI unit**), but it is too large to be useful here. The subdivisions given below are used, each 1/1000 of the unit above.

Unit	Subdivision
1 m	= 1000 millimetres (mm)
1 mm	= 1000 micrometres (µm)
1 µm	= 1000 nanometres (nm)

Another feature of their classification is their diversity in structure. Some are **prokaryotes** (e.g. bacteria) and some are **eukaryotes** (e.g. fungi, **protozoa**). The characteristics of these two types of cell organisation are illustrated in Figure 2.3 and the differences between them are listed in Table 2.1 (page 3). In addition, the **viruses** are not living organisms at all, as we define 'life' in biology (page 18).

Finally, we should recognise at the outset that microorganisms influence **all aspects of life**, with far-reaching effects on the environment, health, disease and decay, the food and drink industries, biotechnological industries old and new, industrial enzymology, and genetic engineering (Figure 1.1).

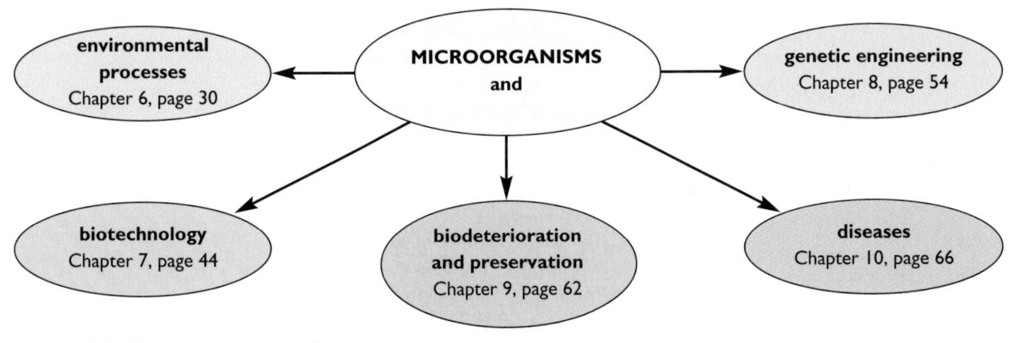

Figure 1.1 Microorganisms and living processes.

Extension: the long history of microbiology – pure and applied

Microbes have been important to human life since our earliest days. Before we knew of their existence, we exploited microbes in cheese manufacture and brewing, and they maintained soil fertility before the mineral nutrition of crops was recognised as an issue. Early microbiologists made progress long before the **electron microscope** (page 18) was on hand to disclose the finest details of structure. In the history of microbiology, some important names to know about include the following.

- **Anthony van Leeuwenhoek, 1632–1723** – a Dutchman with no formal training in science. He made simple microscopes, and was so successful he was elected a Fellow of the Royal Society.
- **Louis Pasteur, 1822–1895** – a Frenchman who made outstanding discoveries. He said 'In experimental science, chance favours the prepared mind'.
- **Robert Koch, 1843–1910** – a German who became the chief founder of medical microbiology, showing the causative agents of anthrax, tuberculosis and cholera.
- **Ronald Ross, 1857–1932** – a British physician who discovered the life cycle of the malarial parasite.

Today, progress in modern microbiology is closely linked with developments in genetics, cell biology, biochemistry, electron microscopy, biotechnology, enzymology, genetic engineering and medicine.

1 How many µm are there in 1.4 mm? Express 660 nm as µm.

2 Bacteria

The structure of *Escherichia coli* is illustrated in Figure 2.1. This organism was named by a bacteriologist, Professor T. Escherich, in 1885. *Escherichia coli* is a permanent resident of the gut, living on our digested food but otherwise doing us no harm. Associations in which one species benefits (*E. coli* in this case) and the other is unharmed are known as **commensalism**. This commensal occurs in huge numbers in the lower intestine of humans and other vertebrates, and is a major component of the faeces. In fact we may benefit from *E. coli*'s presence in subtle ways (e.g. interactions between commensals and parasitic visitors – page 67). However, one strain of this bacterium causes food poisoning.

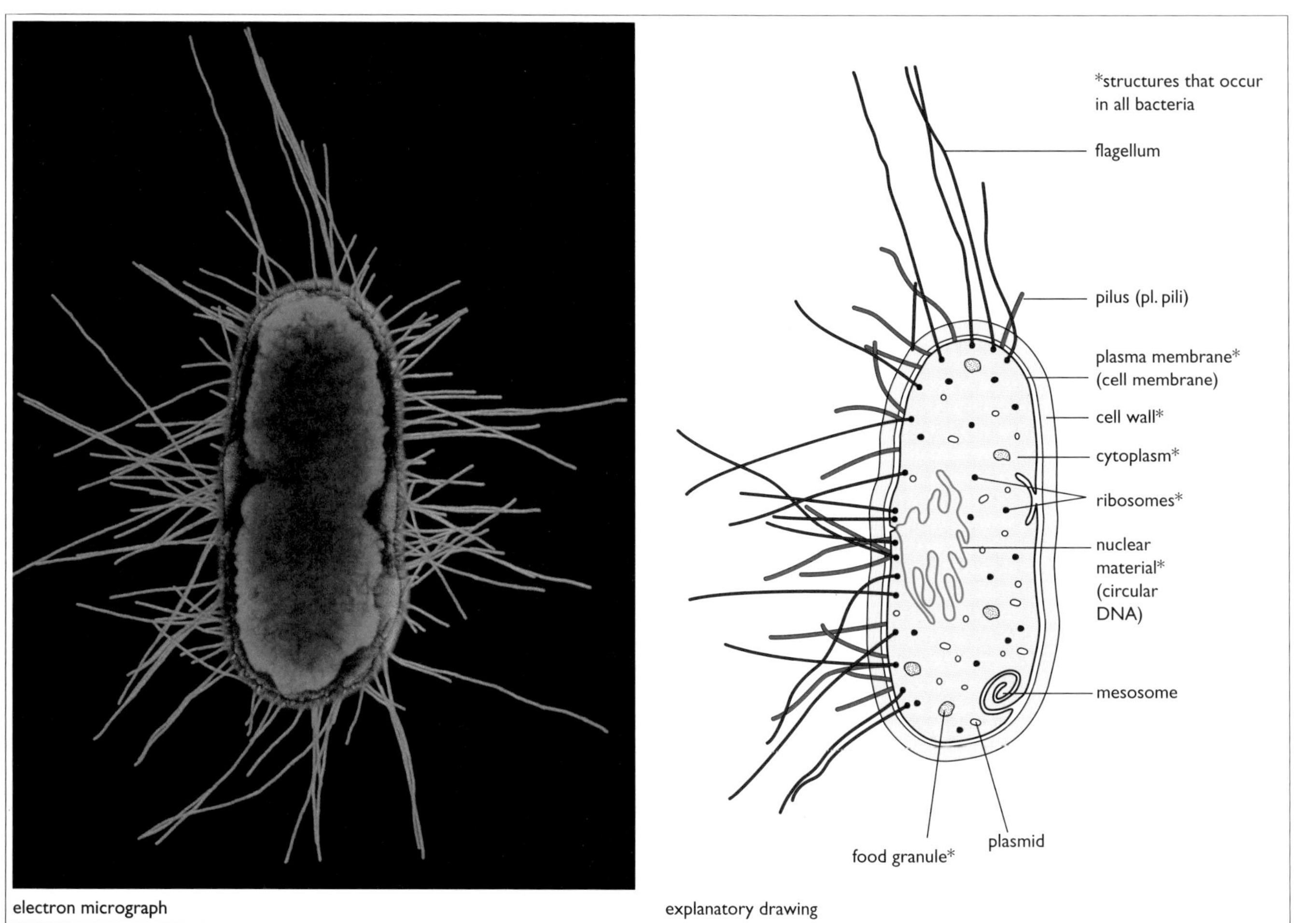

Figure 2.1 Structure of *Escherichia coli*.
Most of the many strains of *E. coli* are totally harmless, but one strain called *E. coli* 0157 releases toxins that destroy gut and kidney cells. It causes a painful, often bloody type of diarrhoea. This is a new pathogen to food poisoning (page 68).

Components of the bacterial cell

- **Cell walls** give shape to cells – and give protection against rupture caused by **osmosis**. They may also help protect against harm caused by other organisms. Bacterial cells have a rigid wall containing giant molecules (polymers) consisting of amino-sugars and peptide units, known as **murein** or peptidoglycan. (This is chemically different from the cellulose of plant cells.) Some bacteria have additional layers on the outer surface of their wall (Gram-negative bacteria, page 15).
- **Cytoplasm** of a bacterial cell is about 75% water, in which are dissolved proteins (mainly enzymes), lipoproteins, sugars, amino acids and fatty acids, inorganic salts, and the waste products of metabolism.

  In the cytoplasm are numerous **ribosomes**. These tiny, spherical organelles of protein and ribonucleic acid (RNA) are the sites of protein synthesis. Here messenger RNA is 'read' and used to dictate the sequence of amino acids in proteins, giving the primary structure of the bacterium's proteins. Bacterial ribosomes are known as 70S ribosomes, because they are smaller than those in the cytoplasm of plant and animal cells and fungi (called 80S ribosomes). The 'S' refers to the rate at which particles sediment in high-speed centrifugation.

  Food stores are common in the cytoplasm of many bacteria. These occur as small granules of lipid or glycogen, held in tiny sacs formed from lipid membrane. These are energy-rich reserves – reservoirs of cell-building materials.

1 What is the length in µm of the bacterium *E. coli* in Figure 2.1, shown magnified ×25 000

- **Plasma membranes** of bacterial cells consist of **phospholipids** and **proteins** arranged as shown in the **fluid mosaic model** (Figure 2.2). Long carbohydrate molecules occur attached to some of the lipid molecules (forming **glycolipids**) and protein molecules (forming **glycoproteins**) on the outer surface of the membrane. This membrane is a barrier across which all nutrients and waste products must pass. Movements are by **diffusion** (including osmosis) or **active transport** using metabolic energy.
- **Genetic material** of a bacterium consists of a single, circular chromosome of a **DNA** helix located in the cytoplasm (i.e. not contained within a **nucleus**), attached to the plasma membrane. The chromosome of *E. coli* has about 4×10^6 base pairs (adenine with thymine, guanine with cytosine) – about 4000 genes. These genes are copied as messenger RNA, as required.
- **Plasmids** are additional hereditary material made up of small rings of DNA. They are present in the cytoplasm of some, but not all, bacteria. Today plasmids are often exploited as vectors in genetic engineering (page 54).

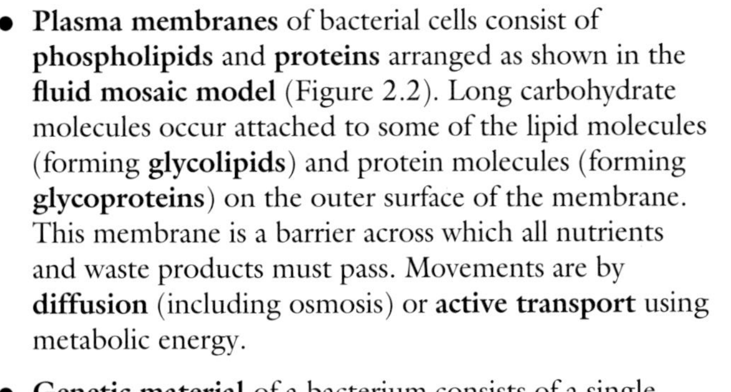

Figure 2.2 Fluid mosaic model of the plasma membrane of a bacterium.

- **Mesosomes** are infoldings of the plasma membrane found in some bacterial cells. In the photosynthetic bacteria (page 16), they are where the photosynthetic pigments are housed.
- **Flagella** are quite rigid protein strands that arise from basal bodies in the plasma membrane in some bacteria. These flagella bring about movement by rotating from their base, driven by the basal body.
- **Pili (or fimbiae)** are tiny tubular structures that arise from the cell membrane of some bacteria. They enable bacteria to attach to surfaces and to other bacteria.
- **Slime layer and capsule** are additional materials that are laid down on the outer surface of the wall of some species. Capsules are firmly attached, whereas slime layers may diffuse into the surrounding medium.

Prokaryotic and eukaryotic organisation compared

Bacteria are known as prokaryotes because they do not have a nucleus. Prokaryotes have a structure fundamentally different from that of plant and animal (eukaryote) cells. These differences are illustrated in Table 2.1 and Figure 2.3.

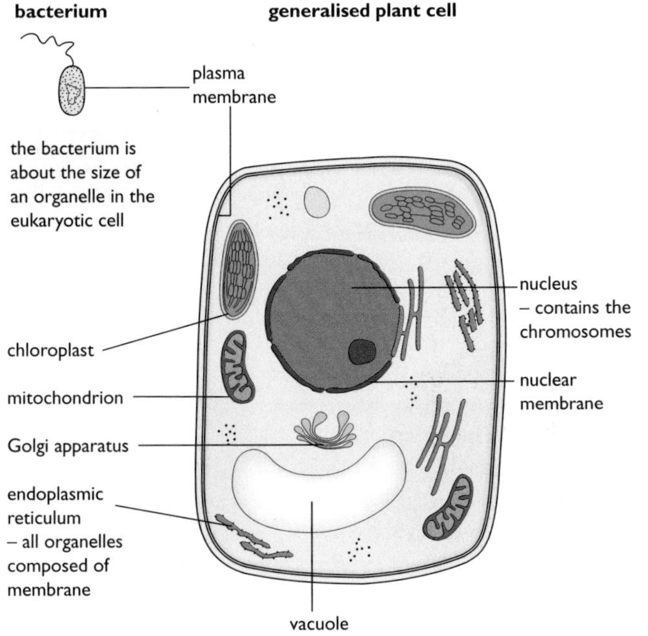

Figure 2.3 Prokaryote and eukaryote cells, drawn to the same scale.

Table 2.1 Prokaryote and eukaryote cells compared

Prokaryotes, e.g. bacteria	Eukaryotes, e.g. animals, plants, fungi
Cells are small, typically 5–10 µm	Cells are larger, typically 50–200 µm
Nucleus absent, single circular chromosome in cytoplasm	Nucleus with nuclear membrane containing linear chromosomes present
Few organelles present	Many organelles present, most bound by membranes
Some have flagella (of a single protein strand)	Some cells have cilia or flagella (made of microtubules)

Cell division and reproduction in bacteria

The cell cycle

Bacterial cells grow to full size and then divide into two by a process called **binary fission**. The complete cycle of growth, from new cell to the point of division, may take as little as 20 minutes, provided the necessary conditions (page 7) are maintained. *Escherichia coli* is one of many species that can reproduce at this rate, at least initially. This **growth rate cannot be maintained** for long, but it does help explain why bacteria are so numerous. For example, it is estimated that a gram of garden soil contains about 1000 million living bacteria – and an average square centimetre of human skin has a mere 10 million individual bacteria on it.

During growth, the cell contents increase so that, after division, each daughter cell has sufficient cytoplasm to metabolise and grow. Prior to division, the single circular chromosome, present in the form of a circular strand of DNA helix, divides. The copying process, known as **replication**, starts at a particular sequences of bases (a gene) that codes for the enzyme triggering the replication process (Figure 2.4). After division of the chromosome, a wall is laid down, dividing the cell into two. Daughter cells each have a copy of the chromosome (Figure 2.5).

2 Tabulate the differences between a chromosome of a eukaryotic cell, e.g. yeast, and the chromosome of a bacterium.

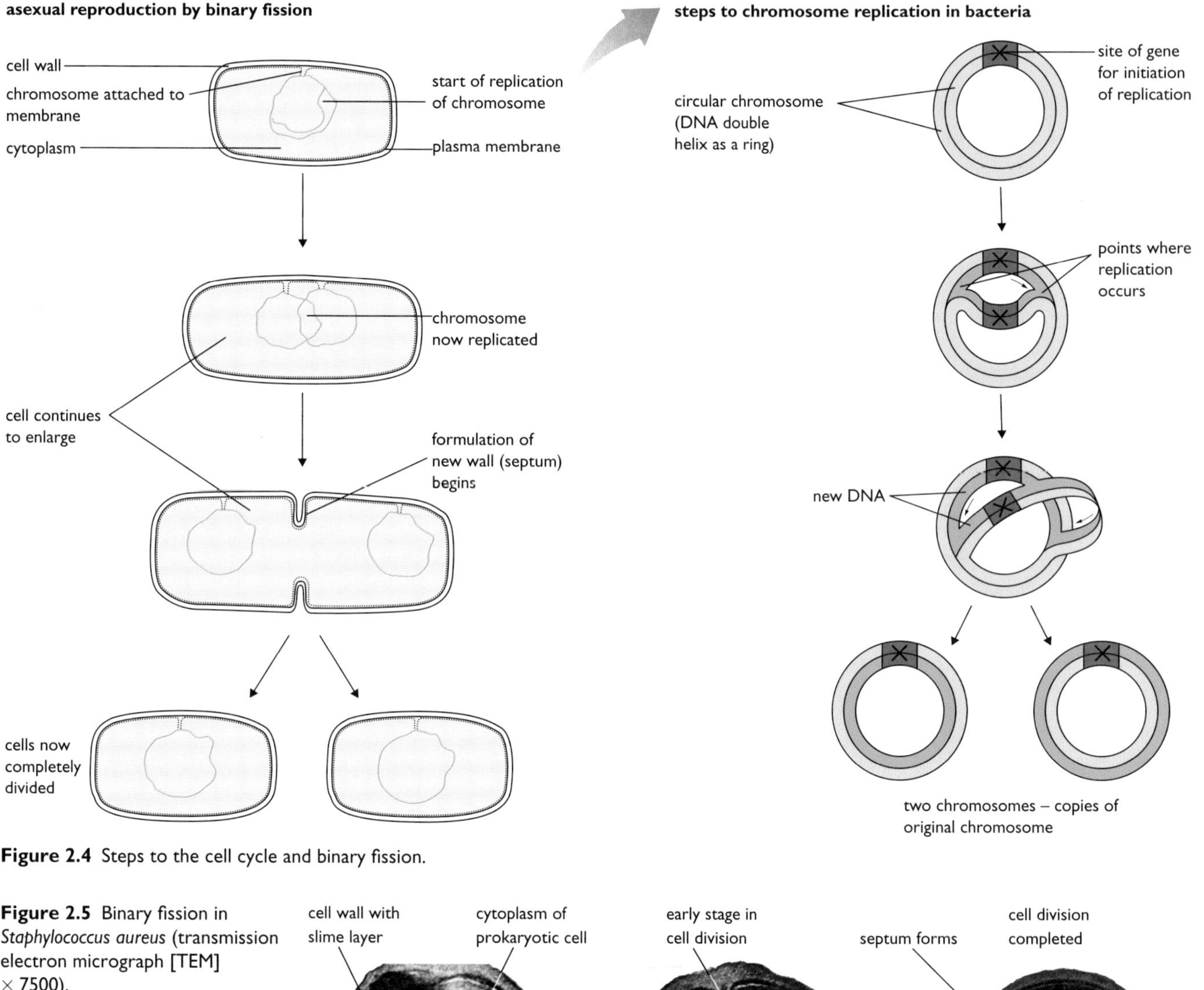

Figure 2.4 Steps to the cell cycle and binary fission.

cell wall with slime layer
cytoplasm of prokaryotic cell
early stage in cell division
septum forms
cell division completed

Figure 2.5 Binary fission in *Staphylococcus aureus* (transmission electron micrograph [TEM] × 7500).

This bacterium is a normal member of the microflora of the nose, conjunctiva, intestine and skin. It is an example of a bacterium with a slime layer on the outer surface of the wall (which consequently appears to be very thick).

Straphylococcus aureus may become a pathogen, e.g. in the tissues of hair follicles (causing boils) and if it enters sterile body tissues during an operation (surgical wound infections)

Sexual reproduction in bacteria?

Binary fission is a good example of **asexual reproduction**. It results in an increase in numbers of individuals, and in new bacteria **identical** to the bacterium from which they arose. On the other hand, **sexual** processes involve the production of new individuals **similar**, but not identical to, the parent(s). In bacteria, sexual reproduction by the production and fusion of gametes does not occur. However, there are exchanges between some bacteria by which new genetic information is acquired.

Conjugation

Conjugation is a form of mating in which genetic material is passed from one bacterium (donor) to another (recipient) down a conjugation tube formed between them. The genetic material may be a plasmid, present in the donor bacterium, a copy of which is transferred to the recipient. Alternatively, part of the circular chromosome may be copied and passed to the recipient, where it is built into the chromosome (Figure 2.6). Conjugation is genetically controlled – it occurs only between bacteria with certain genes. However, conjugation may involve bacteria of different species or the same species.

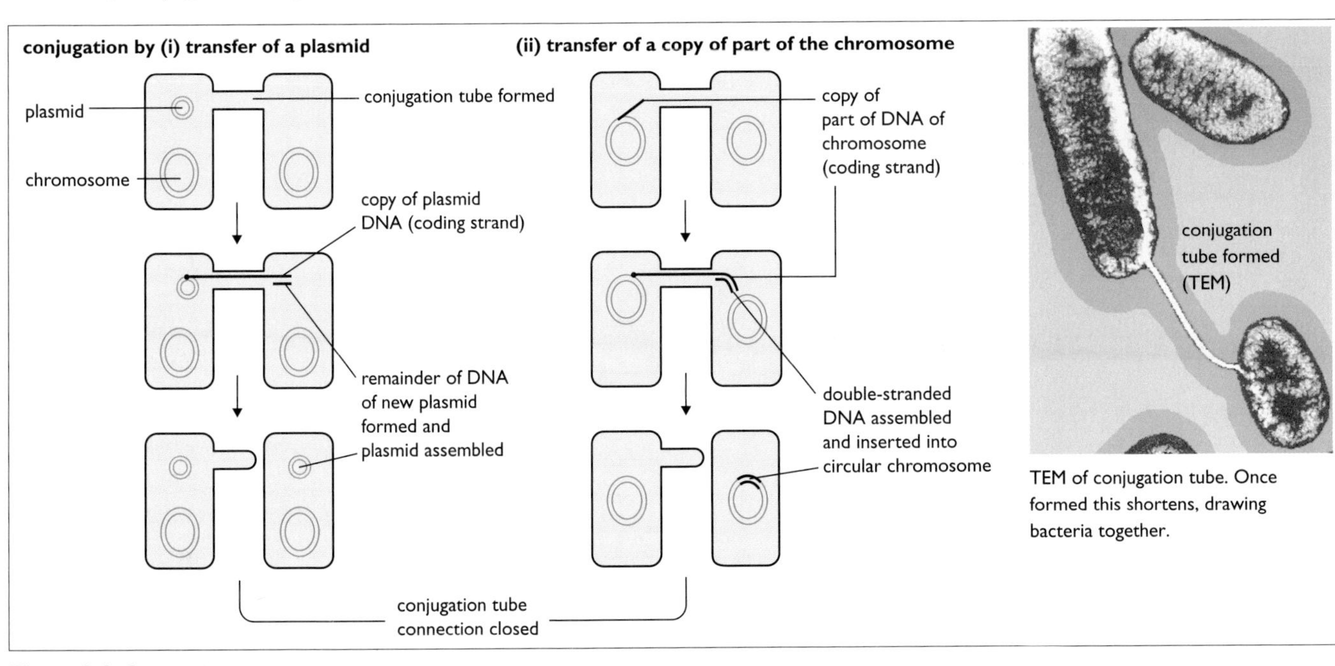

Figure 2.6 Conjugation.

Transformation

Transformation occurs when DNA fragments present outside the bacterium enter the living cell (through the wall) and combine into the DNA of the chromosome. The DNA fragments may be released naturally into the environment from dead cells, or may have been placed there in genetic engineering experiments. Some special treatments of bacteria increase the porosity of the walls (Figure 8.6, page 56).

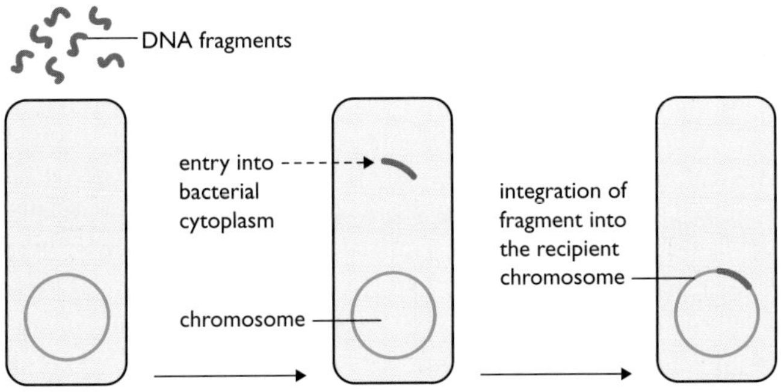

DNA fragments
entry into bacterial cytoplasm
chromosome
integration of fragment into the recipient chromosome

Figure 2.7 Transformation.

Endospore formation

Endospores are formed by some bacteria when conditions for growth are totally unfavourable. The cell contents become dehydrated, and then become contained within a protective wall. Endospores are resistant to low and high temperatures, extremes of pH, desiccation, and the effects of many harmful chemicals.

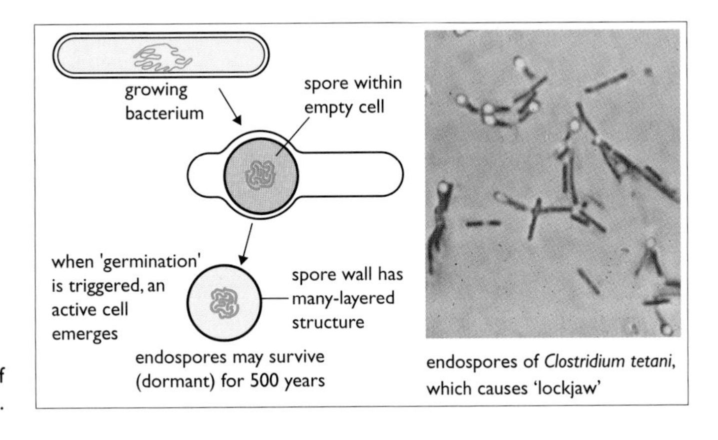

Figure 2.8 Formation of endospores.

Culturing of bacteria – aseptic techniques

Many species of bacteria are harmless to humans. In fact, very many are indispensable – life as we know it could not proceed without them. However, some species are dangerous pathogens, and some strains of certain species are harmful (e.g. page 67). Consequently, bacteriologists handle all cultures as if they were pathogens, using sterile or **aseptic techniques**. The object is to ensure that a bacterial culture is neither contaminated by the environment, nor contaminates the environment in any way.

Consequently, the nutrients used in microbiology, as well as the culture vessels and other equipment, are sterilised both before and after use. Solid or liquid preparations, called **media**, provide the nutrients required for growth. A liquid medium, a **broth**, is typically used for growing microorganisms in bulk. A solid medium is produced by addition of a gelling agent called **agar** (a more-or-less transparent polysaccharide obtained from seaweed) to a solution of nutrients. Agar is used because it is not normally digested by microorganisms. Solid media (e.g. agar plates in petri dishes) are used for holding reference cultures, and also to isolate individual species of bacteria from mixed cultures. Essential bacteriological procedures are illustrated in Figures 2.9, 2.10 and 2.11. The conditions necessary for growth are listed in Table 2.2.

3 How might Pasteur have demonstrated that microbes, freely circulating in the air, are able to contaminate exposed matter?

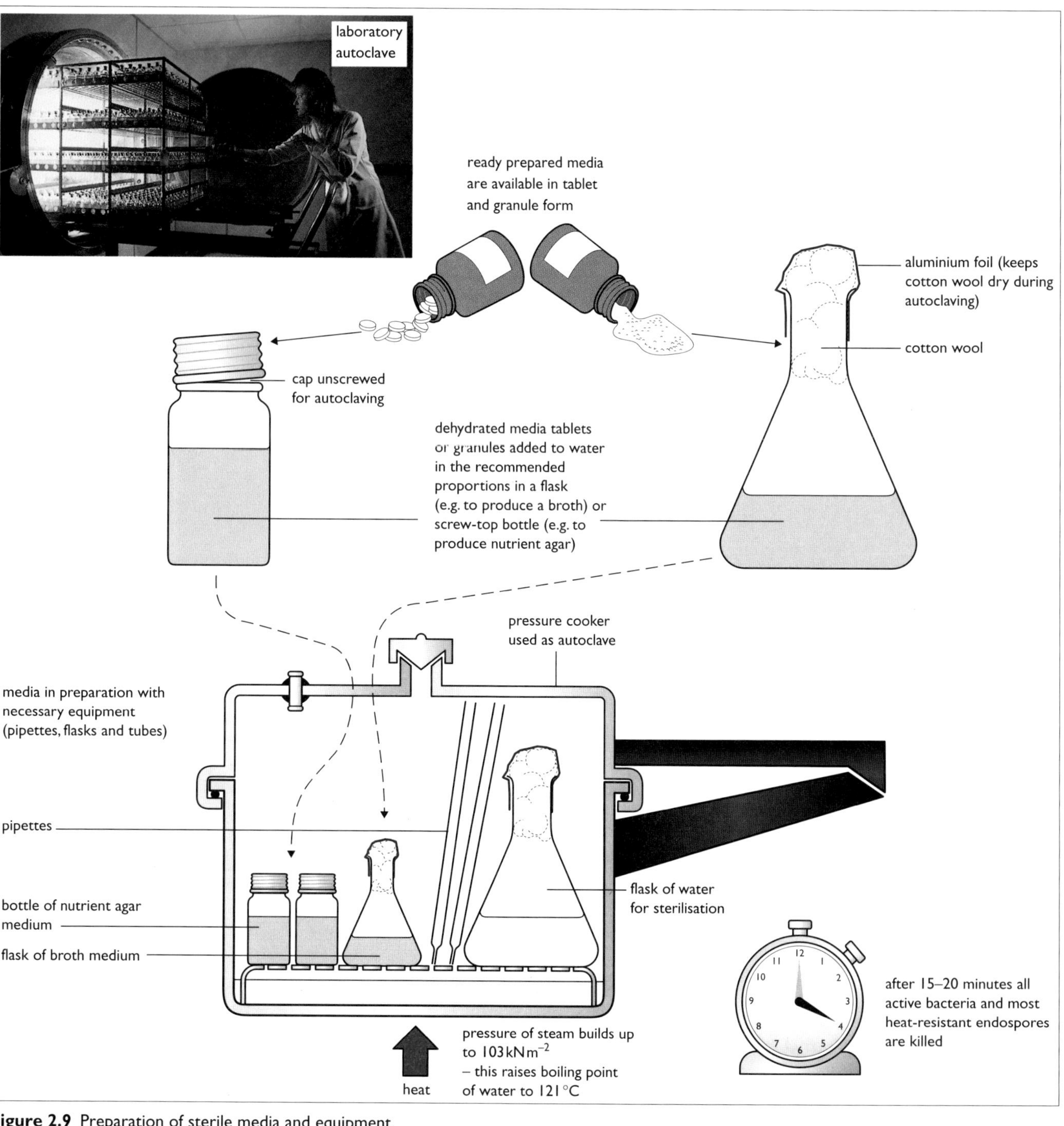

Figure 2.9 Preparation of sterile media and equipment.

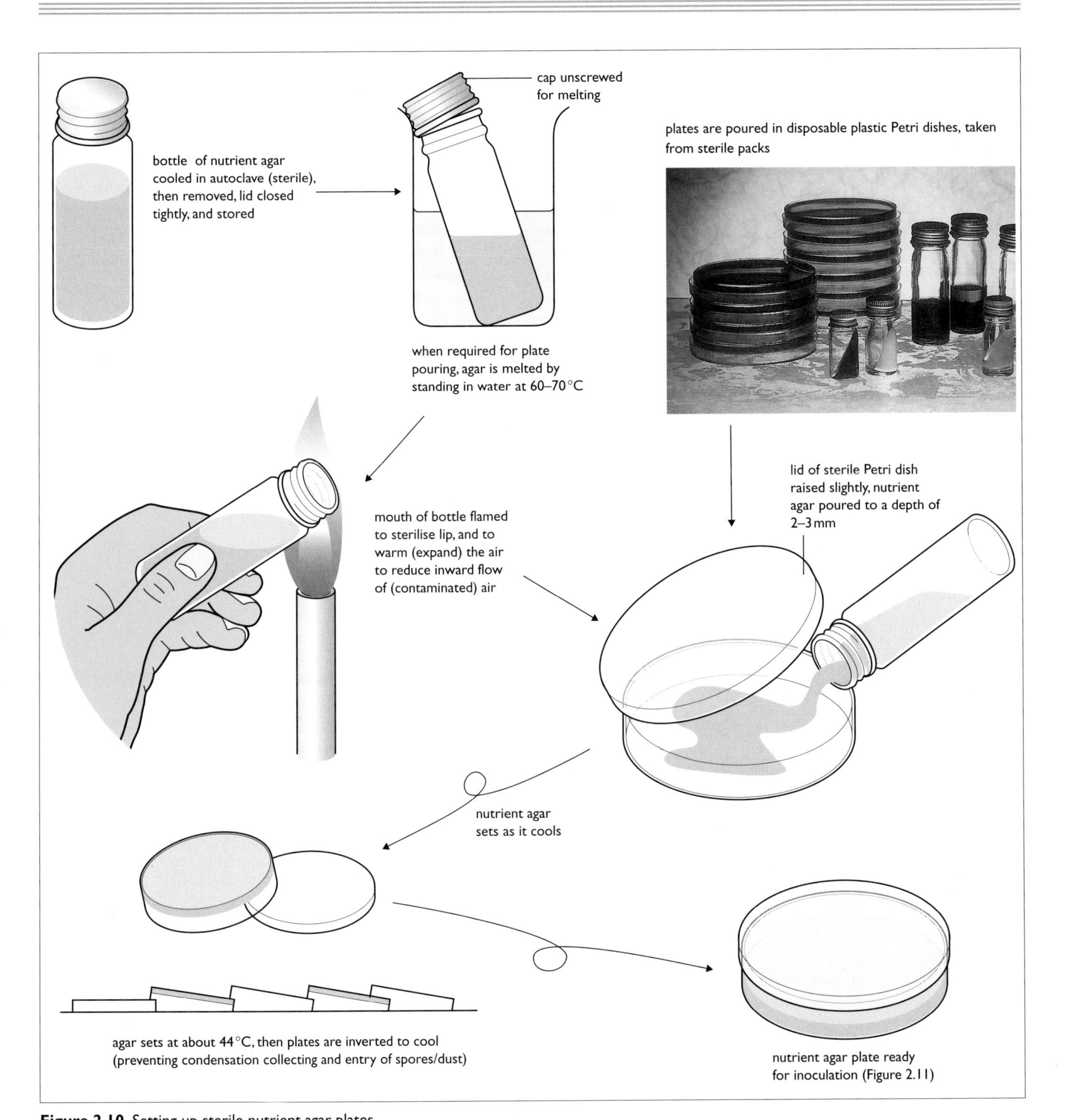

Figure 2.10 Setting up sterile nutrient agar plates.

Table 2.2 Conditions for growth of bacteria

Essential nutrients the medium must supply	Environmental conditions influencing growth
Water	pH – most bacteria are favoured by slightly alkaline conditions (pH 7.4), but some tolerate extremes of alkalinity or acidity. (By contrast, most fungi are favoured by slightly acidic conditions.)
Source of carbon/energy – e.g. carbohydrate such as glucose	Temperature –25–45 °C is favourable for most, but some grow at 0 and some at 80 °C or more (in hot springs)
Nitrogen – many bacteria use inorganic sources, e.g. ammonium ions or nitrate ions; others require nitrogen combined in organic molecules, such as amino acids	Oxygen – most bacteria require air (aerobes) but some can survive in the absence of oxygen if necessary (facultative anaerobes), and others can only grow in the absence of oxygen (obligate anaerobes)
Essential growth factors/vitamins – some bacteria cannot synthesise particular vitamins, e.g. B complex	
Ions, including macronutrients e.g. phosphate, sulphate, potassium, calcium; and micronutrients e.g. manganese, copper	

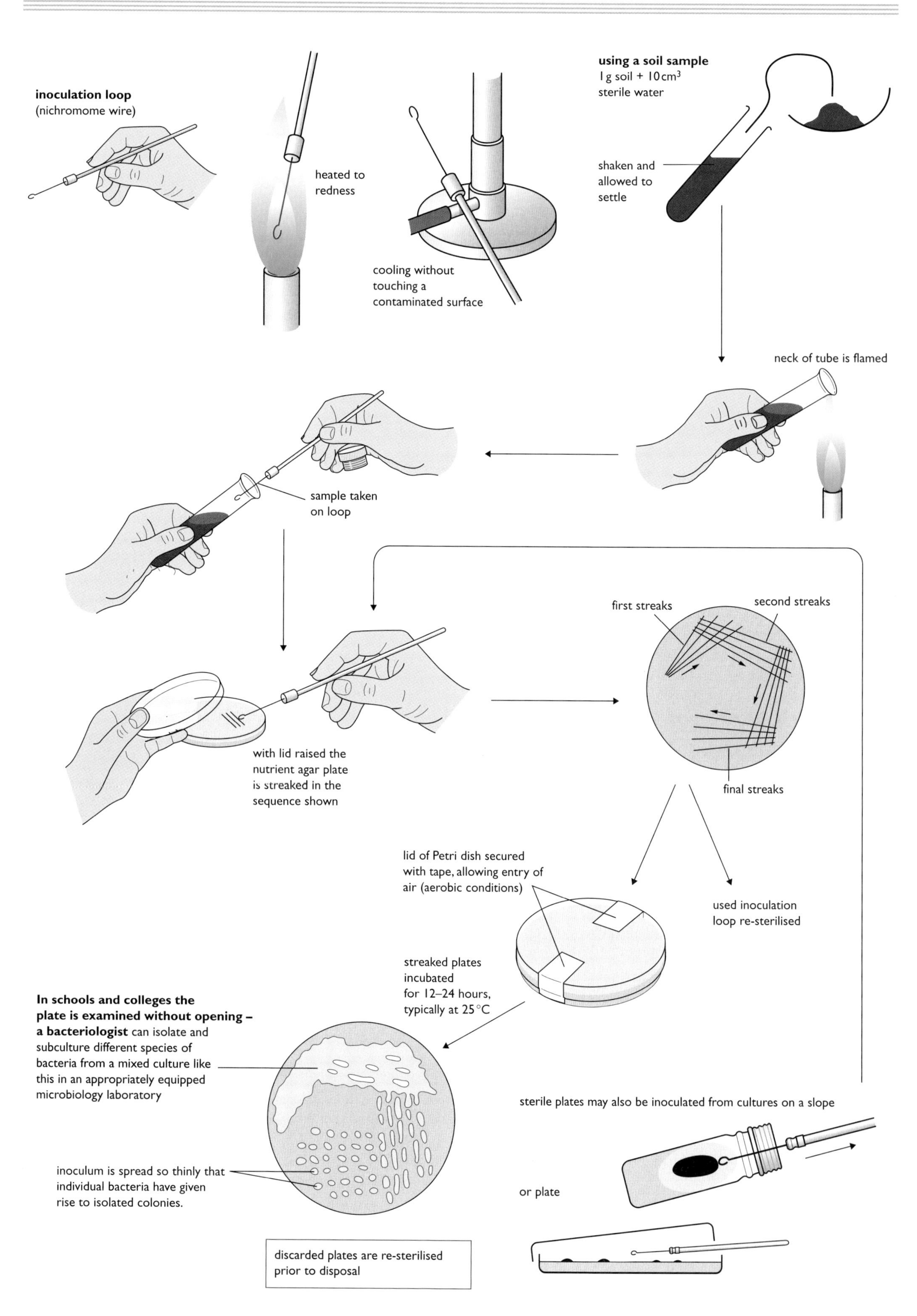

Figure 2.11 Inoculating, culturing and examining cultures.

Monitoring growth of bacteria

It is not practical to examine the growth of an individual bacterium because of its small size. Instead, we follow the growth of a population of bacteria, for example in a liquid culture (broth). The total number of cells in a population of bacteria grown in the culture may be estimated accurately using a small representative sample. The number of cells may be counted indirectly – either by measuring the **change in turbidity** (absorbance of light) of the culture (Figure 2.12), or by the **dilution plate** method (Figure 2.13) – or directly, using a **haemocytometer** (Figure 2.14).

A working microbiological laboratory
Typical investigations here might include the rate of growth of cultures under various conditions, the nutrients and conditions required for that growth, and the steps to bacterial metabolism in particular species (including the waste products of metabolism formed, or the chemical nature of toxins that result).

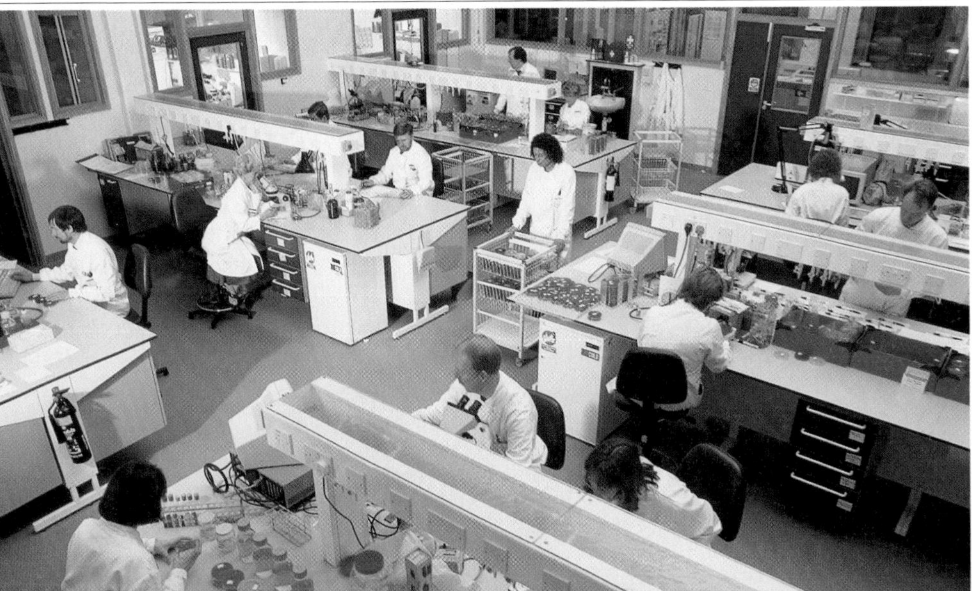

Estimating growth of a bacterial culture by measurement of turbidity change in a liquid culture (broth)
Cloudiness (turbidity) of a culture is due to the bacteria suspended in the broth. Growth of the culture results in an increasing number of cells present. The more cells present, the more light is absorbed by the culture (where absorbance = absorption of light at a particular wavelength).

culturing of bacteria in broth:
culture maintained under constant conditions in an incubator, stirred by magnetic stirrer or a shaker machine

measurement of absorbance of a sample of a bacterial culture:

samples of the culture are withdrawn at regular intervals, and their absorbance measured in the spectrophotometer

lamp
filter
sample chamber
photocell
meter calibrated in absorbance

the changing pattern in turbidity (absorbance) of a growing culture obtained in this study typically shows the distinct phases of growth of a bacterial culture (see Figure 2.15, page 12)

Absorbance
Age of culture

example of a calibration curve

Absorbance
Count (cells/0.1 mm^3)

Making a calibration curve
There is a linear relationship between the number of cells present in a sample of a particular broth culture and the absorbance of light passing through the sample, when measured in a spectrophotometer.

Consequently, a **calibration curve** can be made and used to convert absorbance readings into 'number of bacterial cells present'. This is produced by counting the number of cells present (using haemocytometry – Figure 2.14, page 11) in samples as well as the absorbance. Once produced, the calibration curve can be used to convert absorbance into accurate estimates of the number of cells present (when working with the same species, using the same broth and growing conditions).

Figure 2.12 Estimating growth of bacterial cultures.

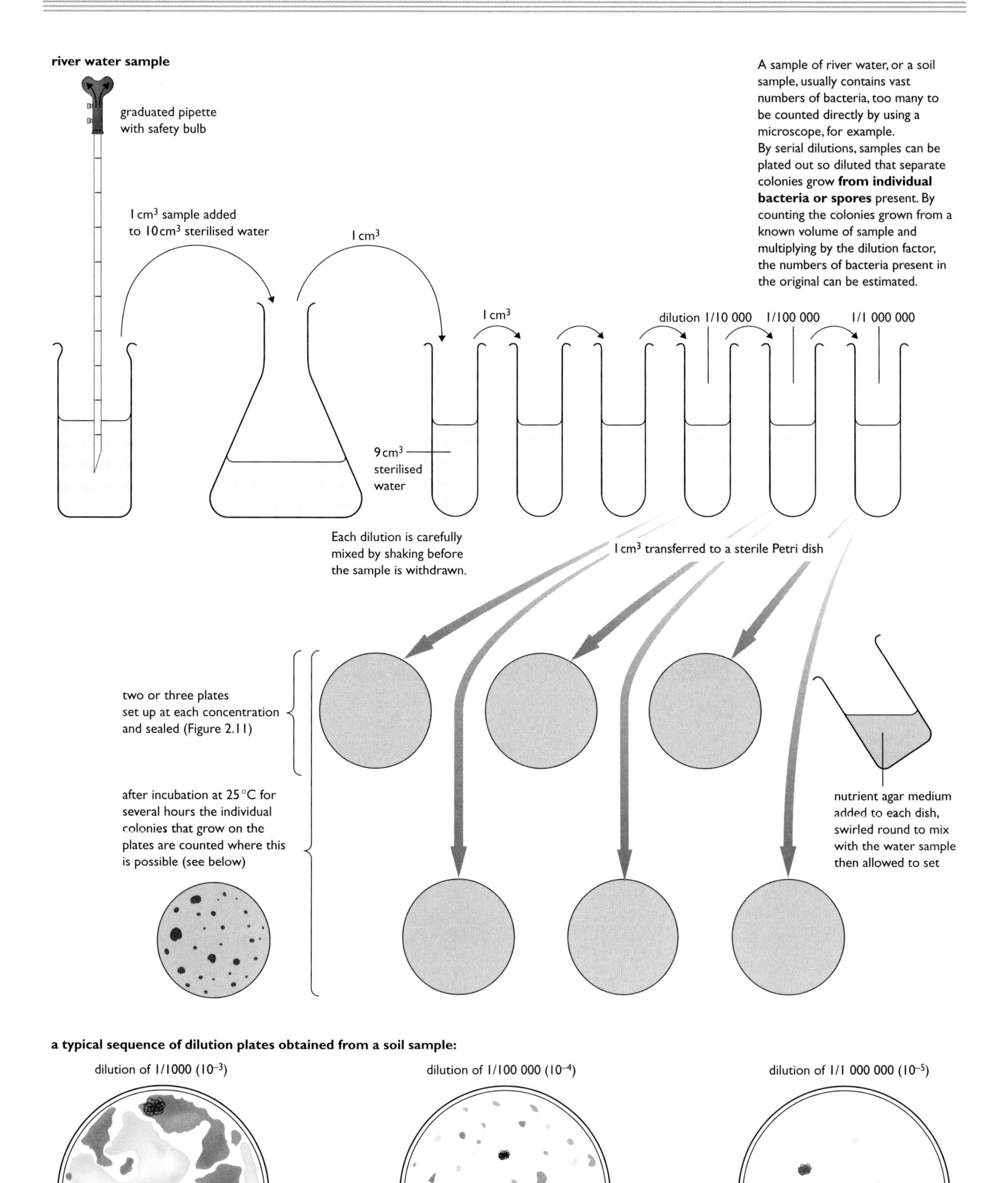

Figure 2.13 Estimating growth of a bacterial culture by the dilution plate method.

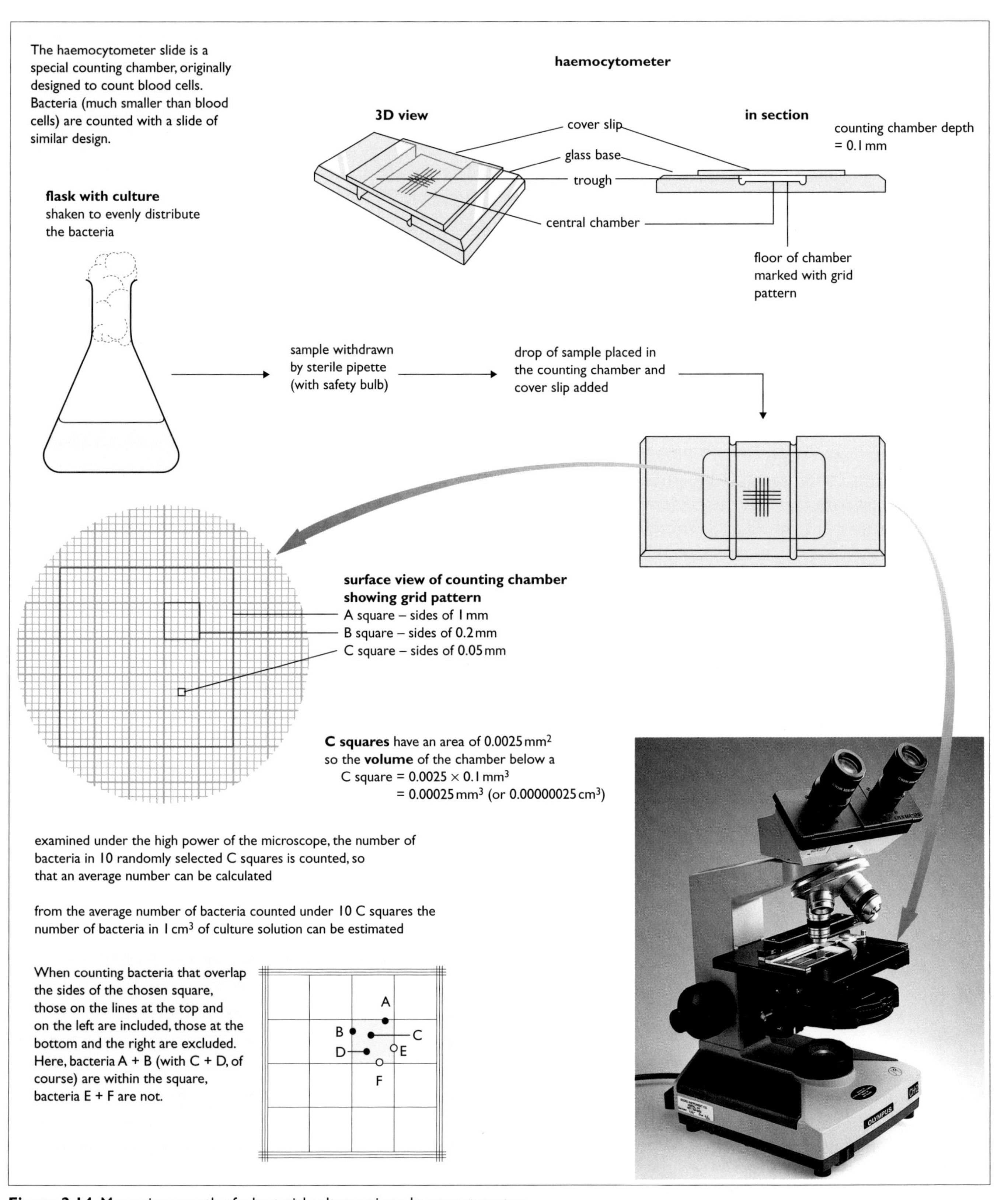

Figure 2.14 Measuring growth of a bacterial culture using a haemocytometer.

Safety note

The guidelines for safe bacteriology in schools and colleges are given in:

ASE (2001), *Topics in Safety*, 3rd edn. Hatfield, UK: Association for Science Education (College Lane, Hatfield, Herts AL10 9AA).

Examples of suitable microbiological experiments, and how to conduct risk assessments, are detailed in:

Clegg, C.J. with Mackean, D.G., Openshaw, P.H. and Reynolds, R.C. (1996) *Advanced Biology Study Guide*. London: John Murray.

4 In ten 'C' squares of the counting chamber, the following numbers of bacteria were counted:

9, 4, 7, 6, 8, 6, 5, 4, 7, 4

Estimate the number of bacteria in 1 mm^3 of the culture being measured.

Growth curve of a population of bacteria grown in broth culture

Growth in a population of bacteria consists of repeated cell division, followed by growth of the daughter cells to full size. This goes on as long as the necessary conditions (Table 2.2, page 7) are maintained. Cell divisions tend to occur simultaneously in all cells of the population (they are synchronised).

For example, if a cell of *E. coli* divides into two cells after 20 minutes (page 4), then 3 hours later (nine more generations) it is likely to have indirectly given rise to more than 1000 *E. coli*. **At this rate**, the cell count would rise to 4.7×10^{24} in only 24 hours. This is an unimaginably large biomass – it could not be supported in a laboratory culture as the available nutrients would soon have been used up. Similarly, the waste products of metabolism could not be removed quickly enough.

Consequently, a characteristic pattern of growth is established in a laboratory culture over a period of about 24 hours (depending on species and temperature, for example). Data on growth of a bacterial culture from the point of inoculation, presented as the number of cells against time (age of culture), are plotted as a **growth curve** (Figure 2.15). Note that the shape of the curve is sigmoid or **S-shaped**, and has distinct phases.

Phases of growth

In the sigmoid curve of growth, the initial period in a new culture medium is a time of adaptation to the conditions (the **lag phase**). This is followed by a period when the cell number doubles with unit time, for there are no constraints on growth (**log phase** of exponential growth). Thereafter, growing conditions start to become unfavourable, and growth is slowed (**stationary phase** of linear growth). Then, increasingly, the occurrences of cell death accelerate, the 'birth rate' is eventually overtaken by the 'death rate', and the culture starts to die (**senescence phase**).

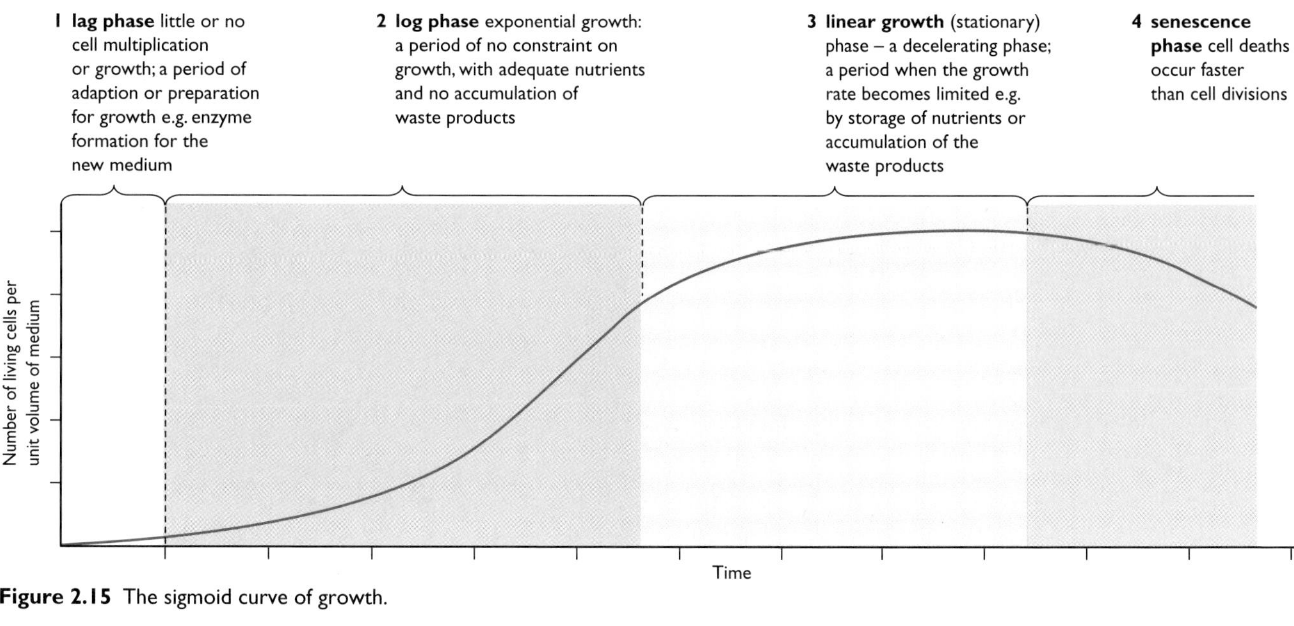

Figure 2.15 The sigmoid curve of growth.

Detecting the phases of growth – total cell counts and viable (living) cell counts

Using a counting chamber similar to the haemocytometer (Figure 2.14), counts of the total number of cells in a representative sample of the culture can be made at intervals. Counts of the viable (living) cells present in the same culture can be done using the dilution plate method (Figure 2.13), as each colony counted on a plate has arisen from a living cell. Typical scores are shown in Table 2.3. Using these data (and log tables or an appropriate calculator), you can construct a graph of the log of the number of cells against time for both total cells and viable cells present.

Table 2.3 Growth of a population of *E. coli* in nutrient broth, showing viable cell numbers and total cell numbers against time

Hours	Number of viable cells / cells per cm³ medium	Total number of cells / cells per cm³ medium
0	20000	20000
2	21900	27000
4	496000	540000
6	5430000	6400000
8	81900000	105760000
12	83400000	126300000
24	80500000	127600000
36	1120000	127900000

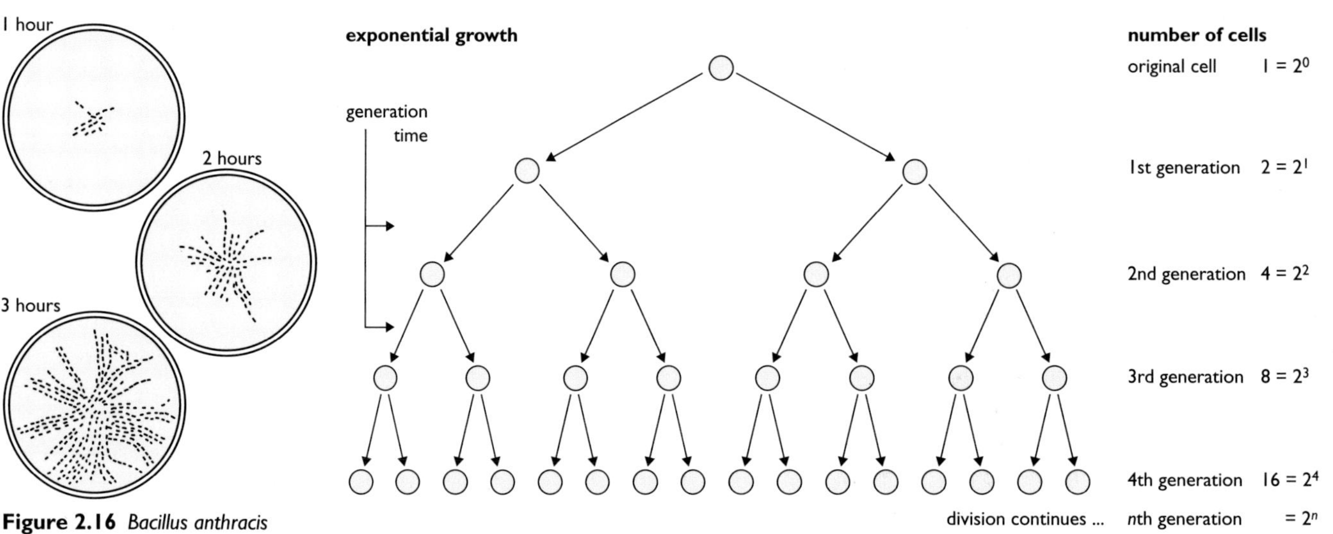

Figure 2.16 *Bacillus anthracis* colonies at 1, 2 and 3 hours.

Figure 2.17 Cell divisions during exponential growth.

Extension: growth rate constants

Because of the geometric progression in population growth during the exponential phase (Figure 2.17), the relationship between the number of cells present initially (N_o) and the number of cells present after n generations of growth (N_n) is given by:

$$N_n = 2^n N_o$$

Taking logarithms:

$$\log N_n = \log N_o + n \log 2$$

This equation can be used to calculate the following.

1 The **number of generations** (n) required to produce a certain number of cells, e.g. from the data in Table 2.3:
the log phase began at the 2 hour stage (N_o = 21 900 cells),
and at the 4 hour stage the cells had increased to (N_n = 496 000 cells,
since

$$n = (\log N_n - \log N_o) / \log 2$$
$$= (\log 496\,000 - \log 21\,900 / \log 2$$
$$= (5.6955 - 4.3404) / 0.3010$$

then

$$n = 4.5$$

2 The **generation time** (g)
In 2 hours (120 minutes), 4.5 generations have been produced, therefore g is given by:

$$120 / 4.5 = 26 \text{ minutes}$$

This equation can also be used to calculate the number of cells expected at a particular time in the log phase.

Diauxic growth

If a culture of *E. coli* is grown in a medium with two sugars, glucose and lactose, as sources of energy, then the glucose is first metabolised before lactose is used. The cells contain the enzymes for glucose metabolism in their cytoplasm. When no more glucose is available, the presence of lactose triggers (induces) the production of enzymes for lactose metabolism. Consequently, the growth curve for *E. coli* cells in this type of nutrient broth shows two phases of exponential growth (Figure 2.18), known as a diauxic growth curve.

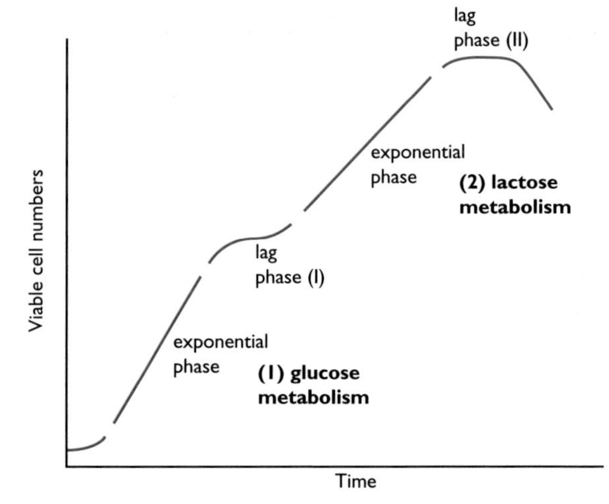

Figure 2.18 Growth curve of an *E. coli* culture in a medium with glucose and lactose as energy sources.

Classification of bacteria

The many thousands of different bacteria that have been isolated, identified and named (genus and species) by bacteriologists have been organised into the hierarchical classification system:

$$\text{Phylum} \leftarrow \text{Class} \leftarrow \text{Order} \leftarrow \text{Family}$$

Biologists also use this scheme of classification with other groups of living things. However, there are alternative, commonly used ways of dividing bacteria into simple groupings, such as **shape** (Figure 2.19), or **nutrition** (Figure 2.20) and **respiration** (Figure 2.21). Also, the **staining properties of the wall** are used to identify bacteria as either Gram-positive or Gram-negative (Figure 2.22). All these methods of classification have applications in the study of bacteria.

Shapes of bacteria

Bacteria are classified into five groups according to their shape, as shown in Figure 2.19. In fact, the majority are rod-shaped or sphere-shaped. Although basically unicellular, bacteria may often appear clumped together. Species showing this form of growth are named accordingly. For example, to those growing in pairs, the generic name is prefixed with 'diplo-'. Where they occur as a chain, the prefix 'strepto-' is used, and where they occur in grape-like clusters, 'staphylo-'.

Figure 2.19 Classification of bacteria by shape.

individual bacteria

rod bacteria (bacilli)
e.g. *Escherichia*
gut-living
Salmonella typhi
causes typhoid

spherical bacteria (cocci)
e.g. *Nitrococcus*
nitrifying bacteria
Methylococcus
methane-oxidising bacteria

comma-shaped bacteria (vibrios)
e.g. *Vibrio cholera*
causes cholera

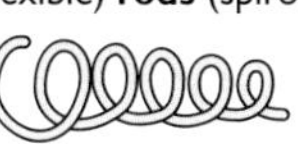
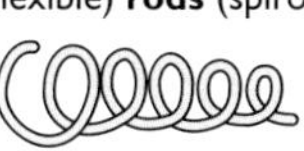

spiral-shaped (rigid) **bacteria** (spirilla)
e.g. *Spirillum minus*
in stagnant water, causes ratbite fever

corkscrew-shaped (long and flexible) **rods** (spirochaetes)
e.g. *Treponema pallidum*
causes syphilis

when found in typical groupings

streptobacilli
e.g. *Bacillus anthrax*
causes anthrax

streptococci
e.g. *Streptococcus pyogenes*
causes sore throats

diplococci
e.g. *Diplococcus pneumoniae*
causes pneumonia

staphylococci
e.g. *Staphylococcus aureus*
causes boils, food poisoning

Nutrition of bacteria

Bacteria are either **autotrophic**, that is, they make their own organic nutrients using an external source of energy, or they are **heterotrophic**, relying on a supply of ready-made complex food substances. Within these groupings there are alternative forms of nutrition which are introduced here (Figure 2.20), and illustrated on later pages.

Figure 2.20 Classification of bacteria by nutrition.

nutrition
provides energy to maintain functions
and matter to build and repair structures

or

autotrophic nutrition
manufacture of sugars (carbohydrates)
and other metabolites from CO_2
and a source of energy

heterotrophic nutrition
the obtaining of nutrients from complex
food molecules (from other organisms and
their products) by the process of digestion

photosynthetic bacteria
use energy from sunlight
to synthesise food substances
(see p.16)

chemosynthetic bacteria
use energy from chemical reactions
to synthesise food substances
(see p.40)

saprotrophic bacteria
secrete enzymes onto external organic
matter (dead organisms or the
products of organisms) and absorb the
soluble products (nutrients)
(see p.30 and 62)

parasitic bacteria (pathogens)
live on or in a host organism and feed
on the organic matter of the host
causing harm (disease)
(see p.68 and 70)

Respiration of bacteria

Respiration takes place in every living cell, using organic nutrients (typically sugar) and making energy available for metabolism, growth and reproduction, for example. For respiration, many bacteria require oxygen (**aerobic respiration**). Others only respire in the absence of oxygen (**anaerobic respiration**). A third group normally respire aerobically but have the facility to switch to anaerobic respiration in the absence of oxygen. The biochemical pathways involved in respiration are summarised in Figure 2.21.

5 Name an intermediate of respiration common to the aerobic and anaerobic pathways.

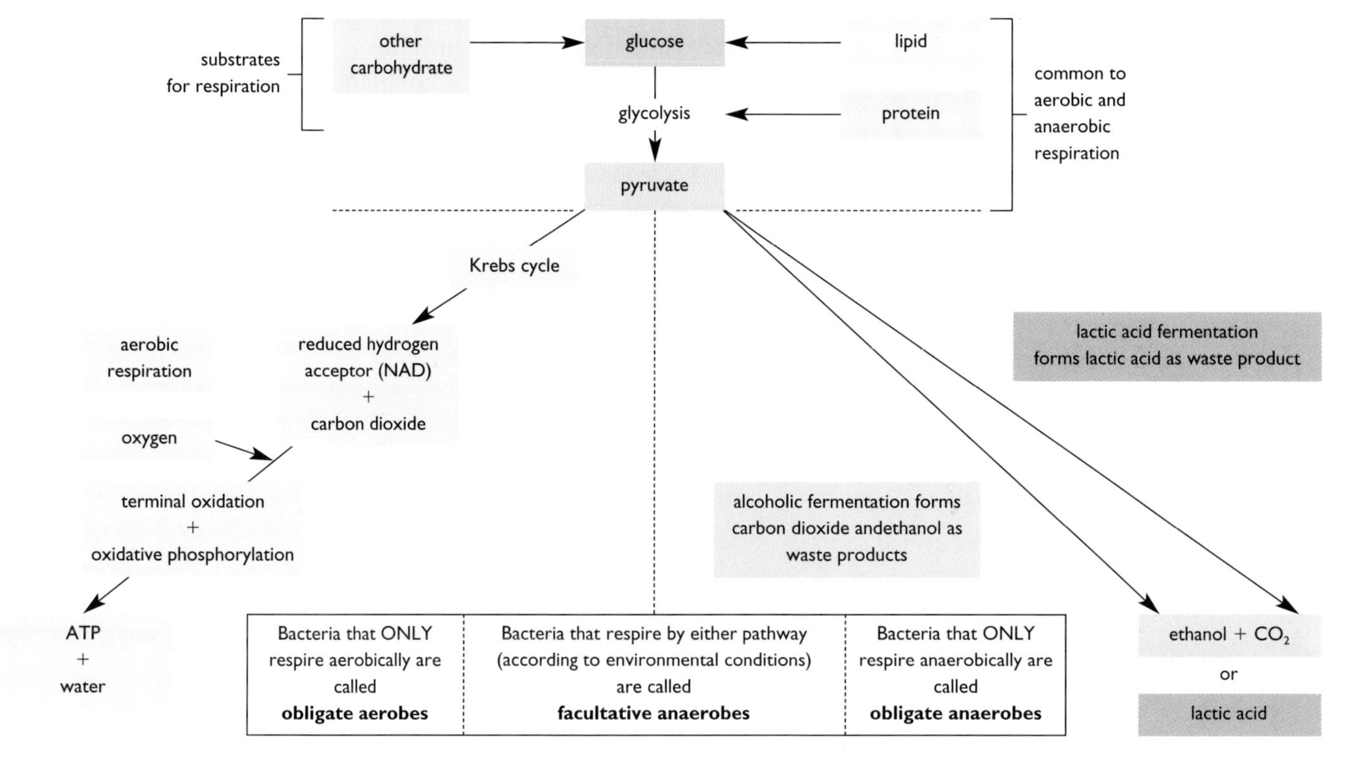

Figure 2.21 Classification of bacteria on their respiration.

Staining properties of the walls of bacteria

Hans Christian Gram, a Dane, showed in 1884 that some bacteria can be stained with crystal violet dye and become purple (Gram-positive species), while others cannot retain this dye in their walls (Gram-negative species). This effect is due to important differences in wall chemistry (Figure 2.22).

stage 1

Bacteria in an air-dried smear on a microscope slide appear colourless.

stage 2

The smear is treated with crystal-violet (a basic stain). All cells appear violet when the stain is washed from the slide.

stage 3

The smear is flooded with Lugol's iodine (a mordant treatment to combine the dye to those bacteria with which it will react).

stage 4

The smear is now treated with a decolorising solution of acetone and alcohol – this removes the violet dye from the cells with which it has not reacted. **Gram-positive bacteria remain purple**.

stage 5

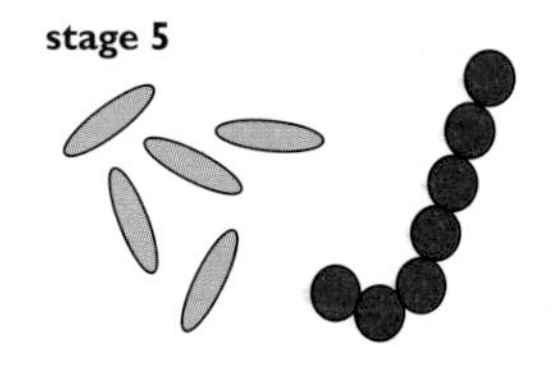

Finally the red dye safranin is briefly added as a counter-stain – it is taken up by the colourless bacteria of the treated smear. **Gram-negative bacteria now appear red**. Gram-positive bacteria remain purple.

peptidoglycan (giant molecules of amino acid sugars and peptides)

plasma membrane

cytoplasm

outer membrane of lipid and polysaccharide (unique to Gram-negative bacteria)

section of walls of Gram-positive and Gram-negative bacteria

Figure 2.22 Gram staining, and the differences between wall structure of Gram-positive and Gram-negative bacteria.

Photosynthetic bacteria – the cyanobacteria

Cyanobacteria are photosynthetic bacteria. They grow in the light, typically on damp, shaded soils, moist rocks and posts, and on the parts of tree trunks and branches where water frequently flows. They also occur in the plankton of fresh water and the sea. Plankton 'blooms', seen in waters polluted by nitrates and phosphates, are usually rich in cyanobacteria. Some of the cyanobacteria are single-celled, but many are colonial forms which grow in clumps or, more commonly, as filaments. *Anabaena*, illustrated here, is a good example.

Cyanobacteria have a typical prokaryote cell structure. Their cell walls have the same composition as those of Gram-negative bacteria, and are often externally coated by a slime layer of polysaccharides. Within the cytoplasm are extensive mesosomes (page 3), where the photosynthetic pigments (chlorophyll *a* and accessory photosynthetic pigment such as phycocyanin) are held. Because of these pigments, cyanobacteria often make their microhabitats a blue-green colour. Also present in the cytoplasm are nutrient reserves (e.g. phosphate granules, lipid granules and glycogen granules), and gas vesicles. Several species of cyanobacteria 'fix' atmospheric nitrogen (page 36).

6 Some microbes can 'fix' atmospheric nitrogen. What is the significance of this?

Figure 2.23 *Anabaena*, a filamentous cyanobacterium.

SEM of *Anabaena* filaments among plankton species (× 400)

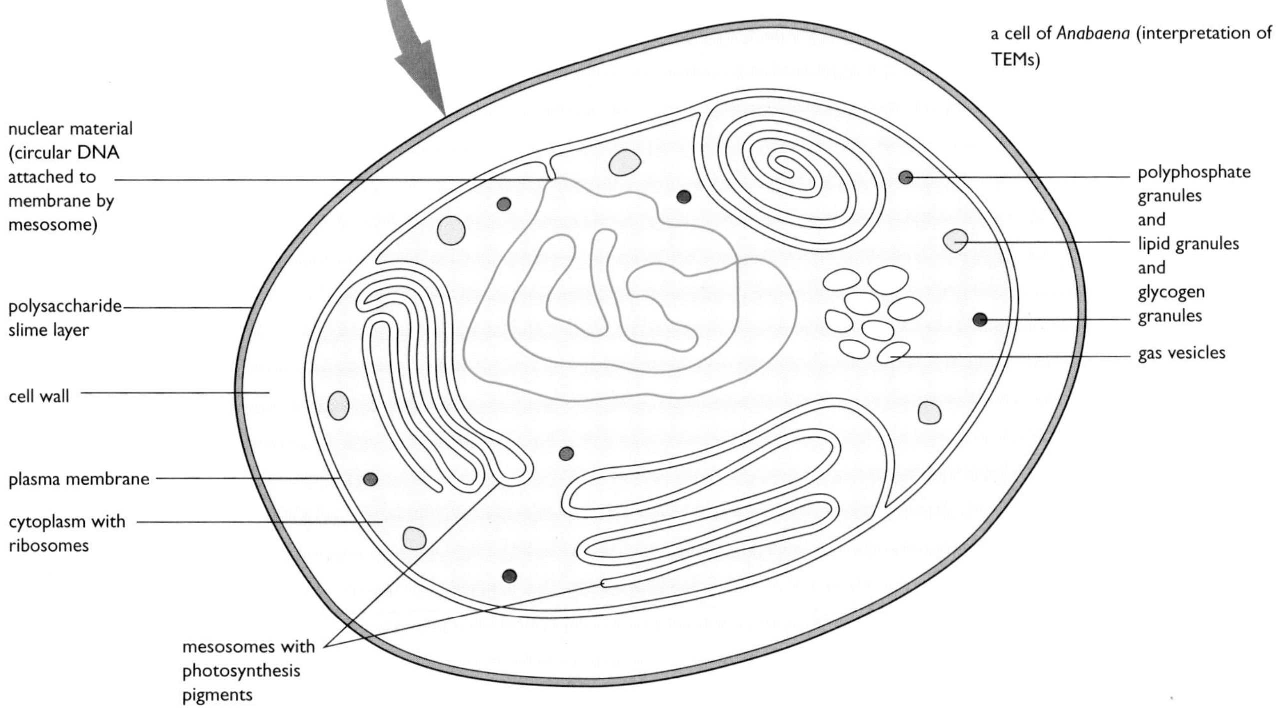

a cell of *Anabaena* (interpretation of TEMs)

nuclear material (circular DNA attached to membrane by mesosome)

polysaccharide slime layer

cell wall

plasma membrane

cytoplasm with ribosomes

mesosomes with photosynthesis pigments

polyphosphate granules and lipid granules and glycogen granules

gas vesicles

Extension: extremophiles

We know that bacteria are on us, in us, and all around us. In addition, some species occur in much more unlikely, and often extremely hostile places. The bacteria of these extreme habitats are called **extremophiles** (*philo-* = liking for). Some examples of extremophiles follow.

- Salt-loving bacteria (**halophiles**), e.g. *Halobacterium halobium*, common in salt lakes and places where sea water becomes trapped and concentrated by evaporation. Here the salt concentration is so high it would dehydrate ordinary bacteria, which are unable to maintain a high salt concentration in their cytoplasm. Similar extreme habitats are soda lakes rich in sodium carbonate. Here the pH is so high that the membrane lipids of ordinary species would dissolve and the proteins unfold.
- Alkali-loving species (**alkalinophiles**), including *Spirulina*, which is of economic importance because of its food value (page 49).
- Species able to withstand extremely acidic conditions (**acidophiles**), e.g. *Thiobacillus ferro-oxidans*, a bacterium exploited in the recovery of metal from low-grade ores (Figure 2.24).

Figure 2.24 Copper extraction from spent ore by microbial mining.

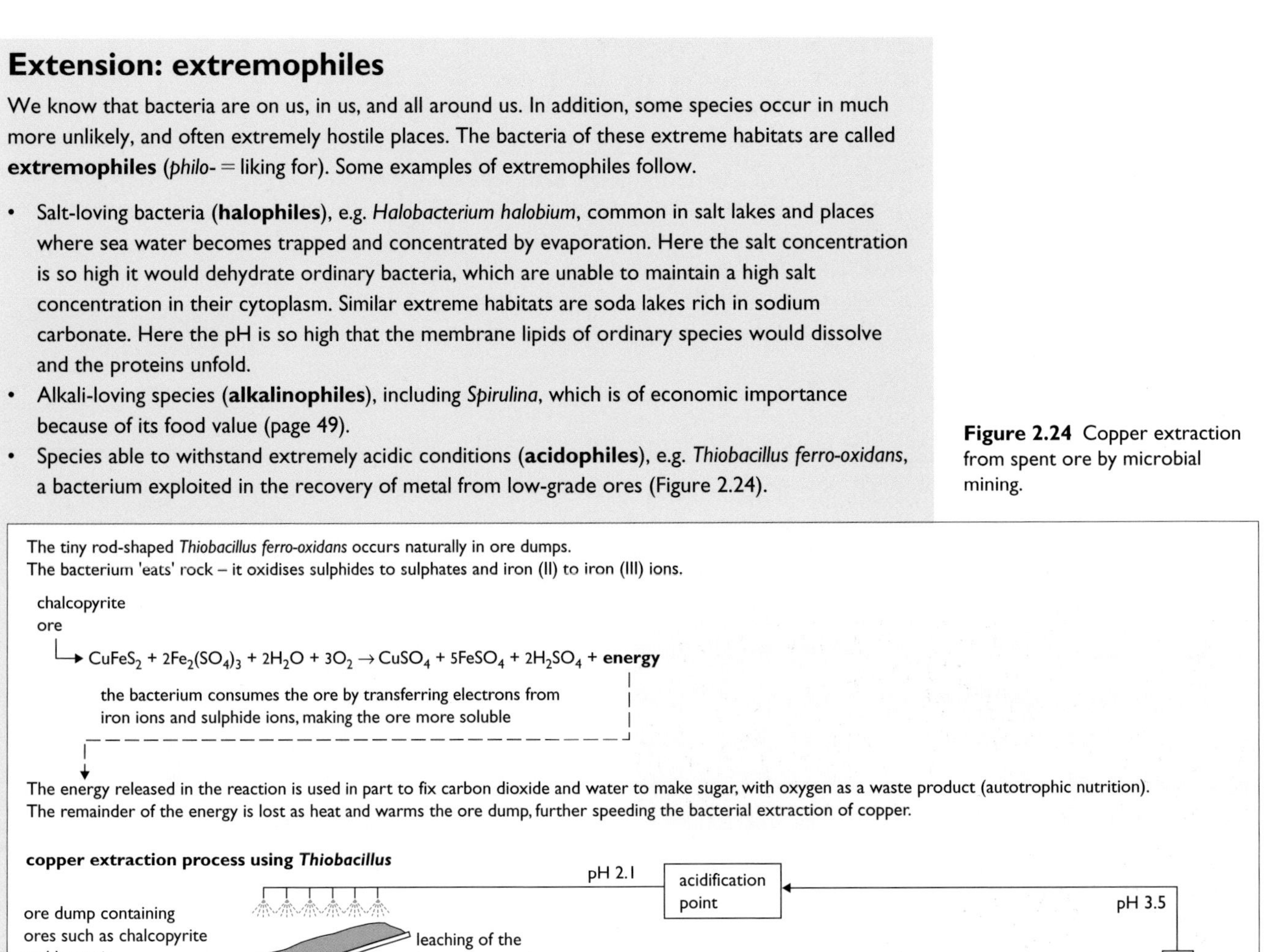

- Heat-loving bacteria (**thermophiles**) occur in hot springs where temperatures of about 70 °C are normal. Some heat-loving bacteria have adapted to survive at extremely high temperatures, e.g. 100–115 °C, and these species never occur in habitats below 70 °C. These are known as **hyperthermophiles**, and are found in deep-sea hydrothermal vents where hot volcanic gases rich in sulphur bubble out at high temperatures, due to geothermal energy (Figure 2.25).
- A few species flourish in sub-zero temperatures (**cryophiles**), and are common at temperatures of −10 °C, as in the ice of the poles where salt depresses the freezing point of water.

Figure 2.25 Volcanic vent of the ocean floor.

Extension: a division within the prokaryotes?

Biochemical studies of the nucleic acids of organisms are used in modern classification studies. For example, in the prokaryotes, analysis of the RNA of ribosomes (page 2) suggests there is a separate, distinctive group within the prokaryotes, now called the **ancient bacteria**. Many of the extremophiles belong to this group of ancient bacteria, including most of the hyperthermophiles.

This study has also shown that the ribosomal RNA of all eukaryotes (plants, animals, fungi, etc.) is similar to that of the ancient bacteria. This discovery is thought to support speculation that life may have arisen in marine deep-sea volcanic vents, and that the earliest life forms were hyperthermophiles.

3 Viruses

Viruses are disease-causing agents, rather than 'organisms'. The distinctive features of viruses are:

- they are not cellular structures, but consist of a core of **nucleic acid** surrounded by a protein coat called a **capsid**
- in some viruses there is an additional **external envelope** of membrane made of lipids and proteins (e.g. HIV, Figure 3.5, page 20; influenza virus, Figure 10.26, page 78)
- they are **extremely small** when compared with bacteria – most viruses are in the range 20–400 nm (0.02–0.4 μm) – and can only be seen using an electron microscope (Figure 3.1)
- they can reproduce only inside specific living cells, so viruses function as **endoparasites** in their **host** organism
- they have to be **transported** in some way between hosts
- they are highly **specific** to particular host species, some to plant species, some to animal species and some to bacteria
- they are classified by the type of nucleic acid they contain, either DNA or RNA, and whether they have a single or double strand of nucleic acid (Figure 3.2).

1 How does resolution differ from magnification?

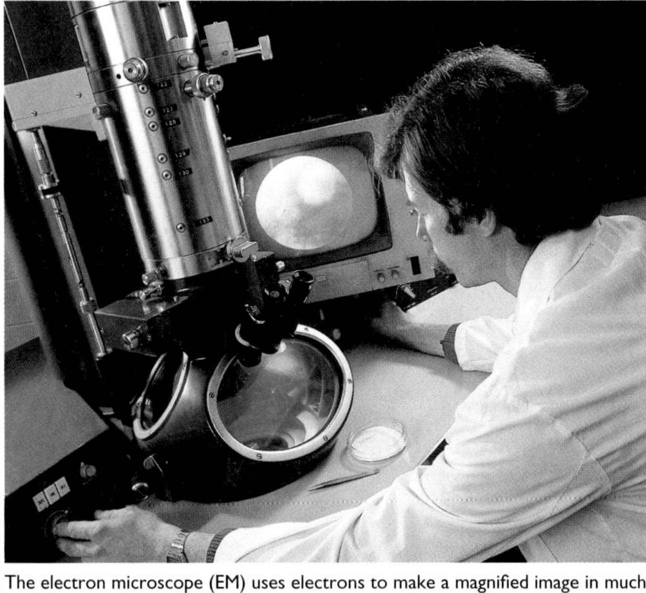

The electron microscope (EM) uses electrons to make a magnified image in much the same way as the optical microscope uses light. The electron beam has a much shorter wavelength, however, so its **resolving power** is much greater. With biological materials such as viruses, the EM has a resolution limit of about 5 nm, so some details of virus structure can be observed.

The electrons travel at high speed but at very low energy. Consequently, inside the microscope must be a vacuum, as air molecules would deflect the beam. So living material must be killed, 'fixed' in a life-like condition, then dehydrated. Specimens must also be very thin They may have to be stained with electron-dense substances to increase the contrast.

electron gun – emits an accelerated electron beam

condenser – electromagnetic lens focuses the electron beam onto specimen

specimen position

vacuum pump

air lock/specimen port – the specimen is introduced without the loss of vacuum

objective – electromagnet lens that focuses the first image (according to voltage)

projector – electromagnetic lens that magnifies a part of the first image

viewing port with binocular viewer

fluorescent screen coated with electron-sensitive compound

camera chamber – allows a black and white photographic image to be made (and the possibility of further magnification)

transmission electron microscope (TEM)

Figure 3.1 The electron microscope and the study of virus structure.

Are viruses 'living' at any stage?

Viruses are an assembly of complex molecules, rather than a form of life. When isolated from their host cell they are inactive, and are best described as crystalline. Within susceptible host cells they are highly active genetic programmes that will take over the biochemical machinery of host cells. Their component chemicals are synthesised and then assembled to form new viruses. On breakdown (lysis) of the host cell, viruses are released and may cause fresh infections. So viruses are not living organisms, but they may become active components of host cells.

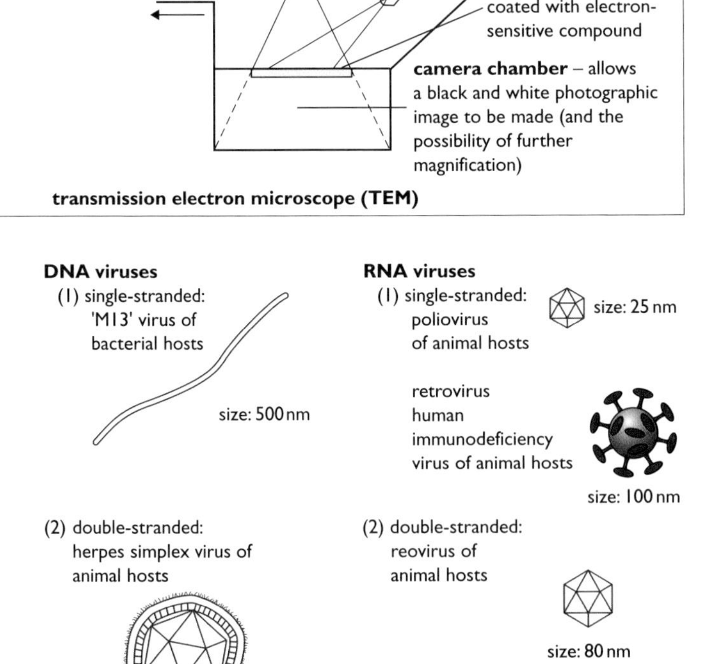

Figure 3.2 A classification of viruses.

Tobacco mosaic virus (TMV)

TMV is a potential infective agent of several species of crop plants, not only of the tobacco plant.

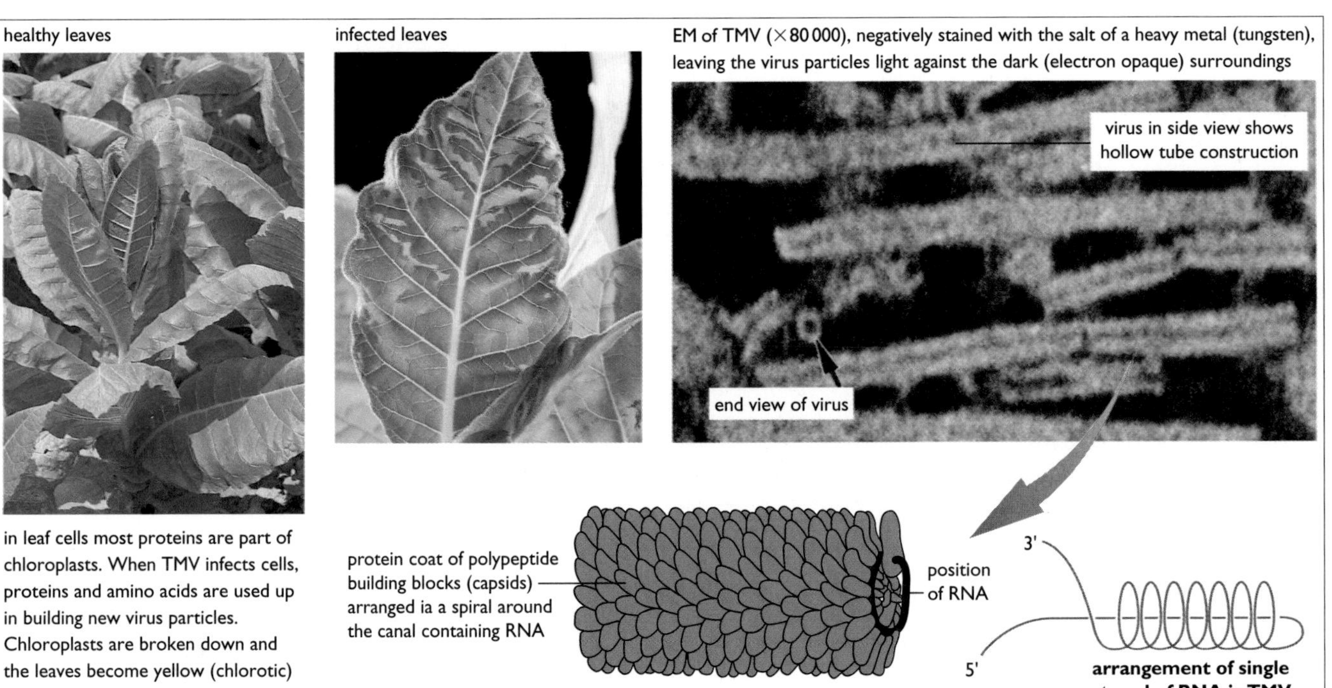

Figure 3.3 Tobacco mosaic virus.

Infection of the host and replication of the virus

TMV may infect healthy cells that have been damaged by aphids or by leaves rubbing together. In host cells, the virus particle unpacks and the RNA strand is copied repeatedly (**replicated**) by host enzymes. Then the viral RNA functions as messenger RNA for synthesis of virus proteins. New virus particles are assembled, and escape when their host cell breaks down. A special transport protein, also coded for by the viral RNA, helps movement of viral particles between cell of the host, spreading the infection there.

Treatment of infected plants

Plants have no immune response to infections (as mammals do). Some plants are naturally resistant to a virus, and crop plants may be genetically engineered to have that resistance. Otherwise, viral infections may be treated by killing aphids (the **vector**) with insecticide, and by removing and destroying infected plants or parts of plants early in an infection, before it spreads.

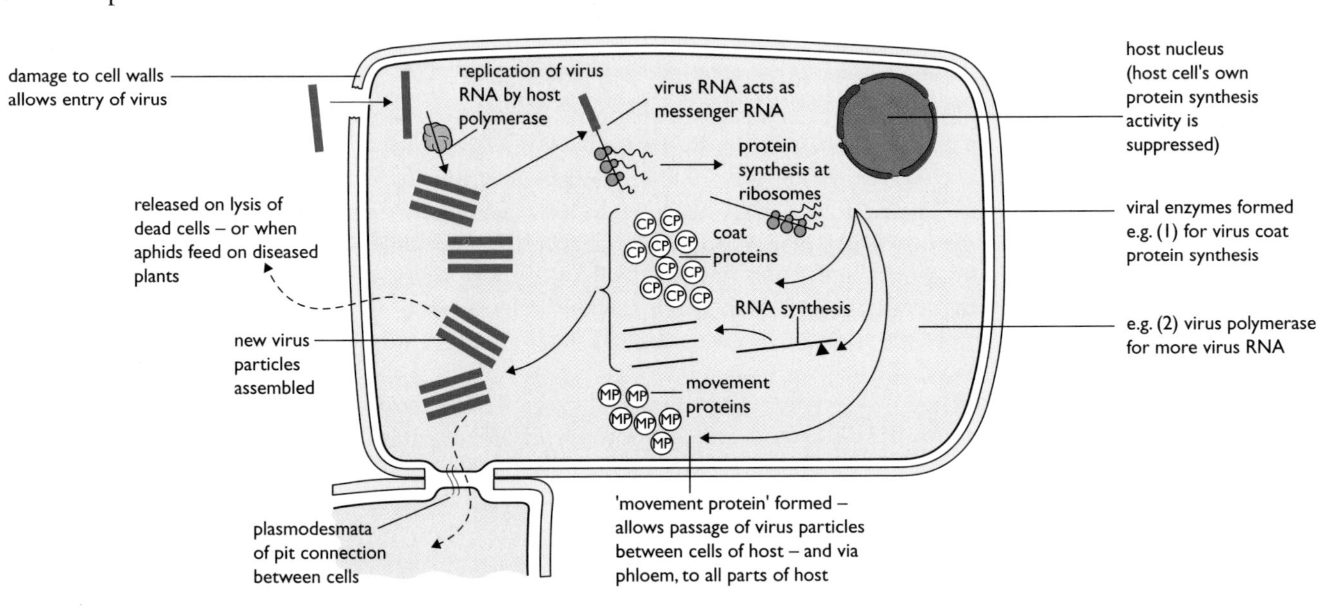

Figure 3.4 Stages of TMV replication in the host cell.

Human immune deficiency virus (HIV)

HIV was first identified in 1983 as the cause of a disease of the human immune system known as **autoimmune deficiency syndrome** (AIDS). HIV is a tiny virus, less than 0.1 µm in diameter (Figure 3.5). It consists of single strands of RNA together with enzymes, enclosed by a protein coat. It is encapsulated with a membrane derived from the human host cell in which it was formed.

The virus first appeared in central Africa in the 1950s, perhaps as a mutation of a similar virus present in African green monkeys (but other theories about the origin of HIV exist). From Africa, HIV was spread to the Caribbean and later to the USA and Europe. Now AIDS occurs worldwide. It is probably already the greatest threat to public health because it kills people in the most productive stage of their lives.

2 Compare the structure and behaviour of the nucleic acid of HIV and TMV.

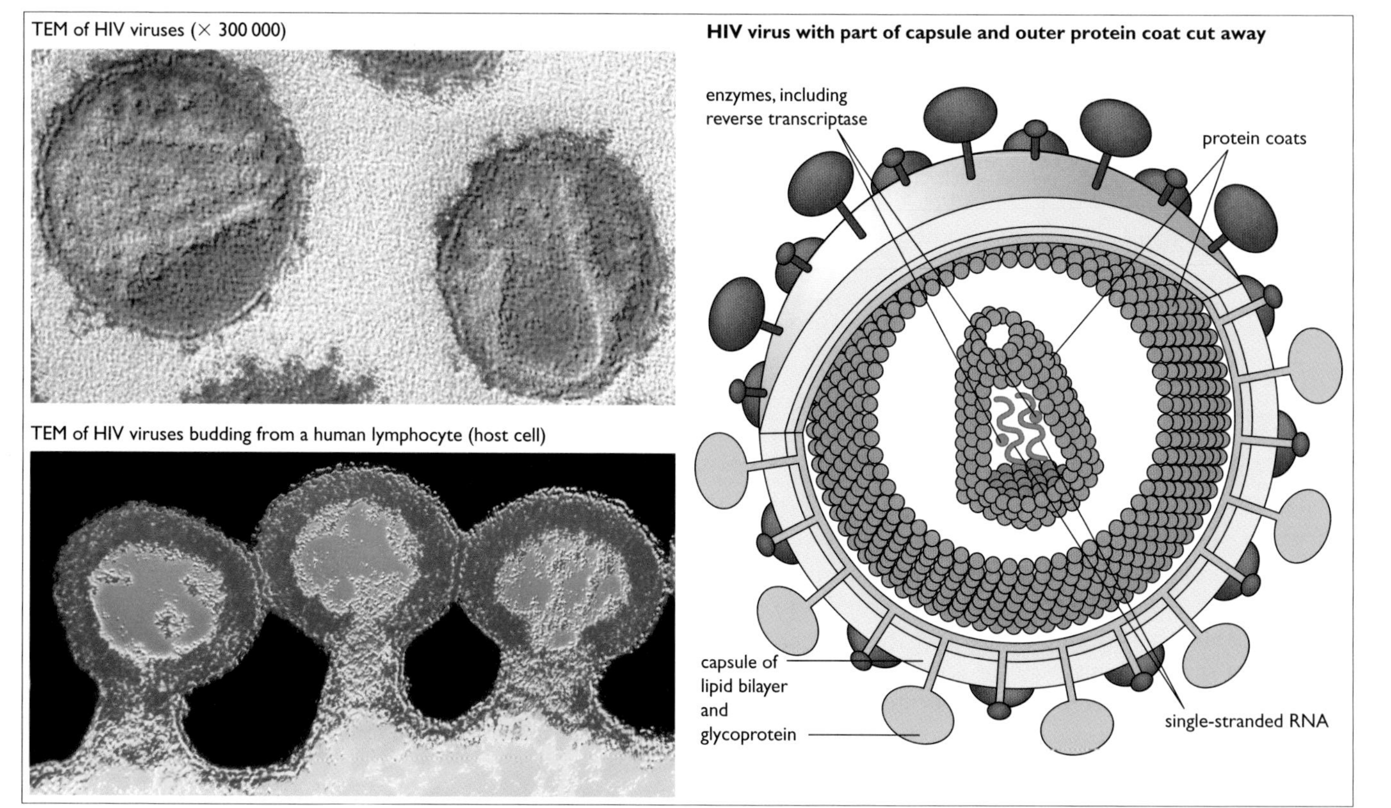

Figure 3.5 The human immune deficiency virus (HIV). HIV is 9–10 nm in diameter.

HIV is a retrovirus

Retroviruses reverse the flow of genetic information, which is normally from DNA of genes to messenger RNA in the cytoplasm. That information always flows in this direction in cells is called the **central dogma** of cell biology. However, in retroviruses the information in RNA in the cytoplasm is translated into DNA of a chromosome in the host's nucleus.

How does a retrovirus work?

Taking HIV as an example, the virus binds to a host cell (lymphocyte) membrane (Figure 10.32, page 80), and the core of the virus passes inside. In the host cell the RNA and virus enzymes are released. One enzyme from the virus, called **reverse transcriptase**, copies the genetic code of the RNA strand into a DNA helix. This DNA then enters the host nucleus and is spliced into the host DNA of a chromosome (Figure 10.33, page 80). Here it may be replicated with the host's genes every time the host cell divides. In these cases the viral genes remain **latent**, giving no sign of their presence in the host cells.

The onset of AIDS

At a later stage, some event in the patient's body activates the HIV genes. The outcome is AIDS. The average interval between HIV infection and the onset of AIDS is about 8–10 years. The result is the synthesis of viral messenger RNA which passes out into the cytoplasm, there coding for viral proteins (enzymes + protein coat) at the ribosomes. Viral RNA (single-stranded), enzymes and coat protein are formed into viral cores. These move against the cell membrane and bud off new viruses (Figure 10.34, page 81). The new viruses infect more lymphocytes, and the cycle is rapidly repeated. The body's reserve of lymphocytes dwindles and eventually no infection, however trivial, can be resisted.

Bacteriophages – viruses specific to bacteria

Viruses whose hosts are bacteria are called bacteriophages (abbreviated to **phages**). There are many different phages, and most bacteria are vulnerable to attack by a specific phage. On entry into the host cell, some phages take over the metabolism of their host immediately, converting it to a 'factory' for production of phage components. These are then assembled, the host cell wall breaks down (lysis), and the new viruses escape and may cause fresh infections. This cycle is a case of a **lytic infection**, and viruses of this type are described as **virulent**. The host cell dies quickly.

Alternatively, some phages insert their genes into the chromosomes of the host and these are reproduced each time the host cell divides. The bacteria carry the virus infection for many generations, but behave normally. This cycle is a case of a **lysogenic infection**, and these viruses are described as **temperate**. Eventually, harmful environmental conditions (such as exposure to UV light) may trigger a switch from a lysogenic cycle to a lytic one (Figure 3.6).

Lambda (λ) virus

The λ virus (Figure 3.7) is an example of a virus that may be temperate or virulent depending on conditions in the host cell. It is a specific parasite of a strain (K12) of the common gut bacterium *E. coli* (page 2). λ has double-stranded DNA surrounded by a protein coat in the form of an isohedral head and tail, with a thin tail fibre.

Figure 3.7 Structure of the λ virus.

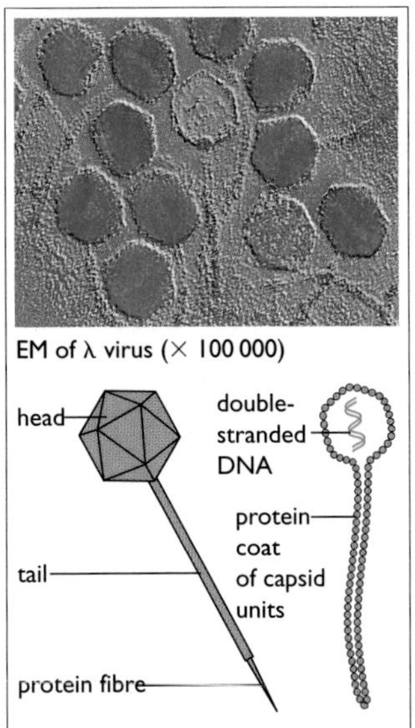

EM of λ virus (× 100 000)

head
double-stranded DNA
protein coat of capsid units
tail
protein fibre

external appearance (left) and internal structure (right)

Figure 3.6 Lytic and lysogenic infection of *E. coli* by λ virus.

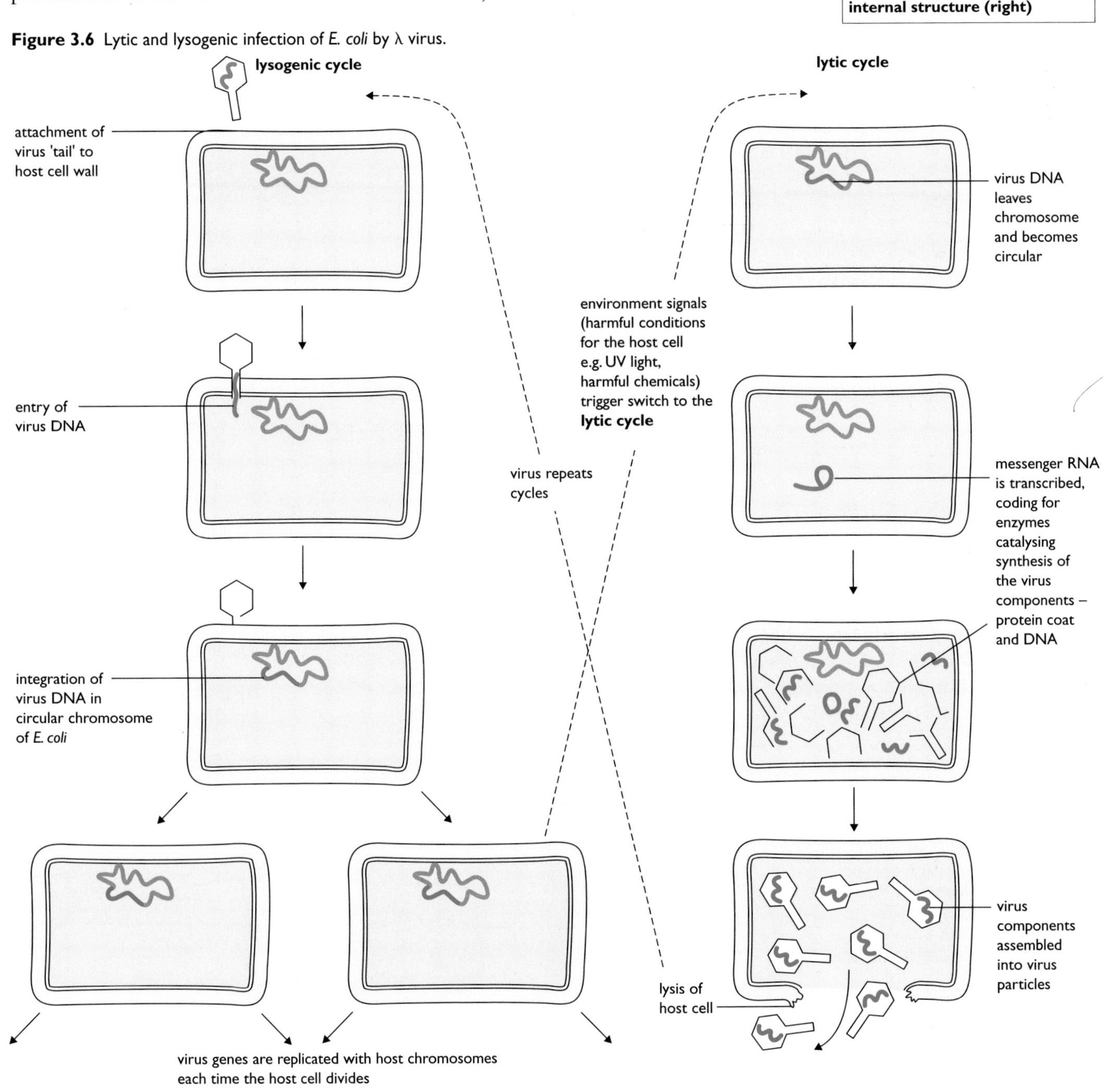

Culturing viruses

Viruses have to be grown for the production of **vaccines** against virus diseases (page 76). Reference collections of viruses are maintained to assist in comparison and identification when new outbreaks of virus diseases are investigated. Unlike bacteria and fungi, viruses cannot be cultured on plates or in broth outside their host cells. Remember, a virus replicates and multiplies only within cells of a host organism.

Culturing animal and human viruses

Initially, human viruses were mostly grown in laboratory-maintained white mice, as cells of mice proved to be susceptible to many human viral diseases. Now cell biologists have perfected the technique of **tissue culture**. As a result, many animal viruses can be cultured in mammalian cell cultures grown in sterile nutrient media *in vitro* (Figure 3.8). Animal and human cell cultures are used. Reference cultures of viruses are maintained in small samples of these cell cultures, held at low temperatures.

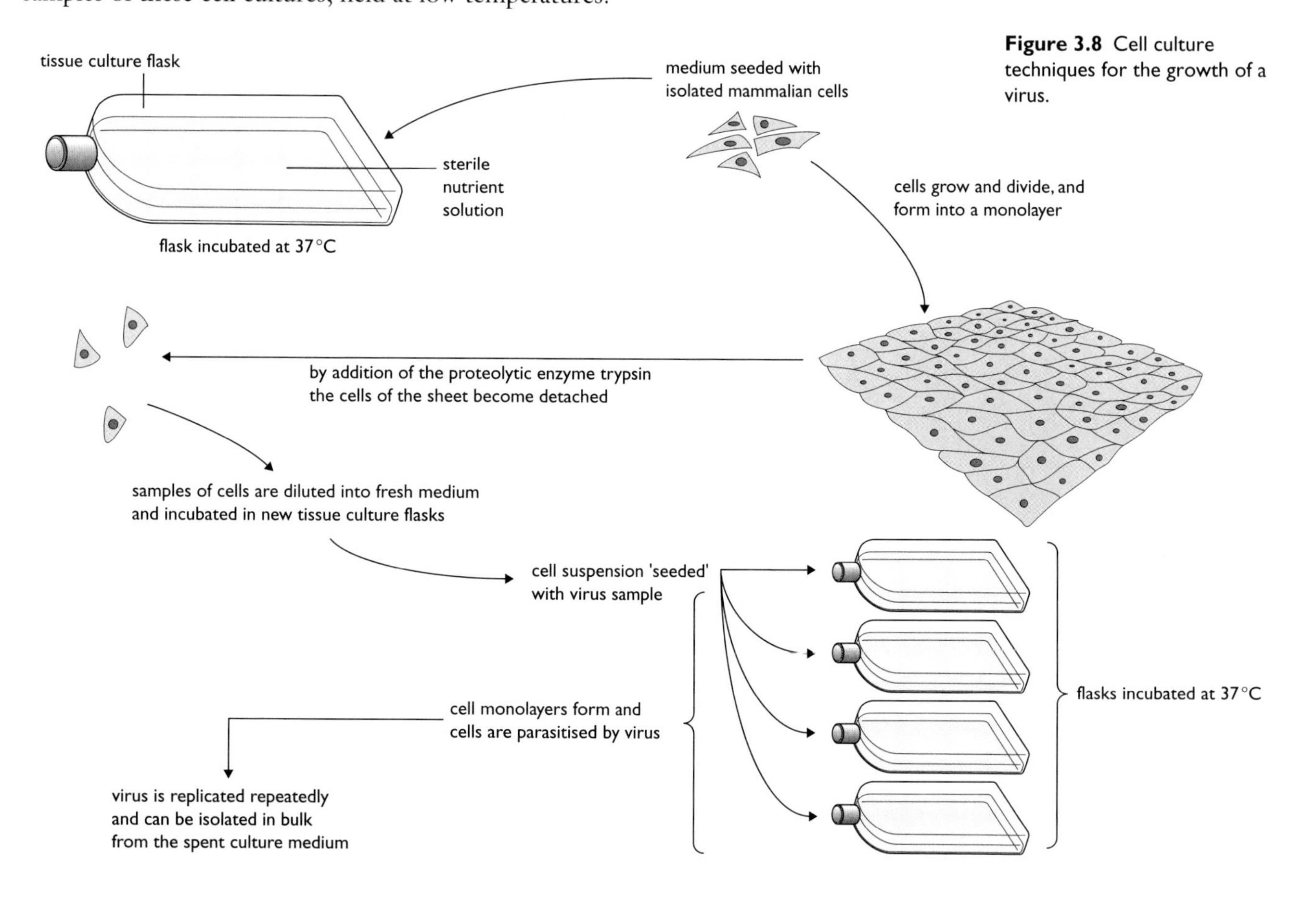

Figure 3.8 Cell culture techniques for the growth of a virus.

An alternative venue for the culture of certain animal viruses is **fertilised hens' eggs** (Figure 3.9). The membranes of these eggs are extremely susceptible to viruses such as influenza, mumps, typhus, psittacosis, herpes, encephalitis (and smallpox). Large quantities of virus are produced.

Figure 3.9 Chick embryo as a venue for virus growth.

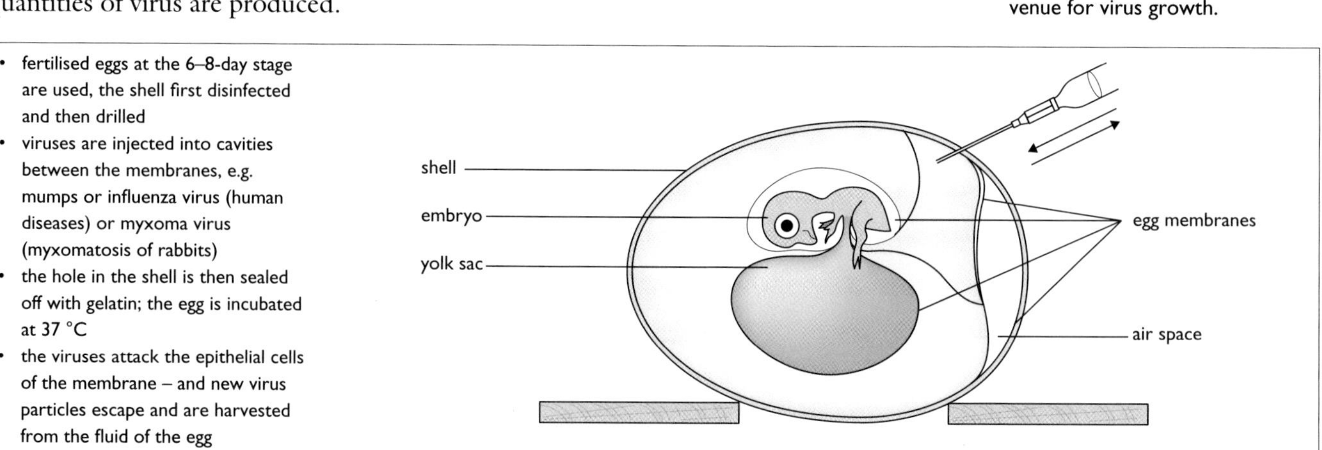

Culturing bacterial viruses (phages)

A colony of bacteria, grown over the entire surface of a nutrient agar plate, is referred to as a 'lawn' of bacteria. When the lawn is inoculated with a virus specific for the bacterium (a phage) and the plate is incubated, clear areas, called **plaques**, develop in the lawn. These are where the host cells have been parasitised and lysis has occurred. Samples of phage can be obtained from plaque regions.

3 Viruses may be cultured *in vivo* (e.g. in a hen's egg) or *in vitro* (in cell cultures). What do these terms mean?

Figure 3.10 Bacterial lawns as venues for phage growth.

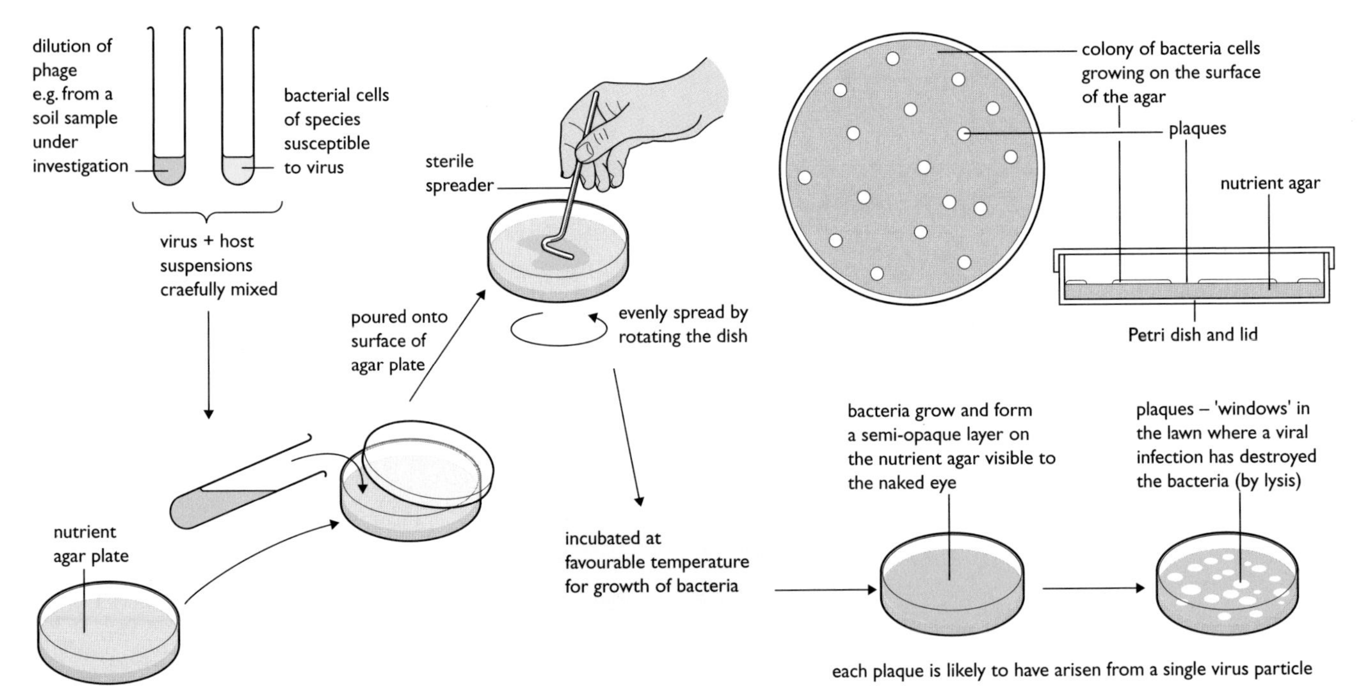

Extension: vaccines by genetic engineering

Foot and mouth is a virus disease of all cloven-hoofed animals (sheep, goats, pigs, cattle and deer). The symptoms are lameness, high temperature, and blisters on the feet, snout, udder, teats and mouth. The virus is easily spread by diseased animals, by stock workers, by frozen carcasses, by birds, and from swill feeds that have not been properly boiled. In the UK, all infected animals and those that have been in contact with them are slaughtered, and the carcasses burnt. In this way, the intention is to keep the national herd free of the disease. Outbreaks arise from time to time; it appears the virus is brought into the country, often on meat or meat products produced from infected animals. Foot and mouth is endemic in some parts of the world.

A vaccine against foot and mouth can be made from genetically engineered bacteria such as *E. coli*. The process of genetic engineering is explained in Chapter 8 (page 54). Steps in the production of a genetically modified (GM) vaccine are summarised in Figure 3.11.

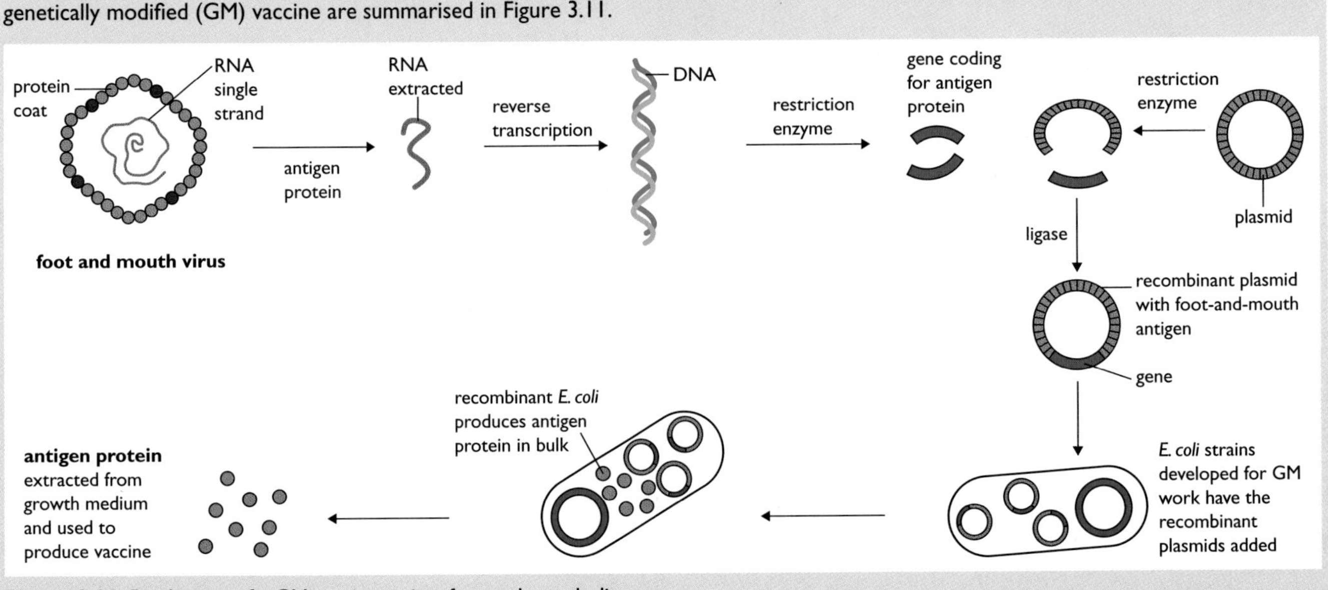

Figure 3.11 Production of a GM vaccine against foot and mouth disease.

Fungi

The fungi include the moulds, yeasts, mildews, mushrooms, puffballs and rusts. The study of fungi is called **mycology**. Mycologists classify fungi on the presence or type of a fruiting body, which **spores** are produced in and dispersed from. Fungi have the following distinctive features:

- they are **eukaryotic organisms** (page 3)
- the fungus body is known as the **mycelium**, and consists of thread-like structures called **hyphae** (Figure 4.1); however, certain fungi are unicellular (Figure 4.5, page 26)
- the hyphae have a wall of **chitin**, an unbranched polymer of glucosamine units (glucose in which one –OH is replaced by –$NHCOOCH_3$), a similar but not identical polymer to cellulose (of plant cells)
- the fungi lack chlorophyll and are unable to manufacture their own food – fungal nutrition is **heterotrophic**: some feed on dead organic matter and are **saprotrophic** (Figure 4.2); others are **parasitic**, mostly with plant hosts (Figure 4.3); some fungi live in mutually beneficial associations with other species (mutualistic organisms, e.g. lichens [algae + fungi], and mycorrhiza, Figure 4.4)
- most fungi can reproduce by both asexual and sexual reproduction, and they produce **spores**.

1 What features of hyphae establish fungi as eukaryotes?

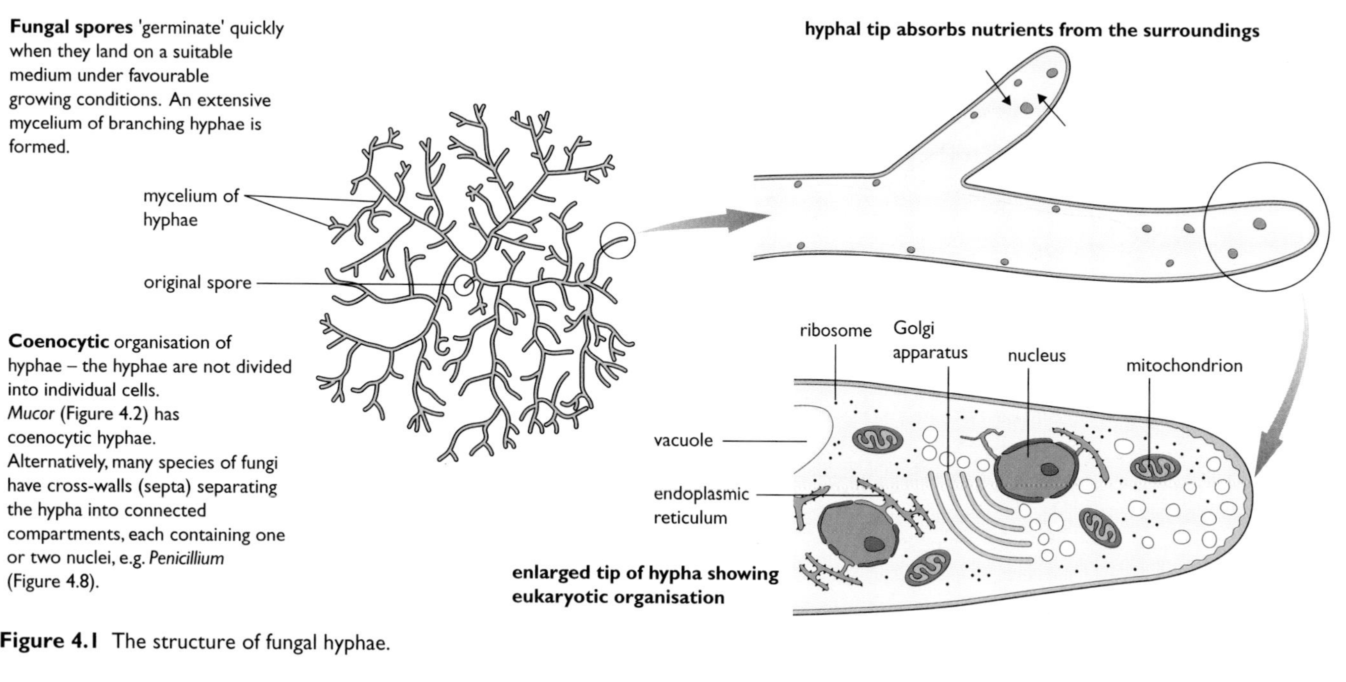

Figure 4.1 The structure of fungal hyphae.

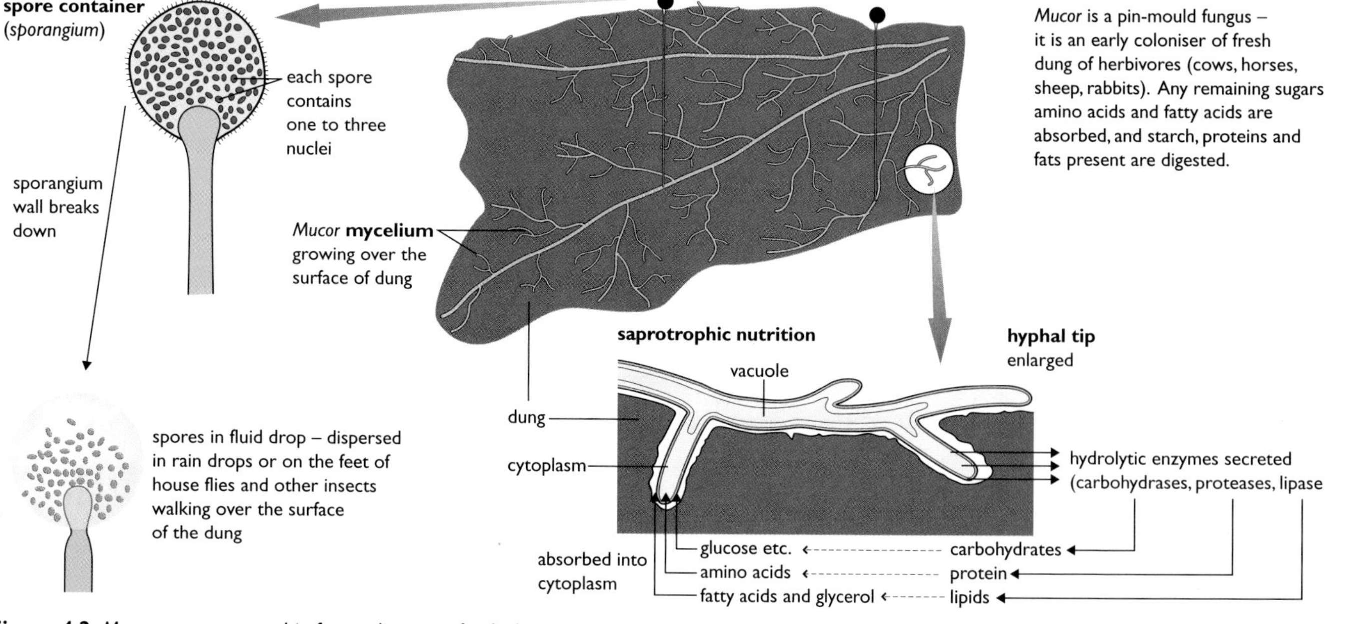

Figure 4.2 *Mucor* – a saprotrophic fungus living on fresh dung.

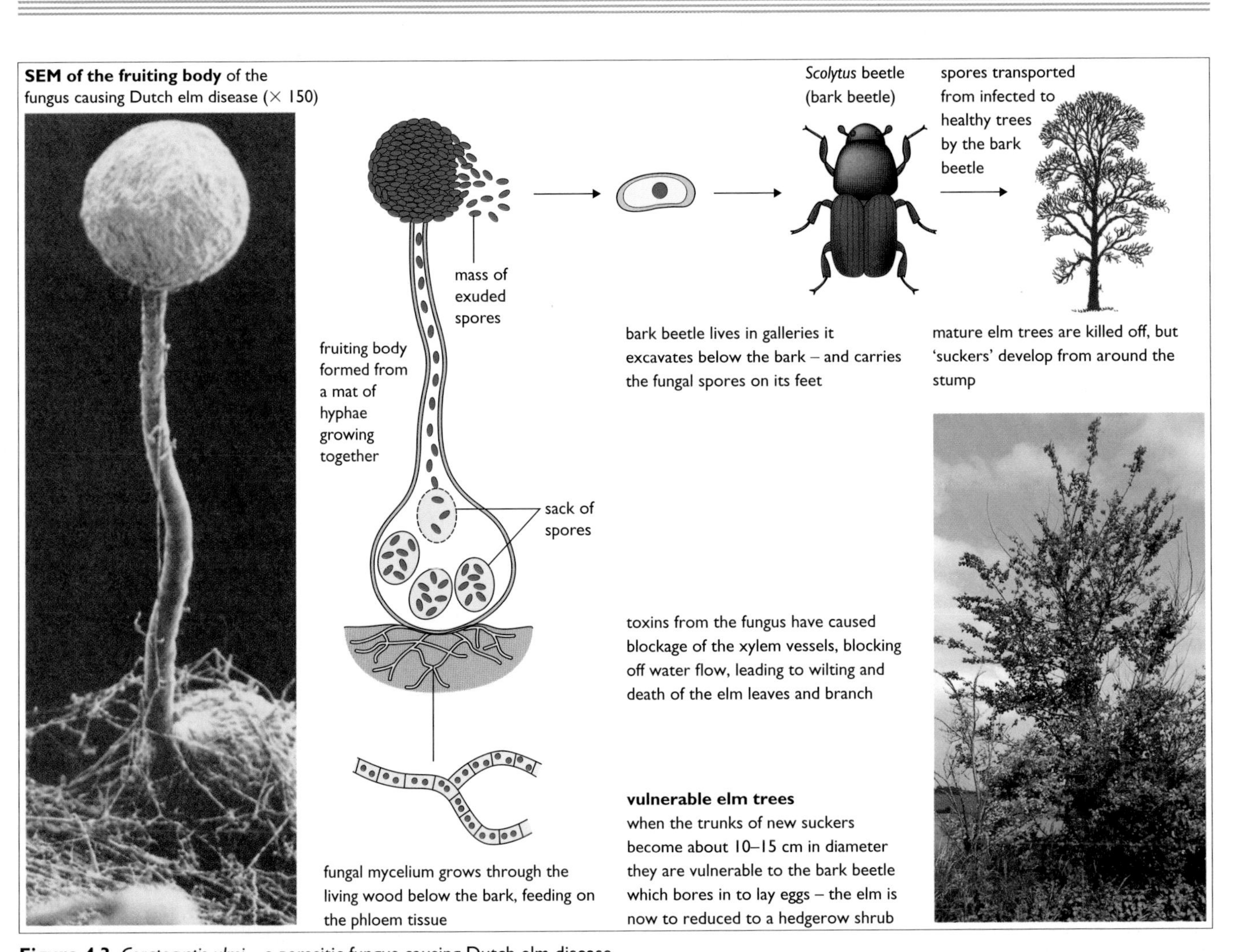

Figure 4.3 *Ceratocystis ulmi* – a parasitic fungus causing Dutch elm disease.

Figure 4.4 *Amanita muscaria* – a mutualistic fungus of birch trees.

Forest trees live in mutualistic association (both organisms benefit) with soil-inhabiting fungi. The fungal hyphae that permeate the soil for a wide area around the tree roots are continuous with those of the compact **sheath** around the roots. This sheath is the site of exchange of nutrients.

this association between a soil-living fungus and tree roots is called a **mycorrhizal relationship**

mycorrhizal root

normal root tips with root hairs

lateral root

mycorrhizal root in section

fungal hypha

mycorrhizal sheath

xylem (water–carrying)

phloem (delivers sugars)

exchange between fungus and root cells occurs here

hyphae in contact with soil particles and soil solution (absorption of ions occurs)

mutually beneficial relationship:

- the tree produces a large excess of sugar and some of this is made available to the mycorrhizal fungus at the sheath
- the fungal hyphae absorb valuable ions ($-NO_3$, $-PO_4$) from a wide area around the tree as they become available, often when the tree's growth is dormant — later the excesses of ions absorbed are supplied to the tree at the sheath in exchange for the sugar

in season (normally autumn) the mycorrhizal fungus forms a mushroom or toadstool fruiting body above the soil, and forms and releases spores from the gills seen on the underside of the cap

Saccharomyces – yeast, a unicellular fungus

Yeasts are saprotrophic fungi that occur everywhere sugar solution is available, such as in flowers, on the surface of fruits, and on leaves and stems where sap is exuded. They are also found in the soil and on animal mucous membranes. The name saccharomyces means 'sugar fungi'.

Yeasts are **unicellular** (Figure 4.5), but under favourable growing conditions the cells reproduce asexually and may divide (bud) so quickly that they temporarily form long, branching chains of connected cells (Figure 4.6).

The yeasts respire by **alcoholic fermentation** and produce carbon dioxide and ethanol as waste products. The ethanol cannot be metabolised further by yeast, and it may accumulate in the growth medium. Yeasts are of great economic importance because:

- waste products of fermentation are exploited in brewing, wine making (**ethanol**, page 44), and in the baking of bread dough (carbon dioxide, Figure 4.7)
- they are easily cultured in liquid media in fermenters (page 48), and so can be useful in modern biotechnology industries
- like many bacteria, they contain **plasmids** in the cytoplasm (very unusual in eukaryote cells), and so yeast cells may be genetically modified to carry and express additional genes (page 59).

2 Yeast shows alcoholic fermentation, whereas some bacteria (page 46) carry out lactic acid fermentation. What do these forms of anaerobic respiration have in common, and how do they differ?

SEM of yeast cell budding (× 4500)

Figure 4.5 Structure of yeast cells.

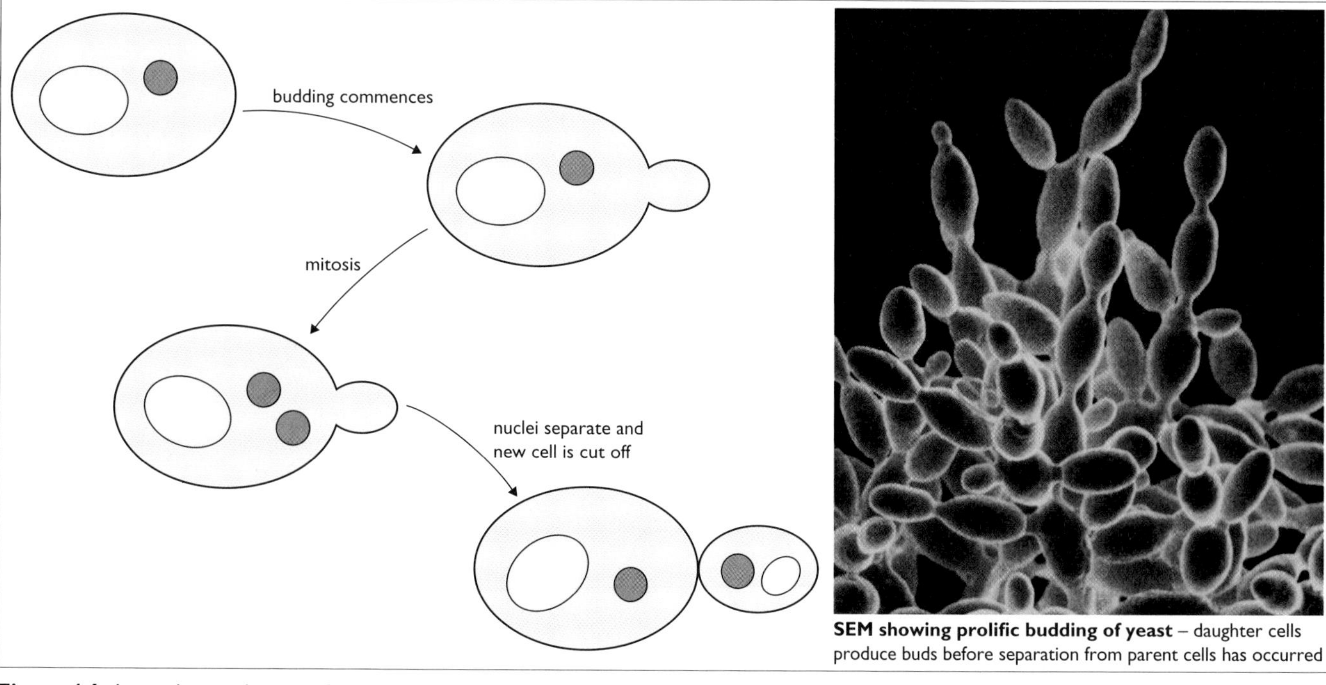

SEM showing prolific budding of yeast – daughter cells produce buds before separation from parent cells has occurred

Figure 4.6 Asexual reproduction of yeast.

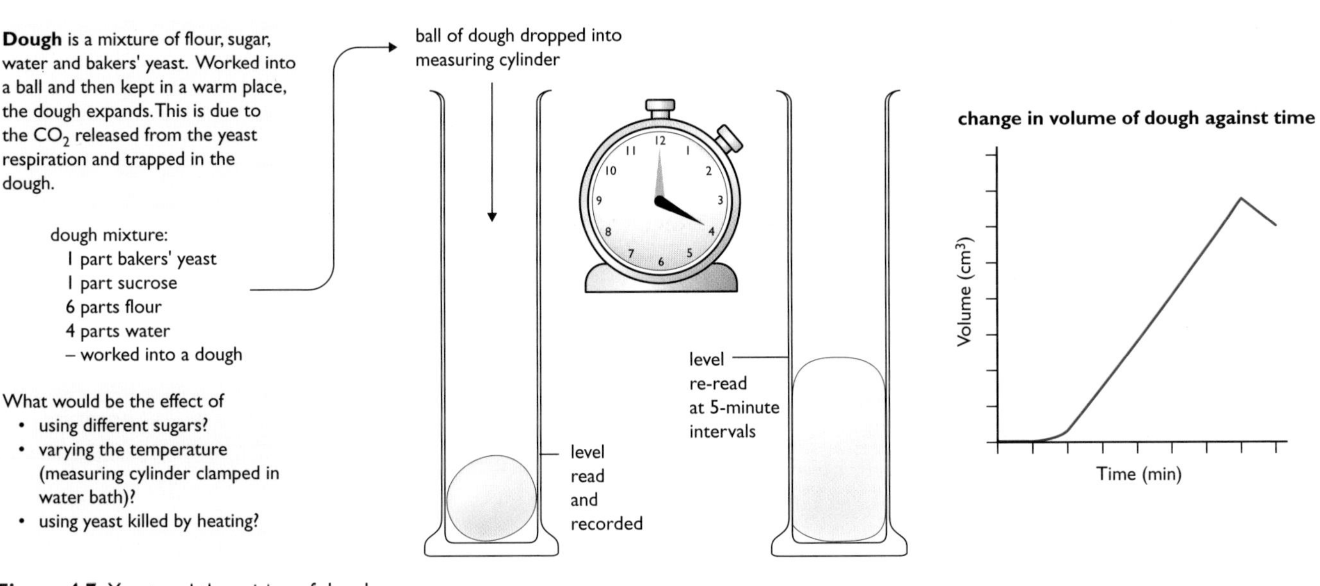

Figure 4.7 Yeast and the raising of dough.

Penicillium, a mould fungus

Penicillium is a very common saprophyte, both of the soil and of decaying organic matter. It is typically seen growing as blue or green moulds on the surface of citrus fruits, which are rapidly destroyed, and on other foodstuffs contaminated by spores. In fact, the spores of *Penicillium* (and a similar species, *Aspergillus*) make up the majority of air-borne fungal spores.

The mycelium of *Penicillium* consists of **hyphae** of multinucleate or binucleate cells, depending on the species. Once established, the fungus produces chains of **spores** on a brush-like head made of branching hyphae (Figure 4.8). The spores are carried away by air currents.

Penicillium species are of great economic importance because:

- many species secrete quantities of citric acid or other valuable **organic acids**, and can be grown in fermenters for commercial production
- species are used in the making of **cheeses**, such as *Penicillium roqueforti* in Roquefort and *Penicillium camemberti* in Camembert. Danish blue and Italian gorgonzola also use *Penicillium* species
- *Penicillium chrysogenum* is used as a source of the antibiotic (page 52) **penicillin** – penicillin works against Gram-positive bacteria by interfering with the biochemistry of their cell wall formation.

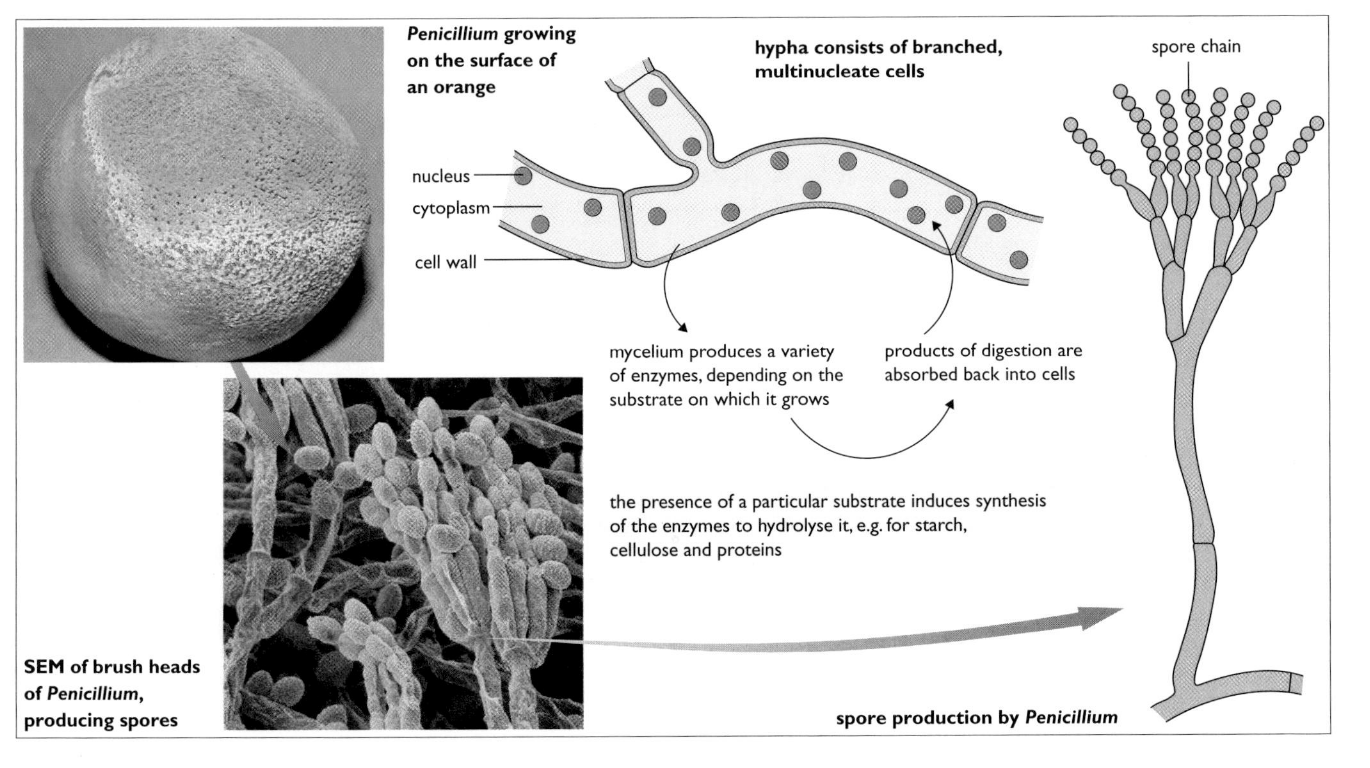

Figure 4.8 Structure of the mould fungus *Penicillium*.

5 Protoctista

The Protoctista are single-celled eukaryotic organisms – namely the **protozoa** or unicellular animals, and the **algae** or simple green plants, together with certain simple multicellular organisms that are related to them. These organisms are classified together because of their relatively **simple level of organisation**, rather than being a natural grouping of closely related organisms. The majority are small enough to be considered as microorganisms.

Protoctista are found in **aquatic** habitats. Many of the Protoctista are of ecological importance in aquatic habitats, or of economic importance in the water supply and purification industries (page 34).

1 Green plants have autotrophic nutrition, but most other organisms are heterotrophs. What feeding strategies are classified as heterotrophic nutrition?

Protozoa

The protozoa typically feed by taking in organic matter and digesting it prior to absorption (holozoic nutrition). However, some protozoa are parasitic and cause diseases of humans (e.g. *Plasmodium* causing malaria, page 82).

Amoeba is a protozoan that lives on the bottom of ponds, feeding on smaller protozoa and algae. Its prey are enclosed in food vacuoles and then digested.

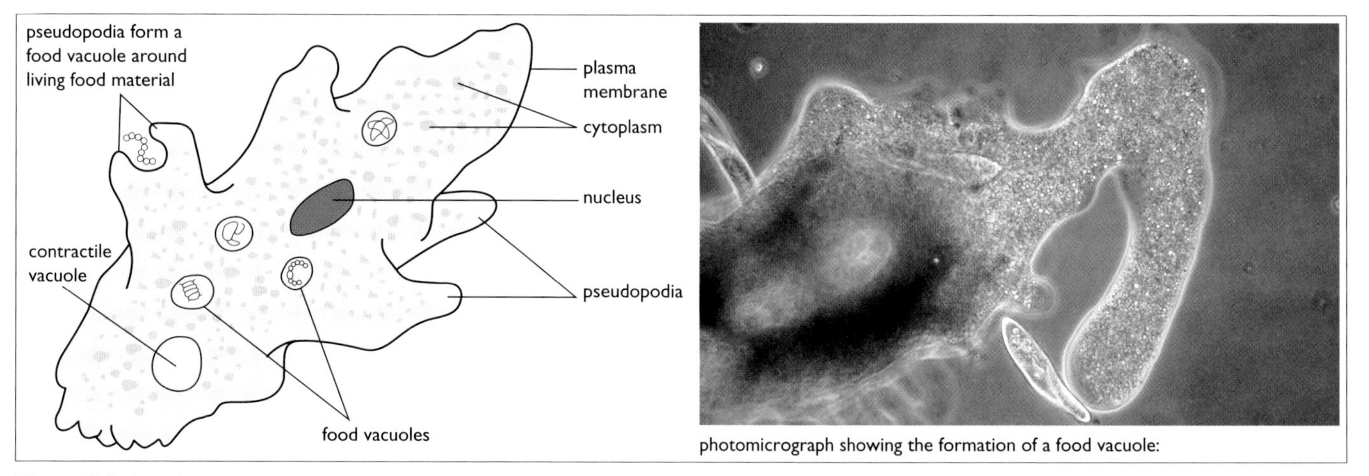

Figure 5.1 *Amoeba.*

Paramecium is a protozoan of ponds that are rich in decaying organic matter and numerous bacteria. These bacteria are the main component of its diet.

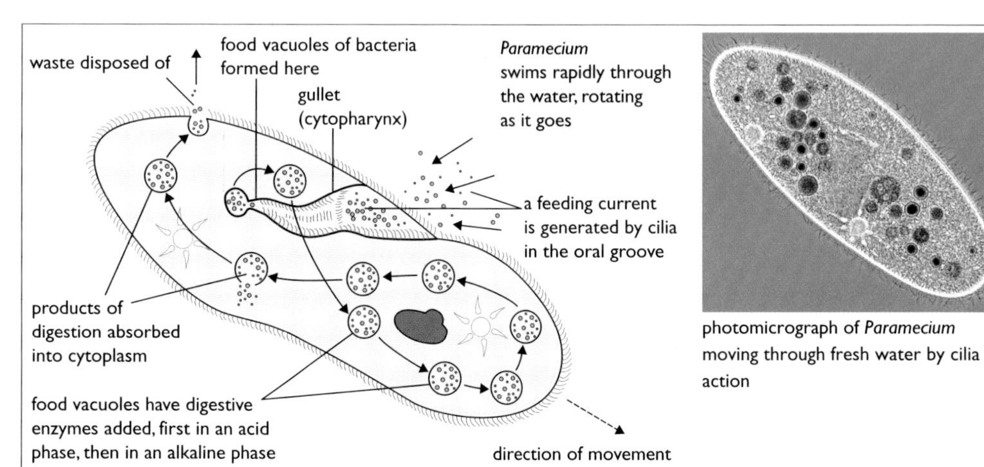

Figure 5.2 *Paramecium.*

Algae

The algae are photosynthetic organisms with a cell structure very similar to that of the green plants, herbaceous and woody (trees), of terrestrial habitats.

Chlorella is a unicellular alga of fresh-water ponds, present at sufficient density to colour the water green. *Chlorella* is a popular organism for use in research, as it is easily cultured in laboratories for the study of green plant metabolism.

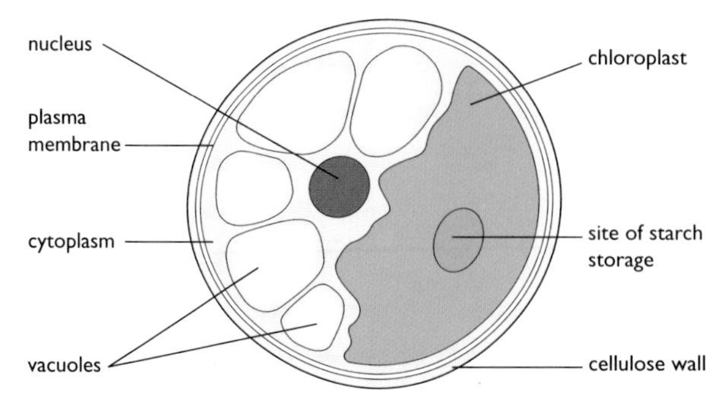

Figure 5.3 *Chlorella.*

Chlamydomonas is a common, motile, unicellular alga of fresh water rich in ammonium ions (e.g. water contaminated by farmyard sewage).

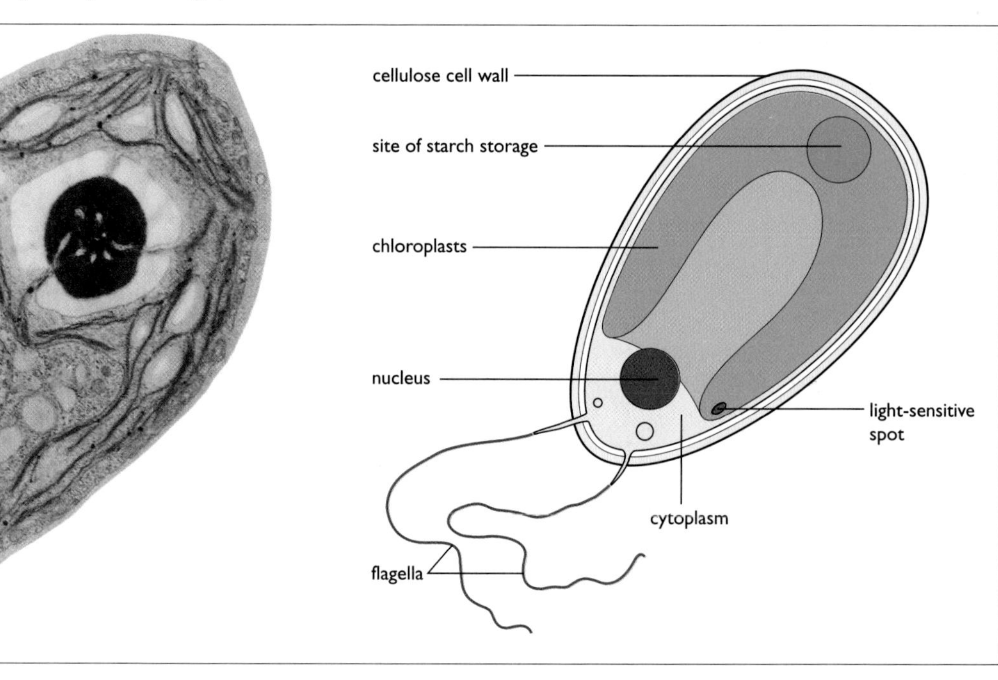

Figure 5.4 *Chlamydomonas*.

Volvox is a **colonial** form – a multicellular alga with the cells arranged in spherical, hollow colonies.

this green, spherical, hollow colony ranges from 5000–60 000 *Chlamydomonas*-like cells supported in a gelatinous matrix. Cells are joined by cytoplasmic strands, and some responses, e.g. movements of the flagella, may be co-ordinated. The colony reproduces by forming smaller colonies within it, and these eventually escape. (× 550)

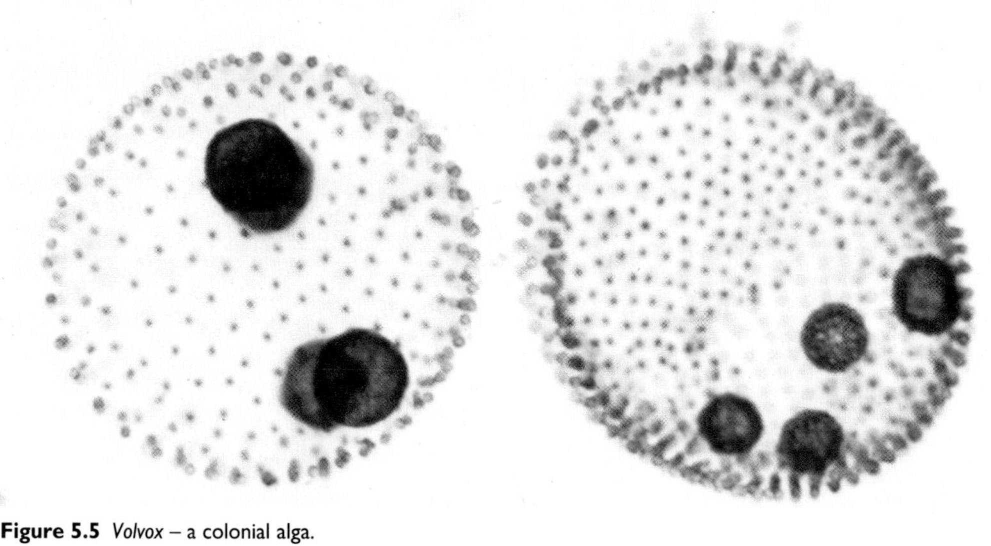

Figure 5.5 *Volvox* – a colonial alga.

Spirogyra is one of three or four commonly occurring, unbranched, **filamentous** algae forming conspicuous floating masses at the surface of ponds at certain times of the year. Bubbles of oxygen gas given off by the algal cell in the light (photosynthesis) are often trapped among the matted algal filaments, making them buoyant.

Spirogyra is a long, unbranched filament. The cellulose cell wall forms a strong cylinder, but the cells can become kinked and broken, and the filament fragmented.

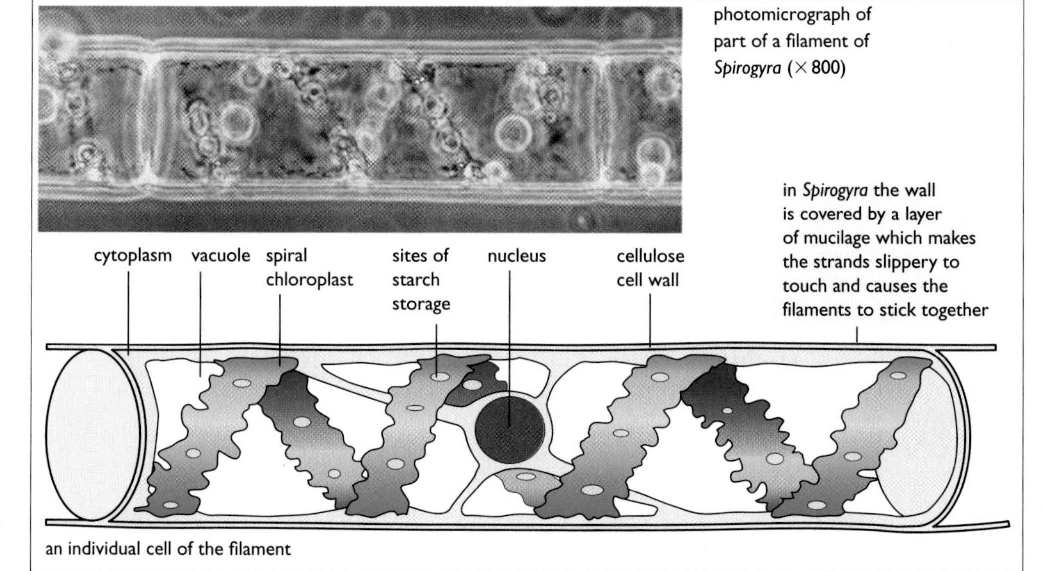

Figure 5.6 *Spirogyra* – a filamentous alga.

6 Microorganisms and the environment

Cycling of nutrients

When organisms die, their bodies are broken down to simpler substances (e.g. CO_2, H_2O, NH_3, various ions, etc.) by a succession of organisms (mostly microorganisms), as illustrated in Figure 6.1. Scavengers and detritivores often begin the process. Ultimately both plants and animals depend on the activities of saprotrophic microorganisms to release matter from dead organisms for re-use. As the supply of chemical elements available to build up nutrients is finite, this recycling is essential for life. Microorganisms are the key movers in the vital recycling that goes on, largely unseen, in the environment (Figure 6.2).

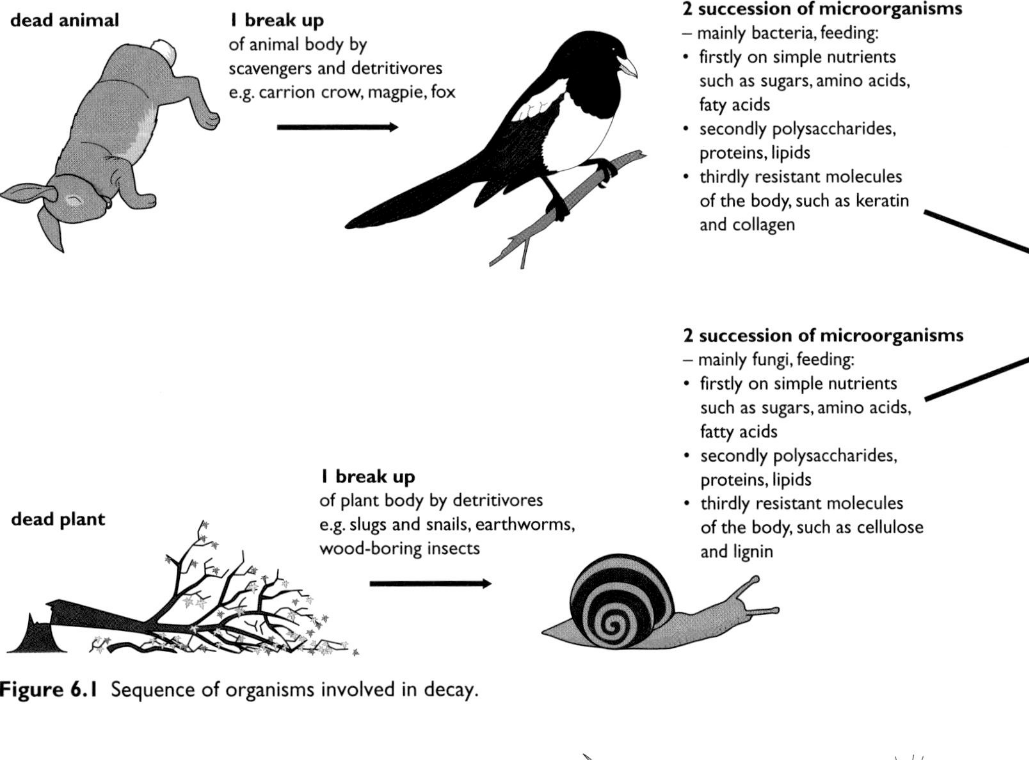

Figure 6.1 Sequence of organisms involved in decay.

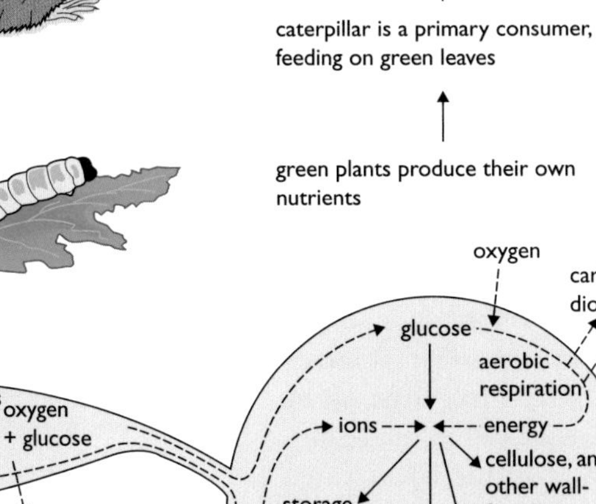

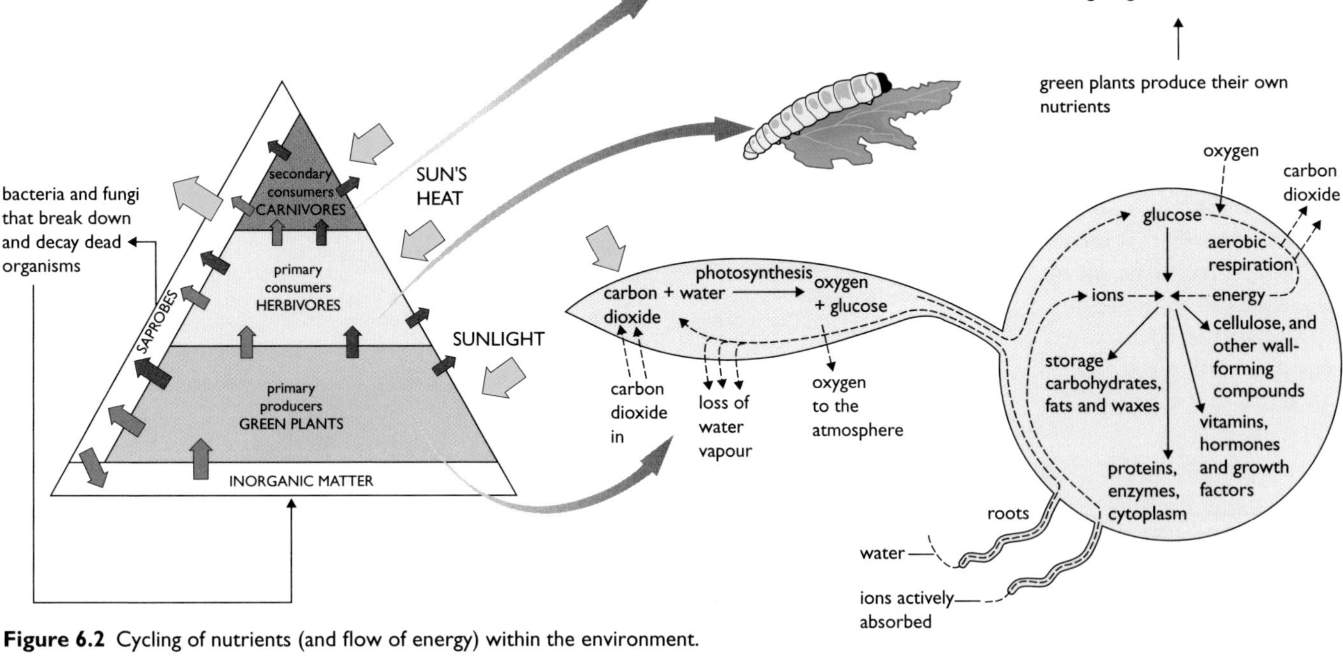

Figure 6.2 Cycling of nutrients (and flow of energy) within the environment.

Decomposition of dung

The faeces of animals are broken down and the elements recycled in much the same way as whole organisms are after death. For example, dung of herbivores is decayed by a succession of fungi (decomposers), many of which are already present as spores in the faeces when discharged from the herbivore. Spores tend to 'germinate' in sequence, according to the most accessible nutrients present in the dung (Figure 6.3). Fresh spores produced on the dung are explosively discharged, or blown or carried away. Many become attached to surrounding grass, and in this position are more likely to be ingested by browsing herbivores, so repeating the cycle.

1 Silage, like compost, is made from plant matter – but silage is a valuable winter food for cows. Use the internet and search under 'silage making' to create a presentation (e.g. a wall 'newspaper' page) on the making and food value of silage.

Figure 6.3 A succession of dung-decomposing fungi on herbivorous dung.

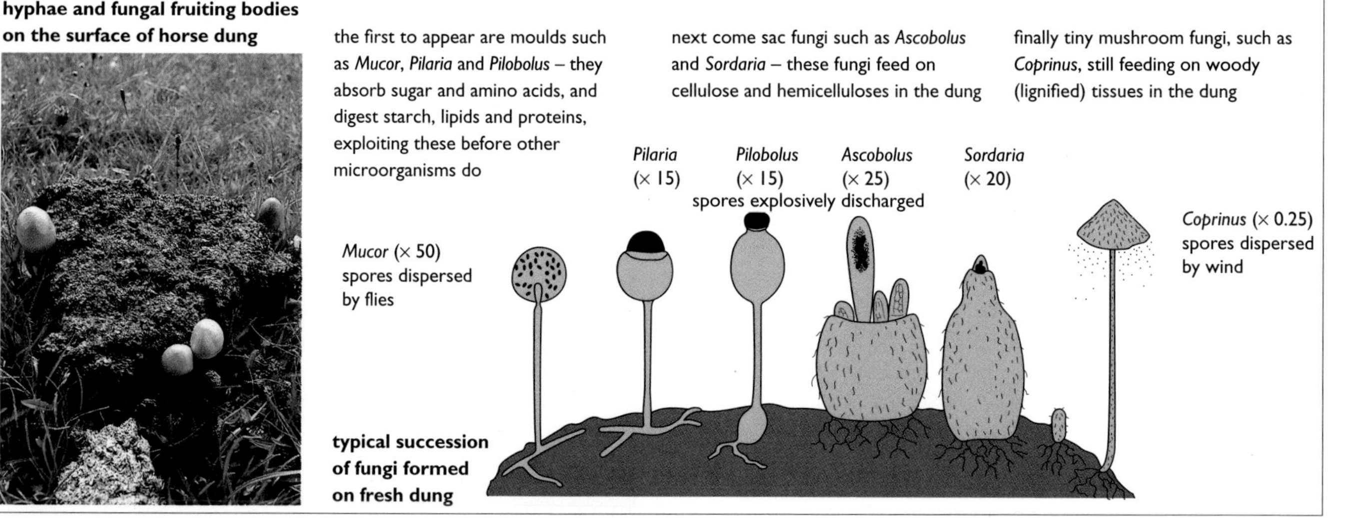

Decomposition of compost

In the soil, decomposition of dead organic matter occurs quite slowly. In a compost heap, however, the process is speeded up by maintaining a compact mass of organic matter held under aerobic conditions. Much of the heat generated is retained. If high enough temperatures occur, the seeds of weeds are killed in the process.

Figure 6.4 A garden compost heap and the sequence of changes as decay occurs.

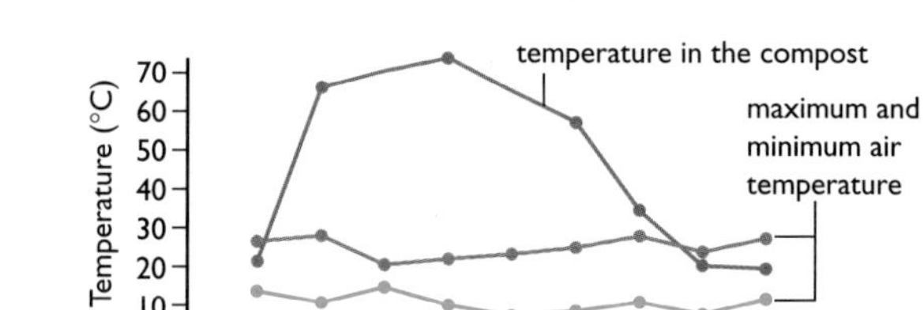

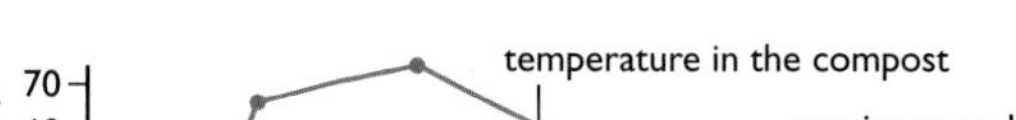

sequence of microorganisms commonly present – microbes that thrive at:

5–30 °C
e.g. *Rhizopus nigrans*
feed on: sugars and starch

10–30 °C
e.g. *Aspergillus fumigans*
feed on: hemicellulose and cellulose

20–60 °C
e.g. *Chaetomium termophile, Humicola langinosa*
feed on: cellulose and 'secondary' sugar (formed) by extra cellular digestion by other microorganisms present

organisms thrive best at different temperatures: **cryophiles** at low temperatures: **mesophiles** at mid-range temperatures; **thermophiles** at high temperatures; **hyperthermophiles** at extremely high temperatures

Summary of carbon recycling

The cyclical processes by which essential elements of living things are released and re-used are called **biogeochemical cycles** (as the changes also involve other living things, and non-living components such as soil, air and water). All essential elements take part in biogeochemical cycles. The carbon cycle is one example.

Figure 6.5 The carbon cycle.

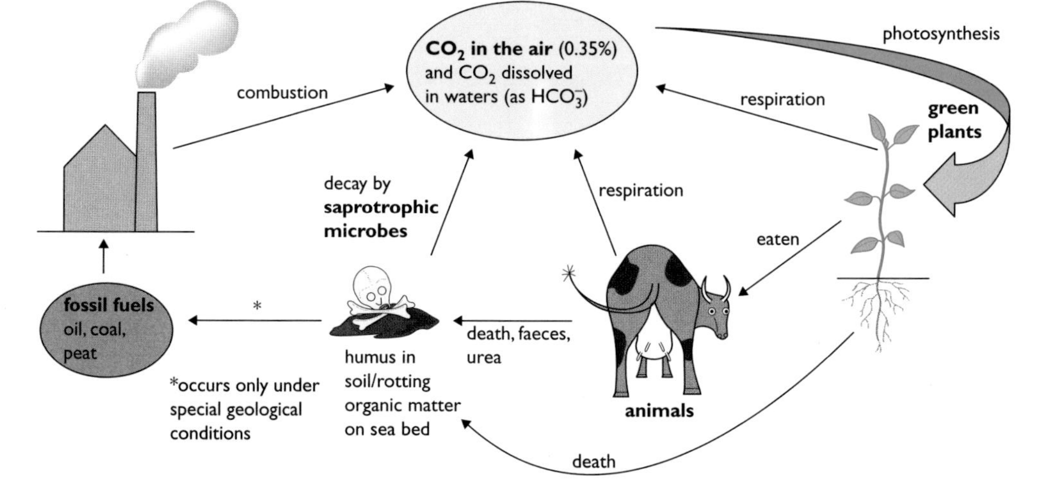

Microorganisms and cellulose breakdown

Cellulose consists of glucose molecules condensed together in long, unbranched, parallel chains, packed in bundles called microfibrils and held together by hydrogen bonds. The whole is an extremely strong structure, and gives great strength to the walls of green plant cells, of which cellulose is the major component.

Herbivores break through plant cell walls to reach the nutrients contained in the cells. Breakdown of cellulose is partly mechanical (e.g. by the grinding action of a herbivorous mammal's teeth, or of the mandibles of some insects). However, breakdown of cellulose by enzymes completes the task and releases soluble sugar molecules (themselves an important nutrient) from which cellulose is built up.

2 Hydrogen bonds are an important component of substances such as cellulose. What are hydrogen bonds, and where are they formed?

cellulase
cellulose — (complex change, not a single step) → glucose

Unfortunately, few animals can produce a cellulase, and certainly no mammals do. However, many herbivores exploit microbes that secrete cellulase. Suitable micro-organisms are found in a part of the herbivore's gut, and many of the products of cellulose digestion by these microorganisms are absorbed into the blood stream of the host organism.

Figure 6.6 Transmission electron micrograph of cellulose fibres of plant cell walls (× 30 000).

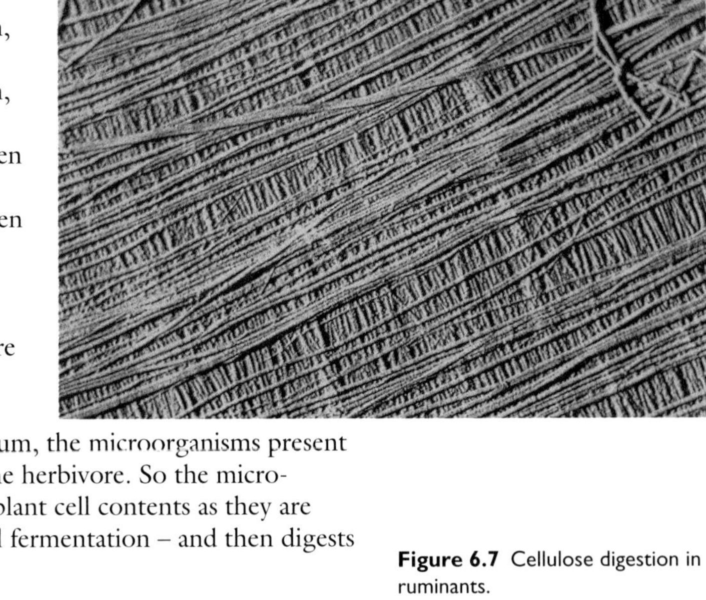

Cellulose digestion by ruminants

Sheep, cows and other ruminants have a four-chambered 'stomach'. The first three chambers, known as the rumen, reticulum and omasum, are a modified part of the lower oesophagus. The final chamber, known as the abomasum, is the true stomach. In the rumen are the bulk of the microorganisms that convert cellulose to glucose, and then glucose to organic acids. Much of the organic acid is immediately absorbed into the animal's blood stream, then used as an energy source, for example in the muscles and the secretory cells of the udder. The rumen is a naturally occurring **fermentation vessel** (page 49). Here other microorganisms synthesise vitamins, and some bacteria are able to produce proteins for themselves, using the ammonium salts present.

Later on, once the digesting food reaches the abomasum, the microorganisms present are themselves digested, and products are absorbed by the herbivore. So the micro-organisms in a ruminant feed on grass. The cow digests plant cell contents as they are released, but also benefits from the products of microbial fermentation – and then digests the microorganisms, too!

Figure 6.7 Cellulose digestion in ruminants.

(3) the fermented grass is passed to the **reticulum** where it is formed into balls known as cud, which is regurgitated to the mouth for further grinding up by the teeth before being swallowed again

(4) in the **omasum** water from the cud is reabsorbed (typically a cow secretes approximately 150 litres of salivia per day) and the more solidified food material is passed to the abomasum

to small intestine
oesophagus
four chambered stomach

(1) grass cropped, ground up by premolars and molars, and mixed with saliva in the **mouth** then swallowed, and passed to the rumen

(2) the **rumen** functions as a 'fermentation vat' under anaerobic conditions – a huge flora of bacteria produce cellulase and other digestive enzymes, and convert the cellulose present to sugars which are fermented to organic acids which are then absorbed into the blood through the rumen wall (waste products, CO_2 water vapour and methane, all greenhouse gases, are belched out)

(5) in the **abomasum** normal gastric secretions begin digestion of the proteins from plant cells, bacteria and ciliates

abomasum
omasum
oesophagus

(6) the chyme now passes to the **duodenum** where the bulk of digestion is completed

duodenum
reticulum
rumen

Cellulose digestion by non-ruminants

In rabbits and horses, the cellulose-digesting bacteria are housed in the caecum and appendix. During the journey through the gut, the digesting food is diverted to the caecum – in by peristalsis, out by reverse peristalsis. Here cellulose is largely converted to organic acids. In the case of the horse, these are absorbed in the remainder of the gut (the colon).

In rabbits, the faecal pellets emerge from the anus overnight and are eaten. They are then re-digested. Nutrients obtained from digestion of the plant cell contents, and the organic acids previously released by bacterial action in the caecum, are absorbed in the small intestine. When the waste is finally eliminated it forms hard, dry faecal pellets, typically seen on grassland where rabbits feed.

the caecum and appendix are a cul-de sac in the gut, between small intestine (ileum) and large intestine (colon)

oesophagus → stomach → duodenum → ileum → colon → rectum
caecum
appendix
colon
appendix
ileum
caecum
site of cellulase-secreting microorganisms, held in anaerobic conditions

Figure 6.8 Cellulose digestion in non-ruminants.

Cellulose breakdown in the soil

Cellulose is the most abundant organic compound in the biosphere; it makes up more than 50% of all organic carbon (CO_2 is inorganic carbon). This cellulose is, in effect, a reservoir of carbon. In tropical countries most carbon is stored in forest vegetation, but elsewhere in the world it is the soil that contains most organic carbon. Cellulose occurs as leaf litter on the soil surface, and as other rotting plant material such as peat in moorland bogs. The carbon of peat is only very slowly released into the atmosphere as CO_2 by the activities of saprotrophic microorganisms able to produce cellulase enzymes (mostly fungi).

Currently, the earth's temperature appears to be warming slightly. One consequence of **global warming** is that microbial metabolism is speeded up. This will enhance the natural release of CO_2. Since CO_2 is a greenhouse gas, more of it in the atmosphere may tend to cause temperatures to rise still further.

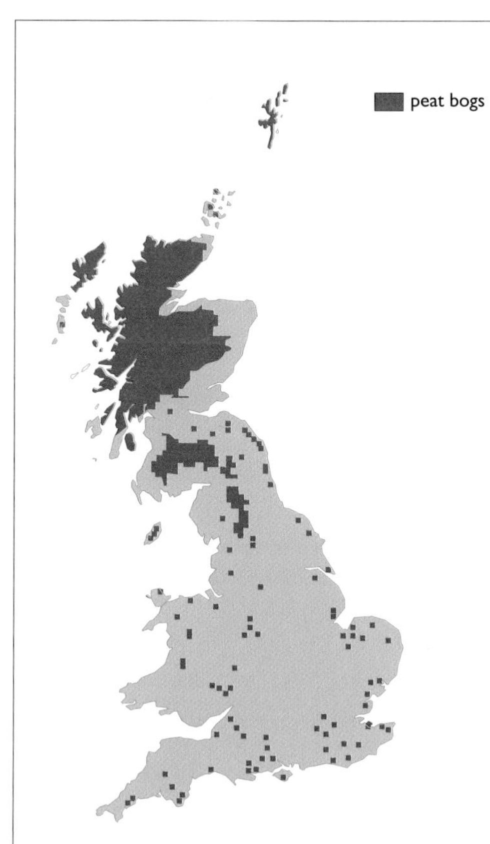

more than 75% of all the carbon locked up in organic matter in British soil and vegetation occurs as peat, most of which is found in Scotland

a mountainside environment in the north of Britain – peat is forming from the moss *Sphagnum*, which accumulates because decay is slowed down by the low pH, high rainfall, low temperatures and poor drainage

Figure 6.9 Peat as an important reservoir of carbon.

Sewage treatment and the production of re-usable water

Sewage treatment

Sewage is the fluid waste of human communities from houses and commercial properties. It is piped to sewage works for processing. Sewage consists largely of used water, faeces and urine, and is rich in organic matter with ammonium, nitrate and other ions present. It contains vast numbers of microorganisms, many harmless saprotrophic ones, but also pathogenic bacteria (page 67).

Treatment of sewage to produce water safe to return to the environment involves the metabolism of vast numbers of microorganisms. During treatment (Figure 6.10), sewage is first **screened** for solid rubbish (grit, timber, car tyres, etc.), and is then separated into **liquid effluent** and **solid matter**. Liquids are cleaned by aerobic metabolism of a mixture of bacteria, fungi and protozoa that are known as activated sludge. Solids are anaerobically digested to break down the organic matter to methane and carbon dioxide.

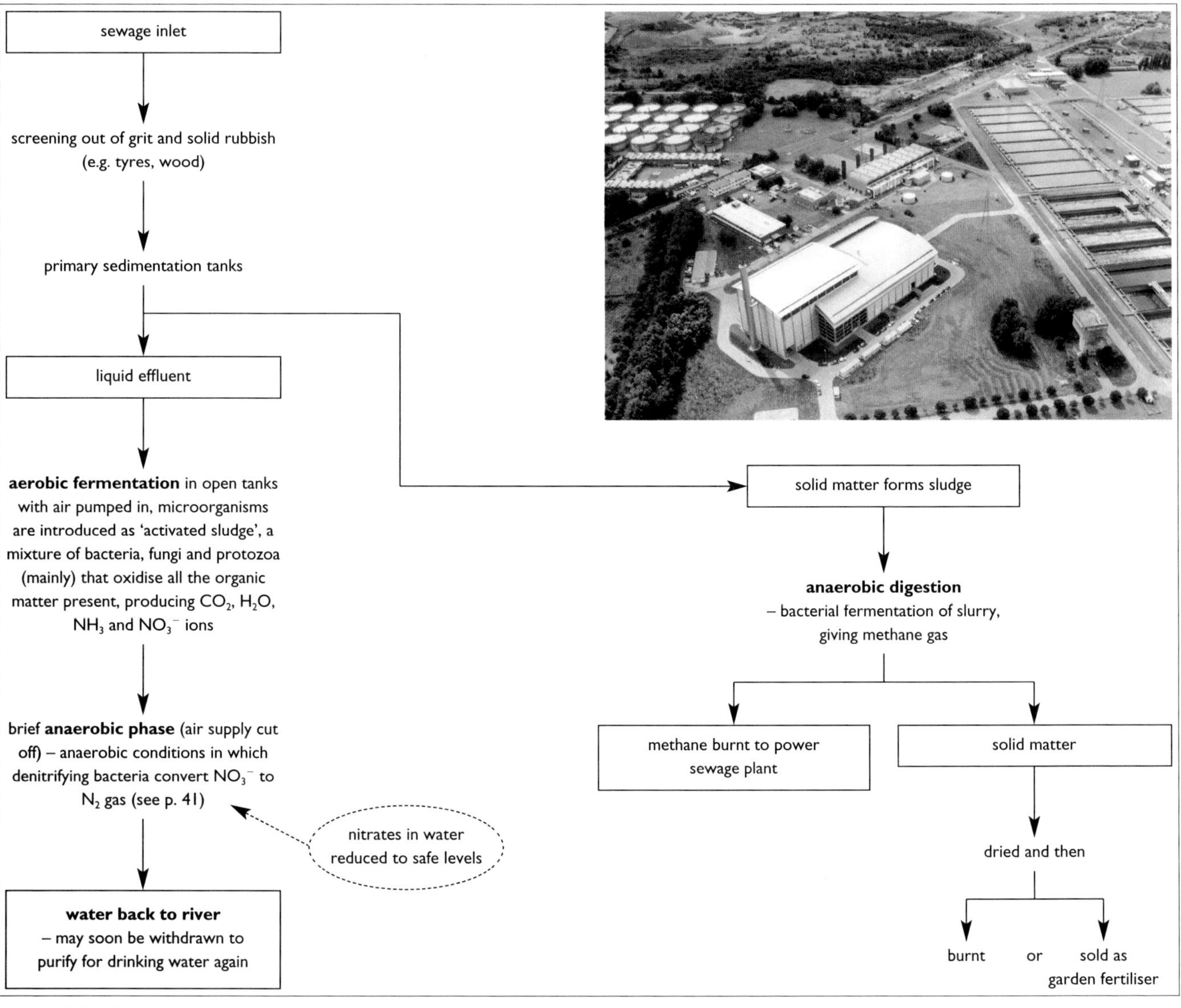

Figure 6.10 Sewage treatment.

Extension: sewage contamination of river water is harmful

A river contaminated with raw sewage is made rich in organic matter. This causes saprotrophic bacteria to flourish in the polluted water (and sewage breakdown starts). However, only a limited amount of oxygen dissolves in water, and the active saprotrophic bacteria quickly use it all up. So, sewage added to a river causes anaerobic (anoxic) conditions there.

The absence of oxygen then causes the death of many aquatic aerobic organisms, such as fish, and anaerobic bacteria start to flourish. Anaerobic decay of organic matter produces other waste products, including hydrogen sulphide. This highly soluble gas is extremely poisonous, and kills much of the flora and fauna of the river. As the water flows on, sewage is slowly broken down, oxygen returns, and eventually the river community recovers.

Water purification for domestic supplies

Water delivered to homes is regarded as pure if:

- it doesn't transmit water-borne diseases such as typhoid or cholera (page 70)
- it is bright and clear
- it is free from chemical poisons
- it is reasonably 'soft', i.e. it forms a lather with soap.

aerial view of London's water supply where it is taken from the River Thames

one of the Elan Valley reservoirs providing water for Birmingham

much of our drinking water is drawn from rivers – often downstream of points where the effluent from sewage works (from towns upstream) has been added

↓

storage reservoir – settling out of fine particles

↓

aeration by passing through fountain or waterfall

↓

filtration by slow flow through fine sand beds

↓

chlorination to kill any residual bacteria, using chlorine at 0.5 p.p.m.

↓

removal of excess chlorine – by SO_2 treatment

↓

temporary storage in service reservoirs close to points of distribution

Figure 6.11 Water supply and the steps to water purification.

Water supply, a major world challenge

The Earth contains an estimated 1400 million cubic kilometres* of water, and yet the water supply to many human communities in parts of the world is inadequate.

The bulk of the world's water occurs in the oceans or as groundwater. Groundwater may originate as rain, and percolate through soil to porous rocks. Such groundwater is a vital human resource. Other groundwater originated in porous rock below the sea, and is saline (not suitable for drinking). Consequently, the treatment of contaminated waters (such as sewage and the waters of accessible rivers and lakes), to make them safe for re-use by human communities, is an important technology. The world's water supplies are represented in Figure 6.12.

* 1 cubic kilometre = 10^{12} litres.

3 In water rich in nitrates and phosphates, algae tend to bloom. What does this mean, and why is it a threat to other organisms?

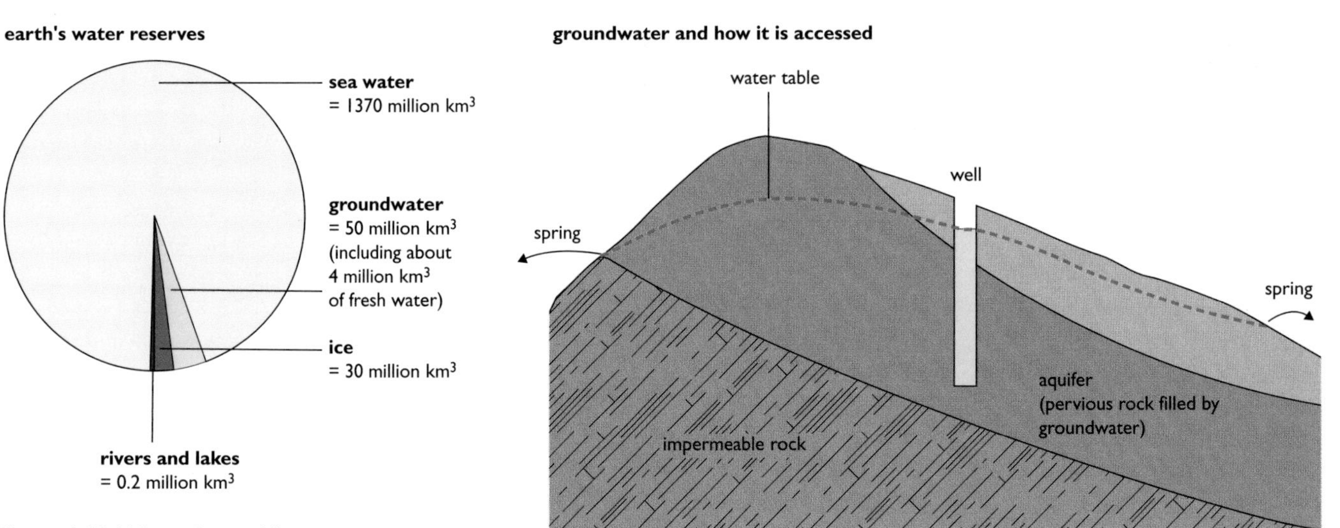

Figure 6.12 Where the world's water occurs.

Microorganisms and nitrogen cycling

The importance of combined nitrogen.

Proteins are essential components of living things, used as enzymes and in the membranes of cells. Proteins are built up from amino acids. Amino acids contain the element **nitrogen** combined with carbon, hydrogen and oxygen. Combined nitrogen also occurs in nucleic acids (DNA and RNA).

Nitrogen abounds in the environment – it makes up about 80% of the atmosphere. But nitrogen gas (N_2) is not available to most organisms to use in metabolism; most require nitrogen to be combined with other elements in compounds (green plants as nitrates or ammonia; animals as amino acids from the digestion of proteins).

However, some microorganisms can fix atmospheric nitrogen to form ammonia, and then combine it with organic acids, forming their own amino acids. This is known as **nitrogen fixation**. Nitrogen fixation requires energy (as ATP) and hydrogen (reducing power) from respiration. The enzyme involved is nitrogenase.

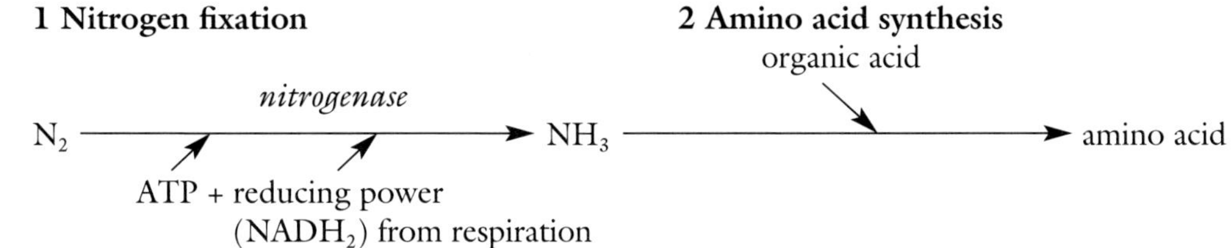

Nitrogen fixation by free-living microorganisms

Nitrogen-fixing microorganisms that are free-living (not occurring in mutualistic association with other organisms; page 25) include certain species of **bacteria**, e.g. *Azotobacter* species, *Clostridium* species, and many species of **cyanobacteria** (photosynthetic bacteria) such as *Nostoc* and *Anabaena* species. Here nitrogen fixation is illustrated in *Anabaena*.

The cyanobacterium Anabaena

Anabaena, a filamentous **cyanobacterium** of pond water and damp soils, is widely distributed on the Earth's surface (page 16). The cells of the filaments are mostly small, and are the site of photosynthesis, but larger, thick-walled cells, called **heterocysts**, also occur. These are the sites of nitrogen fixation.

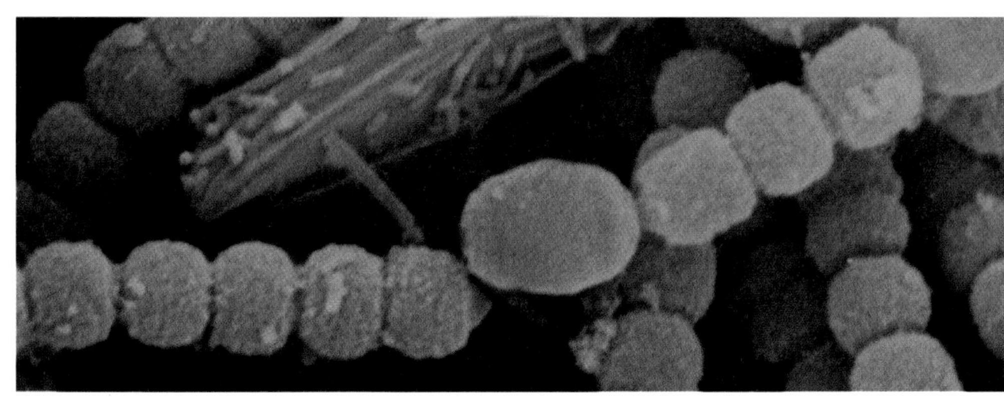

Figure 6.13 SEM of filament *Anabaena* species showing a heterocyst.

How nitrogen is fixed

Sugar is manufactured by photosynthesis in the ordinary cells of the filament. Some of this sugar passes into the heterocysts and is respired, producing ATP and reducing power – the reduced hydrogen acceptor known as $NADPH_2$. Also in the heterocysts, light triggers the formation of additional ATP by a modified form of photosynthesis (known as cyclic photophosphorylation) in which no oxygen gas is formed.

Nitrogen gas diffuses into the heterocysts via the ordinary cells and their cytoplasmic connections. Here the enzyme nitrogenase reduces dinitrogen molecules (N_2) to ammonia. (The enzyme is inhibited by oxygen, and so nitrogen fixation does not occur in the ordinary cells of the filament.) The ammonia reacts with organic acids in the presence of enzymes, forming amino acids, and these are built up into proteins (and into the other nitrogen-containing metabolites the cells require). As a result of nitrogen fixation, *Anabaena* can grow in environments where combined nitrogen ions (nitrate ions and ammonium ions) are absent or in very low supply.

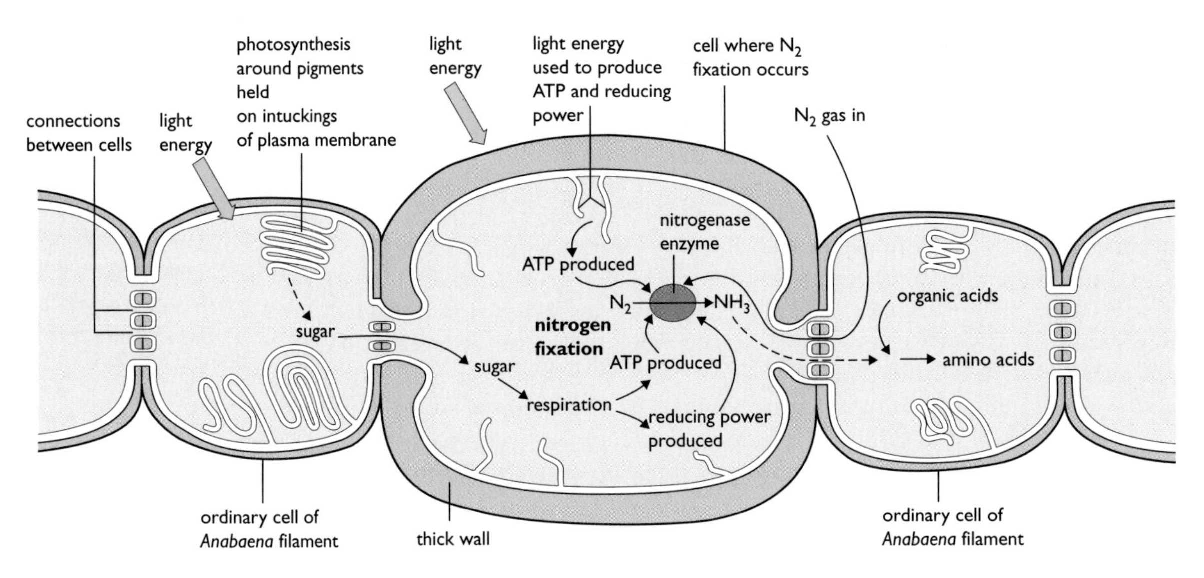

Figure 6.14 The process of nitrogen fixation in *Anabaena* sp.

Economic importance of nitrogen fixation by free-living microorganisms

The combined nitrogen of free-living microorganisms is released into the environment when they die and decay. Nitrates and ammonium ions are then available to other green plants to absorb. So cyanobacteria and bacteria that are nitrogen-fixing species provide a form of green manure or fertiliser for the plants around them.

Cyanobacteria mats in paddy fields where rice is grown

Rice (*Oryza sativa*) is the principal food of about 60% of the human population. It has been grown intensively in paddy fields in tropical and sub-tropical parts of the world for more than 2000 years – long before humans understood about plant metabolism and the importance of fertilisers. It is estimated that about 625 g of combined nitrogen are fixed annually by cyanobacteria per km^2 of the paddy fields of the world. This has permitted repeated harvesting of successful rice crops, as many as the growing seasons permit.

4 How is the combined nitrogen, formed by *Anabaena* in nitrogen fixation, made available for use by plants and animals growing in the same habitat?

Cyanobacteria in subsistence agriculture of mountain terraces

Subsistence farming on poor soils of mountainsides in arid regions is often made possible by terracing and the building of irrigation channels. Typical crops may include *Triticum dicoccum* – durum or macaroni wheat. Limited water supplies are shared between terraces, according to the needs of the crops. Cyanobacteria may also survive on the soil surface.

high terracing – protects the lower slopes in periods of very heavy rain/snow melt

terracing creates small fields – surrounded by irrigation channels

a common crop is durum wheat

river – tapped higher up the valley; channels deliver water to the irrigation system here

Figure 6.15 Subsistence agriculture in the High Atlas Mountain foothills, Morocco.

Mutualistic nitrogen-fixing microorganisms

What is mutualism?

Mutualism is the name for a close **association between two organisms from which both organisms benefit**. An example given earlier is the association between tree roots and some fungi (mycorrhiza: Figure 4.4, page 25).

The bacterium *Rhizobium* commonly occurs in the soil around plant roots, where it feeds saprotrophically on dead organic matter. But it is also able to enter the roots of leguminous plants (clover, peas, beans, soya and many others) growing nearby (Figure 6.16). Here it causes the host tissues to form into a nodule around the cells 'infected' with the bacterium. *Rhizobium* produces the enzyme nitrogenase, and reduces nitrogen gas to ammonia, just as free-living nitrogen-fixing cyanobacteria do (page 36). In the case of *Rhizobium*, energy and reducing power are obtained by respiration of sugars taken from the host plant. Again, the ammonia is combined with organic acids to form amino acids, but here many of the amino acids pass out to the surrounding cells and are used by the host plant.

5 What flowering plant family do the leguminous plants belong to? Name three green vegetables frequently eaten in the UK that are legumes, and three common wild flowers that belong to this family.

Root nodules of leguminous plants – economic importance

Because of their association with the bacterium *Rhizobium*, leguminous plants can grow and flourish in poor soils. As crop plants they are found to be rich in proteins (Table 6.1).

Consequently, leguminous crops are vitally important to vegetarians, and to communities otherwise suffering from shortages of protein in their diets, such as those living by subsistence farming on very poor soil. In fact, rich and poor communities alike, all around the world, make good use of leguminous plants:

- as fodder for farm animals – e.g. grass with clover; grass with lucerne (known as alfalfa in the USA)
- as crops producing food for human consumption – e.g. pea, bean, peanut, soya bean, chickpea.

Table 6.1 Protein content of some important foods of plant origin

Plant	Protein content/%
Non-leguminous	
Whole rice grain	7.5–9.0
Wheat flour	9.5–13.5
Potato (dry weight)	10.0–13.5
Cassava	1.0–2.0
Leguminous	
Chickpea	22–28
Soya bean	33–42
Red kidney bean	20–23
Peanut	25–30

Figure 6.16 Root system of a leguminous plant showing root nodules.

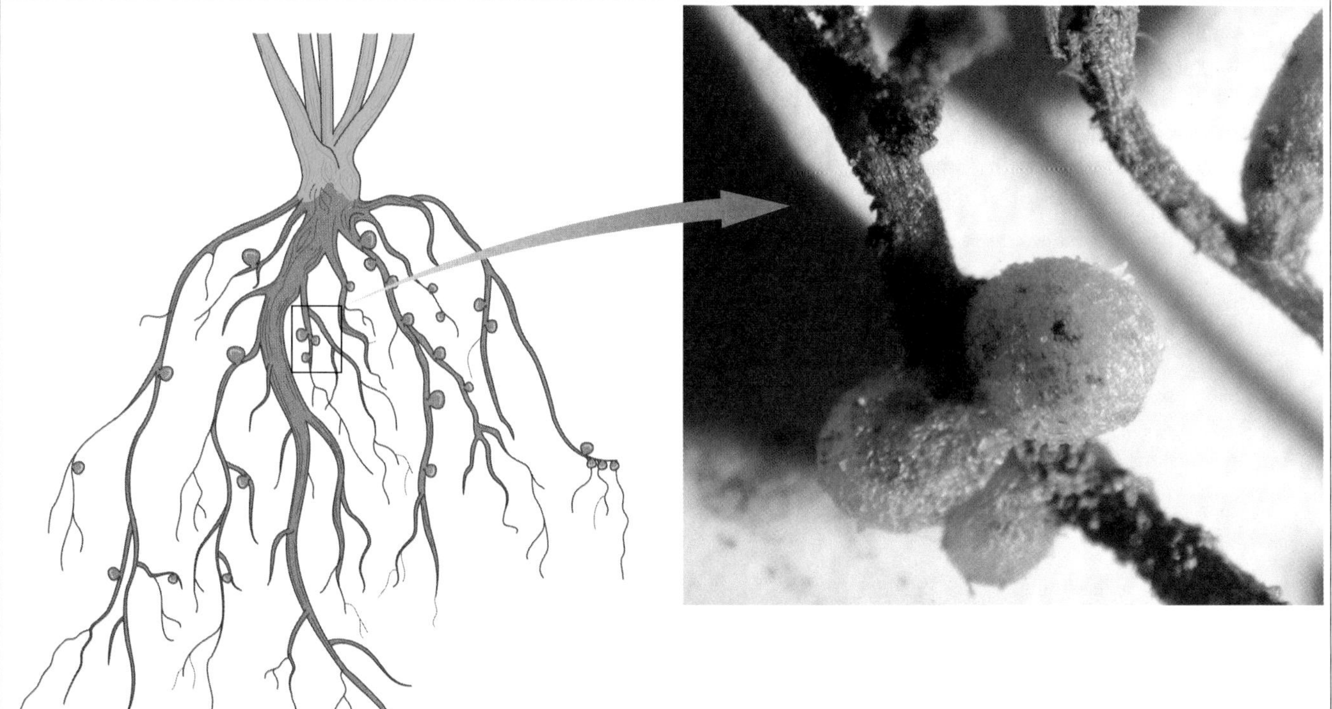

Other species with mutualistic nitrogen-fixing microorganisms

Several families of non-leguminous plants have similar mutualistic associations with root-infecting bacteria. For example, the alder tree (*Alnus glutinosa*) has root nodules which house the nitrogen-fixing microorganism *Frankia*. Alder trees grow well on soils poor in nutrients. They are commonly found on river banks where their root systems resist waterside erosion.

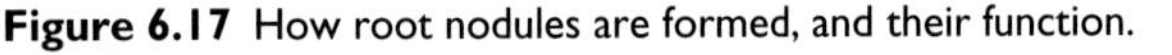

Figure 6.17 How root nodules are formed, and their function.

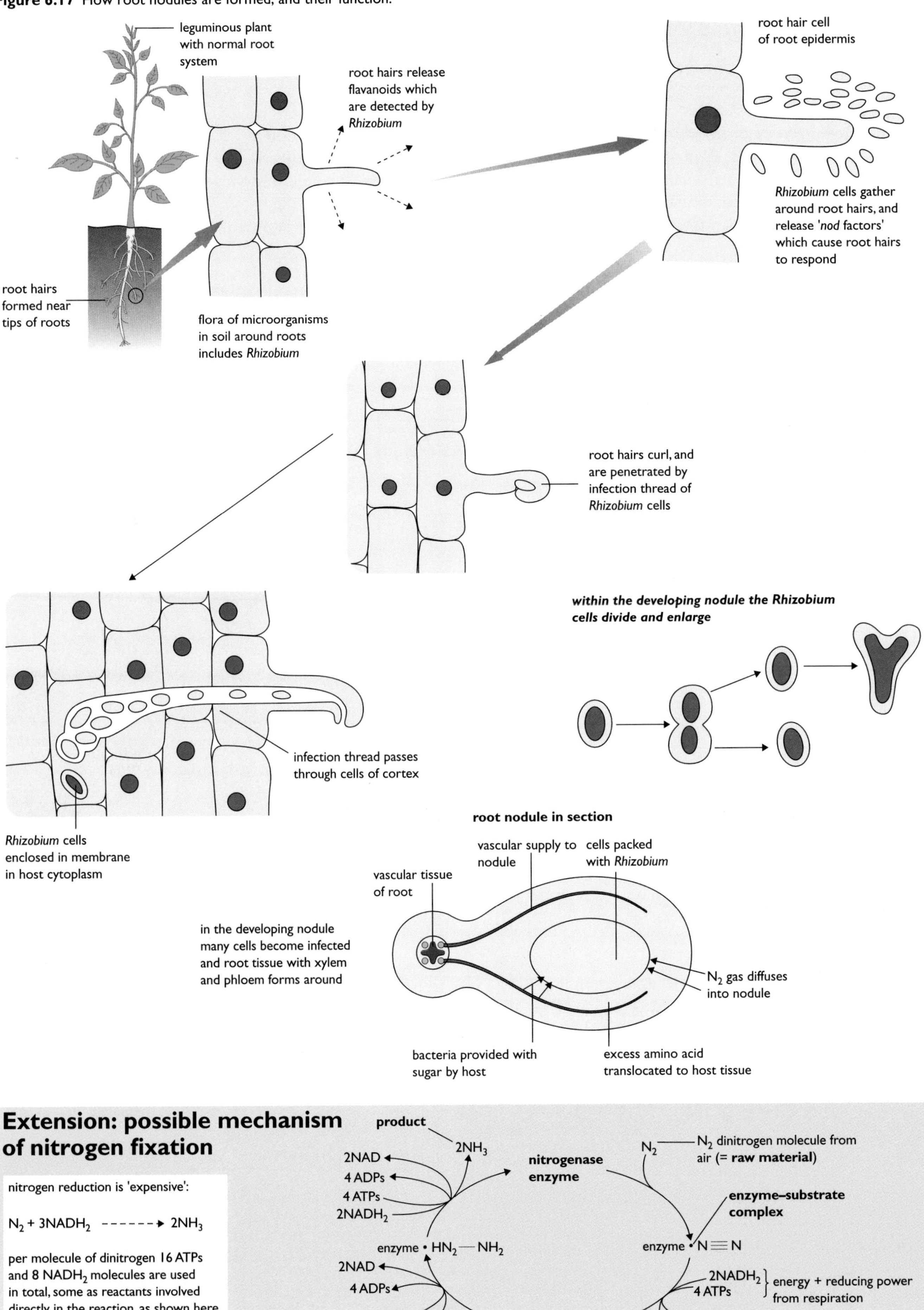

Extension: possible mechanism of nitrogen fixation

nitrogen reduction is 'expensive':

$N_2 + 3NADH_2 \dashrightarrow 2NH_3$

per molecule of dinitrogen 16 ATPs and 8 $NADH_2$ molecules are used in total, some as reactants involved directly in the reaction, as shown here

Maintenance of soil nitrates

Nitrate ions (NO_3^-) are the common form in which combined nitrogen is absorbed from the soil solution by plant roots. Fortunately, nitrates are added to the soil when dead organic matter is broken down aerobically. Many different microorganisms are involved.

Formation of nitrates

A very wide range of saprotrophic bacteria and fungi in the soil community break down organic matter in the soil (dead remains of organisms, waste matter excreted by animals, humus around soil particles, and so on). The final products of breakdown include carbon dioxide, water, and a range of metal ions (e.g. Ca^{2+}, K^+) and non-metal ions (e.g. PO_4^{3-}). Meanwhile, the combined nitrogen present is reduced to ammonia (NH_3). This is known as **ammonification**.

Ammonia occurs as the ammonium ion (NH_4^+), dissolved in the soil solution, particularly in acidic and neutral soils. Because ammonia is volatile some loss can occur from soil, especially in basic soils (such as soil formed from chalk rock). Loss from soils is prevented when rapid nitrification (oxidation of NH_4^+ to NO_3^-) follows the ammonification step.

Nitrification

The **first step** to nitrification is the oxidation of ammonium ions to nitrite ions. This is an aerobic process and an exothermic reaction. It occurs in the soil, carried out by enzymes of bacteria of the *Nitrosomonas* genus.

$$2NH_3 + 3O_2 \xrightarrow{\textit{Nitrosomonas}\text{ species}} 2HNO_2 + 2H_2O + \text{energy}^*$$

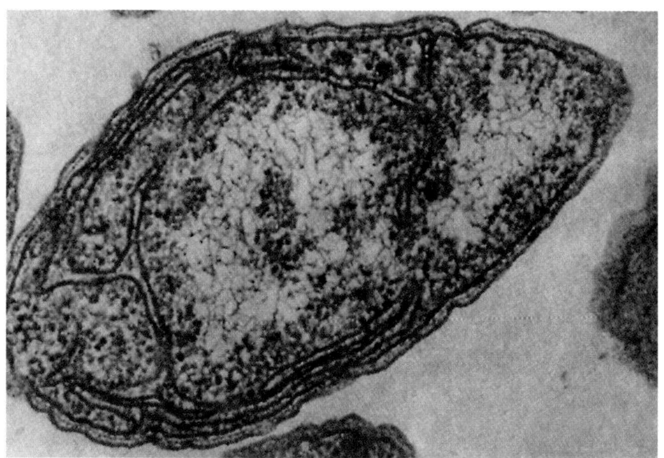

Figure 6.18 Transmission electron micrograph of *Nitrosomonas* (×100 000).

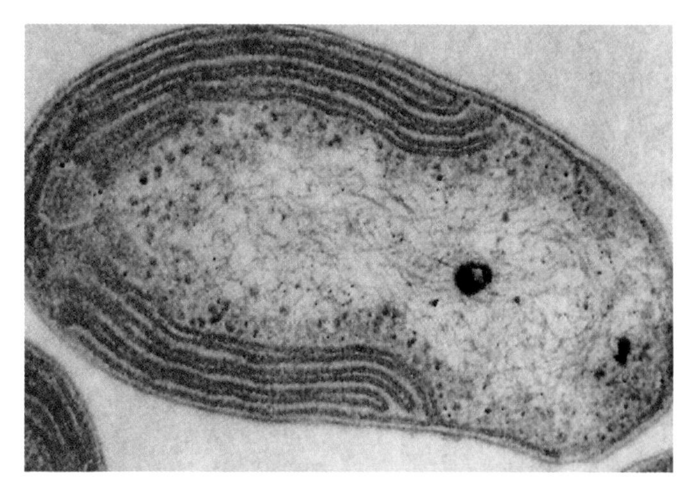

Figure 6.19 Transmission electron micrograph of *Nitrobacter* (×200 000).

The **second step** to nitrification is the oxidation of nitrite ions to nitrate ions. This is also an aerobic process and an exothermic reaction. It occurs in the soil, carried out by enzymes of bacteria of the *Nitrobacter* genus.

$$2HNO_2 + O_2 \xrightarrow{\textit{Nitrobacter}\text{ species}} 2HNO_3 + \text{energy}^*$$

Extension: a note about chemosynthesis

Nitrobacter* and *Nitrosomonas* bacteria use the energy released from the chemical reactions they catalyse in their nutrition. The energy is coupled to the synthesis of sugar from carbon dioxide and water in a process known as **chemosynthesis (Figure 2.20, page 14). Chemosynthetic bacteria are autotrophic, but use energy released from chemical reactions, rather than from light energy (as green plants and photosynthetic bacteria do), to manufacture nutrients.

$$6CO_2 + 6H_2O + \textbf{energy}\text{ from chemical reactions} \longrightarrow C_6H_{12}O_6 + 6O_2$$

Loss of nitrates from soil by bacteriological action – denitrification

Breakdown of soil nitrates to nitrous oxide or nitrogen gas occurs in waterlogged soil. This is due to the metabolism of anaerobic bacteria, particularly *Pseudomonas denitrificans*, but other species carry out denitrification too, including certain species of *Bacillus*. The anaerobic metabolism of these bacteria in soils under waterlogged conditions is harmful to green plants because of this loss of nitrate from the soil.

However, denitrifying bacteria can be useful. In the processing of liquid effluent at the sewage works, these bacteria are exploited to reduce the nitrate concentration of water that may become human drinking water at a later stage (Figure 6.10, page 34).

6 How do electric storms contribute to combined nitrogen in soil?

The nitrogen cycle – a summary

Although the element nitrogen makes up about 80% of the Earth's atmosphere, nitrogen compounds available for use by living things are relatively scarce within the biosphere. The nitrogen cycle summarises the cycling of nitrogen between soil, the atmosphere and living things.

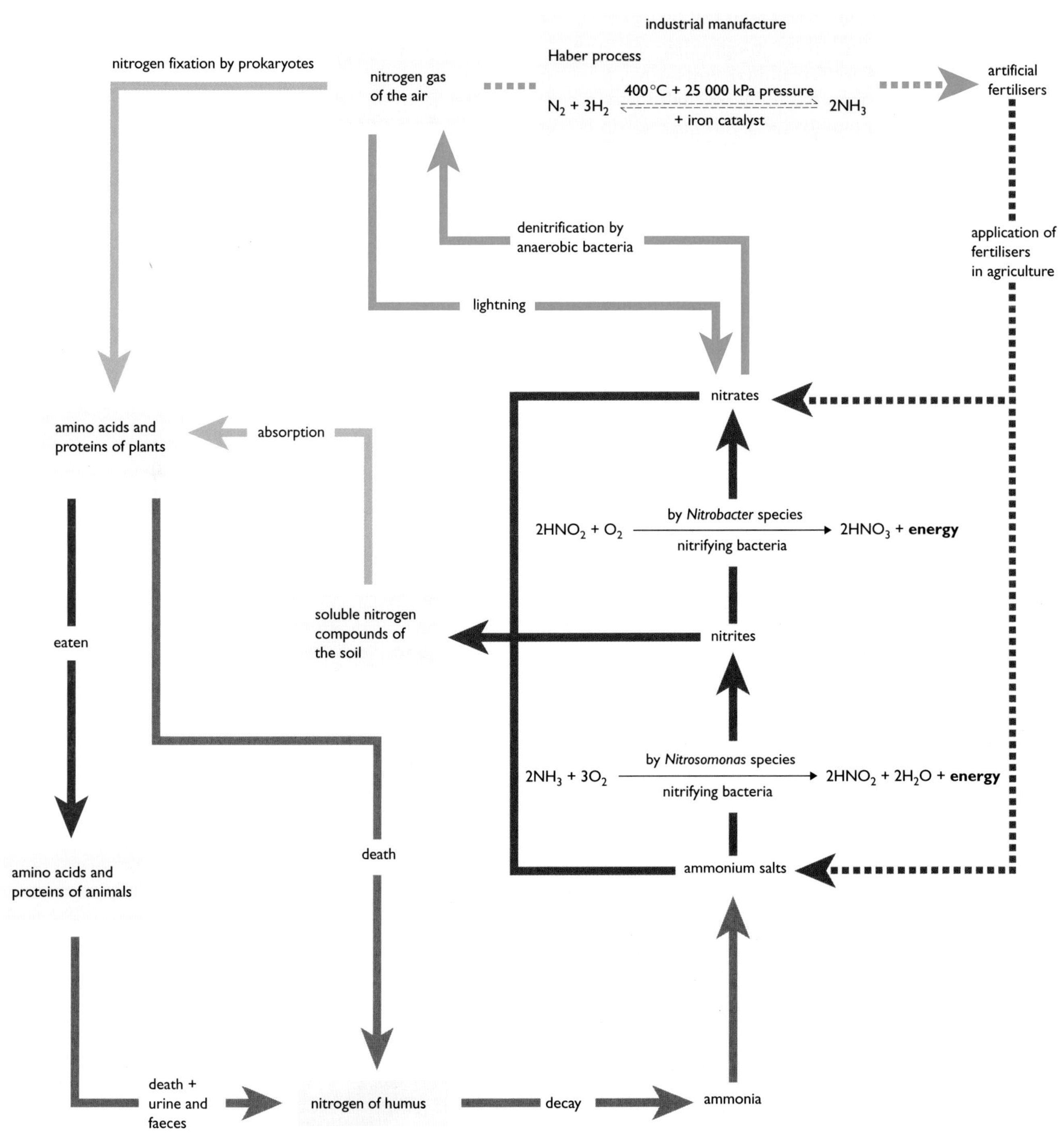

Figure 6.20 The nitrogen cycle.

Fuels from biomass

Oil, gas and coal are known as **fossil fuels**, as these resources were laid down long ago, during the Carboniferous period of Earth history. Dependence on fossil fuels generates problems because:

- they are expensive for poor people when bought in from distant sources
- they are non-renewable sources of energy – stocks will eventually run out
- burning of fossil fuels releases into the atmosphere carbon dioxide gas that has been locked away for over 300 million years. Consequently, the carbon dioxide concentration of the Earth's atmosphere is rising. This gas is one of many so-called greenhouse gases, so when we use fossil fuel we may be contributing to raising the Earth's temperature.

There are often good economic and environmental reasons for developing fuels derived from organic waste matter or from recently grown crops (**biomass**), in preference to using fossil fuels. Microorganisms are often involved in fuel production from biomass.

Methane gas from waste matter in the less-developed world

In parts of rural India and China, **biogas** digesters are in regular use. Into the digester goes animal dung and/or agricultural waste; there it is fermented anaerobically by bacteria such as *Methanobacterium*, known as **methanogens**. Simple digesters need to be operated in climates where the contents can be maintained at or about a temperature of 15 °C if fermentation is to occur satisfactorily. The biogas formed is about 50–80% methane, 15–45% carbon dioxide, and about 5% water vapour. This compares reasonably well with natural gas (from oil and gas fields), which is about 80% methane. Biogas may be used by the local community for cooking, lighting, electricity generators, and sometimes as tractor fuel. These simple digester plants may work well, but remain sensitive to pollutants such as detergents or heavy metal ions. Consequently, the gas supply may be erratic. The plant may need labour-intensive and unpredictable maintenance work.

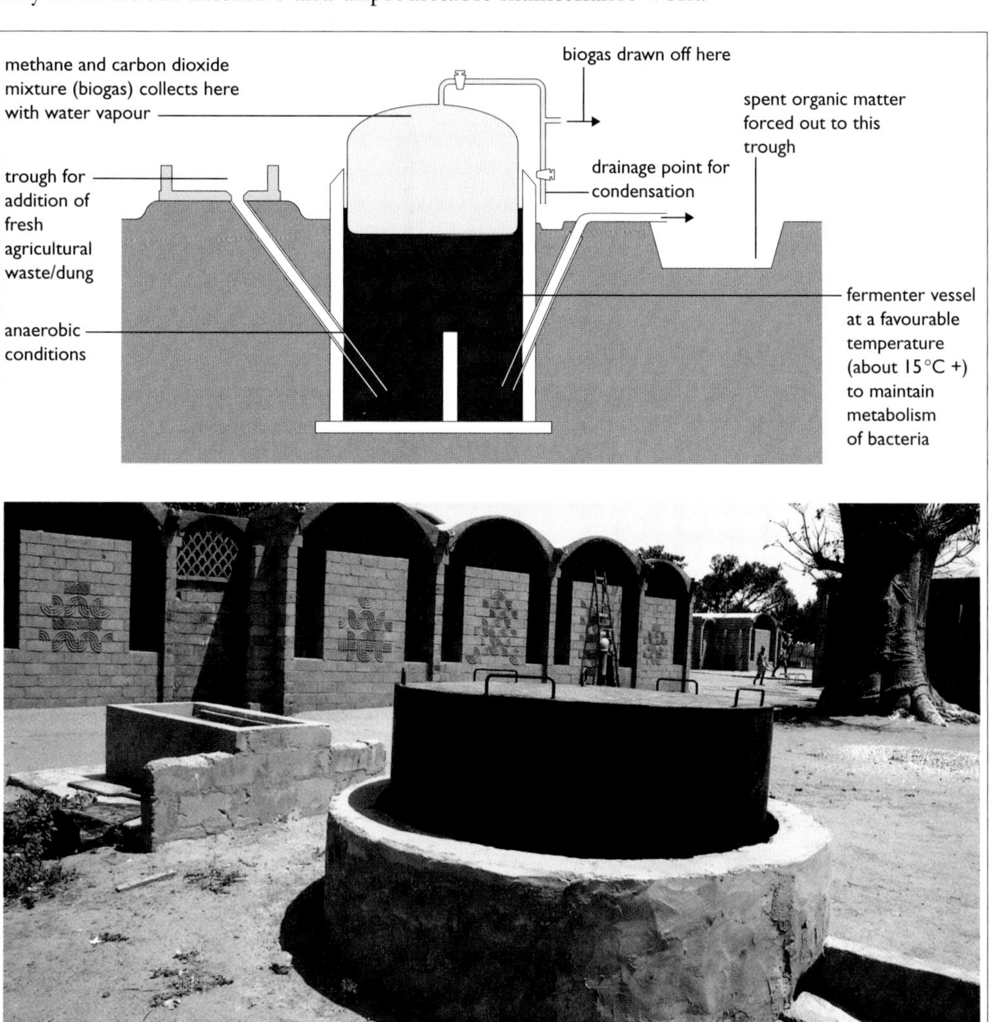

Figure 6.21 Biogas plants.

Methane gas from waste matter in the developed world

Urban communities in the developed world typically arrange to dump their domestic waste, much of which is of organic matter, into disused pits or quarries. These are called landfill sites. The rubbish is naturally compressed, anaerobic conditions quickly develop, and a **cocktail of microorganisms** break down the organic matter, giving off the gaseous waste products CO_2 and methane. Completed landfill sites have to be vented so that these gases can safely escape into the atmosphere. However, methane is a more powerful greenhouse gas than CO_2, and so it is a unwelcome addition to the atmosphere.

Schemes now operate to tap biogas mixture and pipe it to nearby sites where it can be usefully burnt to generate heat and/or power, wherever this is practical. (At sewage works, a similar mixture is used in the power house to help run turbines that generate electricity to run the sewage plant's air pumps and other equipment; page 34.)

7 How may greenhouse gases such as methane contribute to global warming?

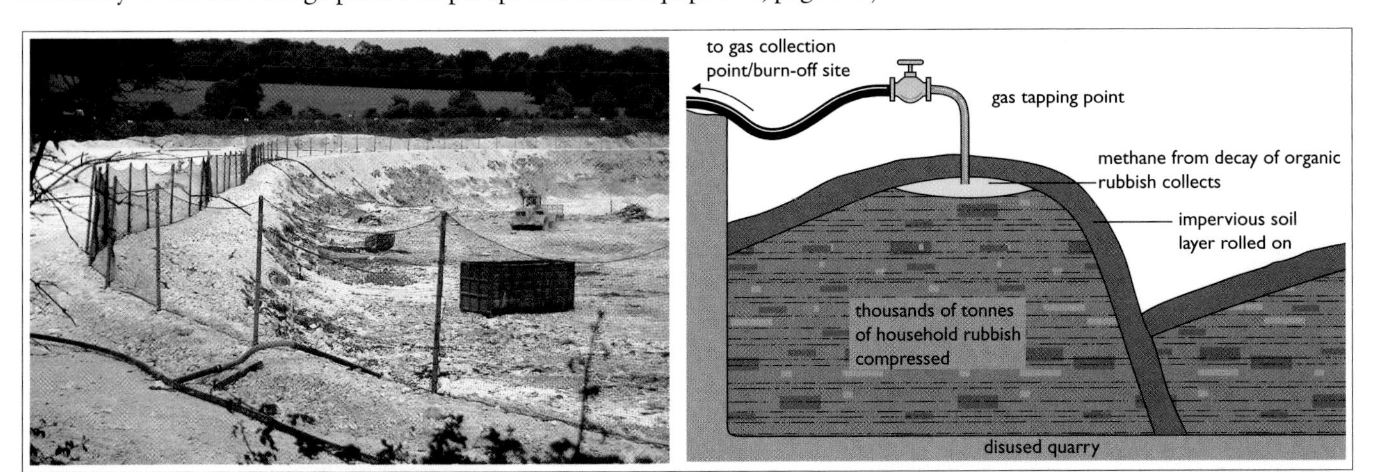

Figure 6.22 Collection of biogas from the decaying compressed rubbish at an urban landfill site.

Alcohol from biomass – for the internal combustion engine

'Energy farming' of green plant crops for the production of ethanol as a fuel is another way that energy is obtained from biomass. For example, in Brazil a scheme was initiated in the 1970s, when the price of fossil fuels rose steeply and that of sugar from sugar cane (a major export from Brazil) fell disastrously. Some of the sugar extracted from sugar cane was no longer offered for export. Rather, it was fermented to ethanol by **yeast** (page 26).

A 15% aqueous ethanol solution results, and has to be distilled to form a solution pure enough (containing only 4.4% water) to power cars adapted to burn ethanol. The scheme continues to run, but it has to be subsidised at times when crude oil is cheap enough for conventional petrol to be less expensive than Brazilian Álcool, as the fuel is known.

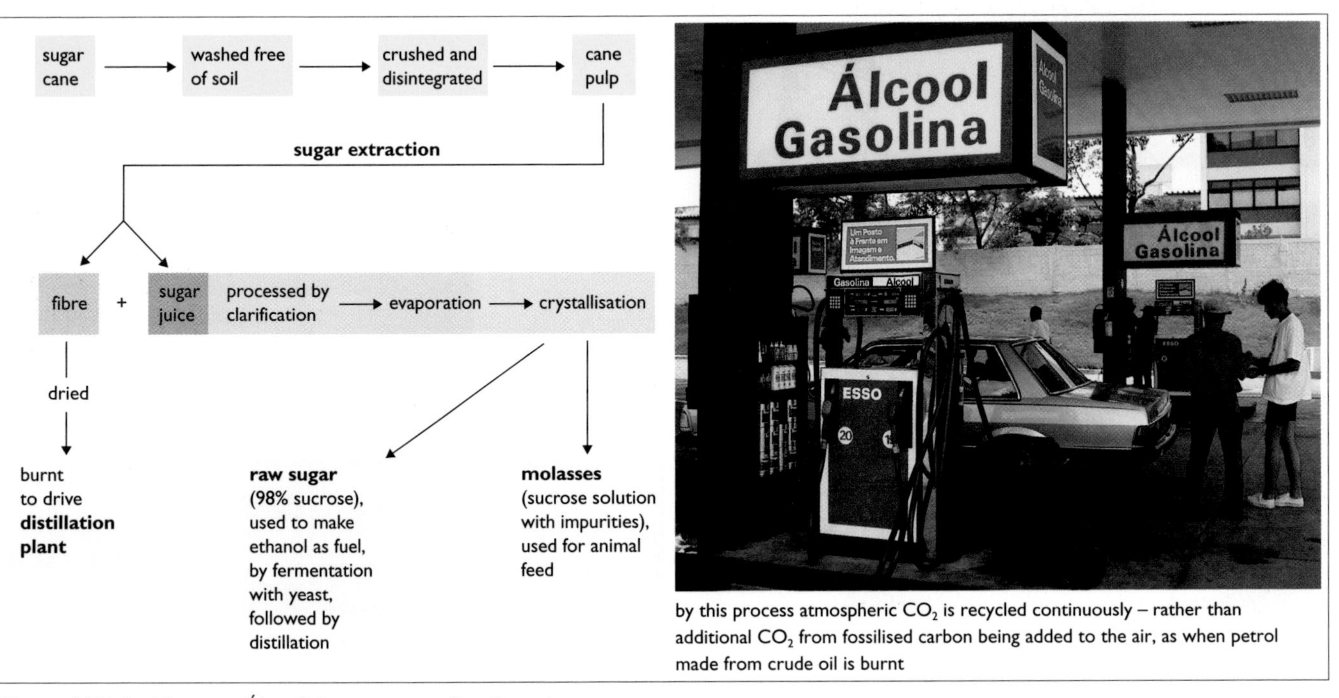

Figure 6.23 Fuel for cars (Álcool) from sugar – a Brazilian scheme.

7 Microorganisms and biotechnology

Biotechnology is the industrial and commercial application of biological science, particularly of microbiology, enzymology and genetics. Modern biotechnologies are enormously important to industry, medicine and research. These technologies are often presented as a recent development, but in fact their origins go back several millennia. Some early biotechnologies are illustrated first.

Wine making

Wine production begins with the harvesting of ripe grapes, followed by crushing the fruit to produce a 'must' (slurry of pulp, skins and pips). Today the must is treated with sulphur dioxide to kill the flora of wild yeasts and bacteria that were living on the skins of the ripening fruit. Then a chosen strain of yeast is added to ferment the sugar. There are many variations to the basic process of wine production (Figure 7.1), by which the distinctive flavours and aromas (known as bouquet) of different wines are produced.

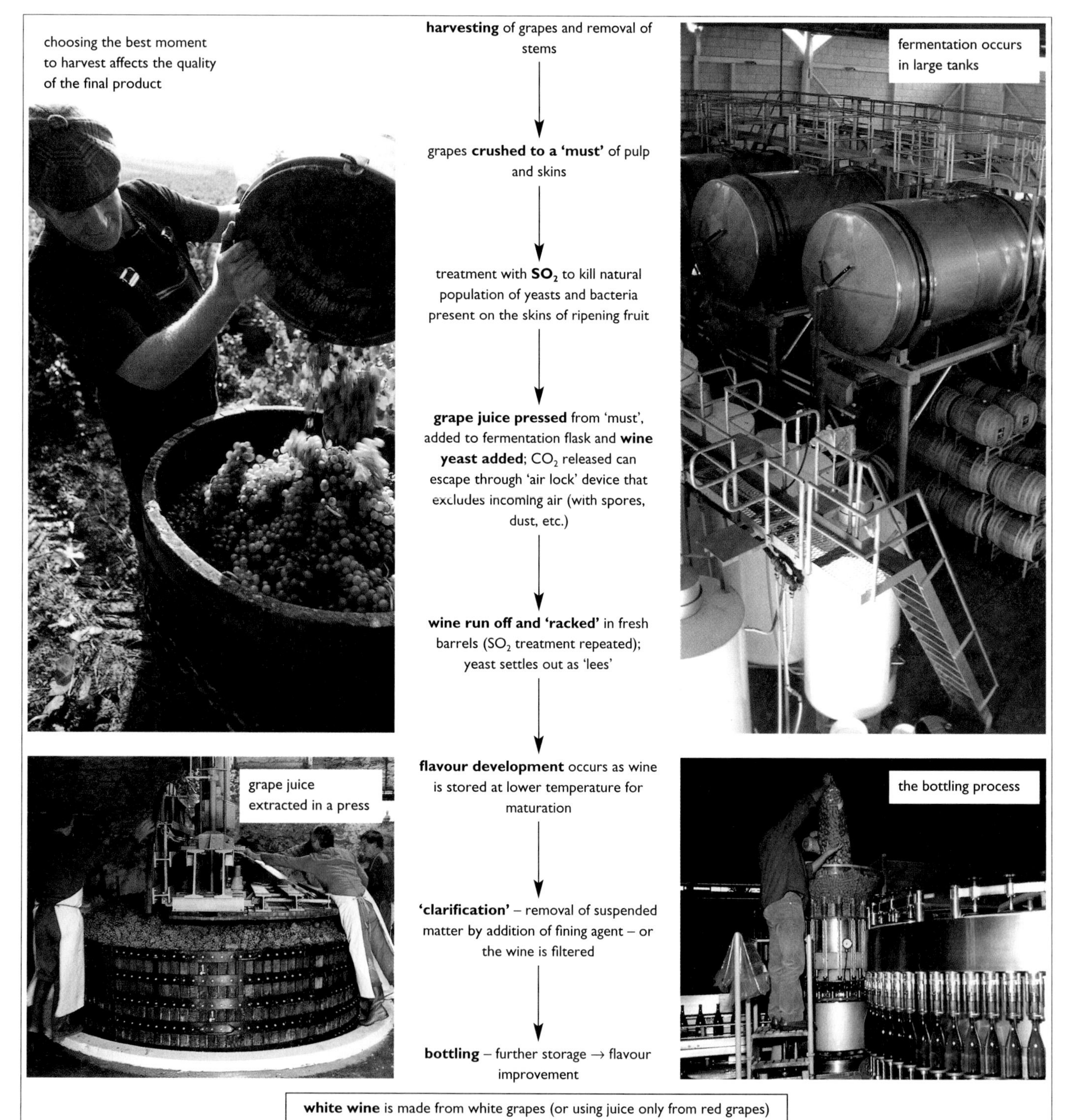

Figure 7.1 Red and white wine production.

Brewing

Beer is made from barley grains (or rice or maize), which are first allowed to germinate. By this first step, the enzymes of the grains are used to convert the stored starch and proteins to sugars and amino acids. Yeast is added to the resulting sugar solution and produces alcohol (Figure 7.2).

Flavours of particular brews are due to the ingredients selected (e.g. malt, water, hops and yeast), the proportions used, and the final cask conditioning of live beer allowed by different brewers. The differences between lagers and beers are due to the yeasts used. Market variety is also due to the industrial development of keg beers – processed, pasteurised beers served using nitrogen and carbon dioxide gas mixtures. These keg beers are viewed unfavourably by real ale enthusiasts.

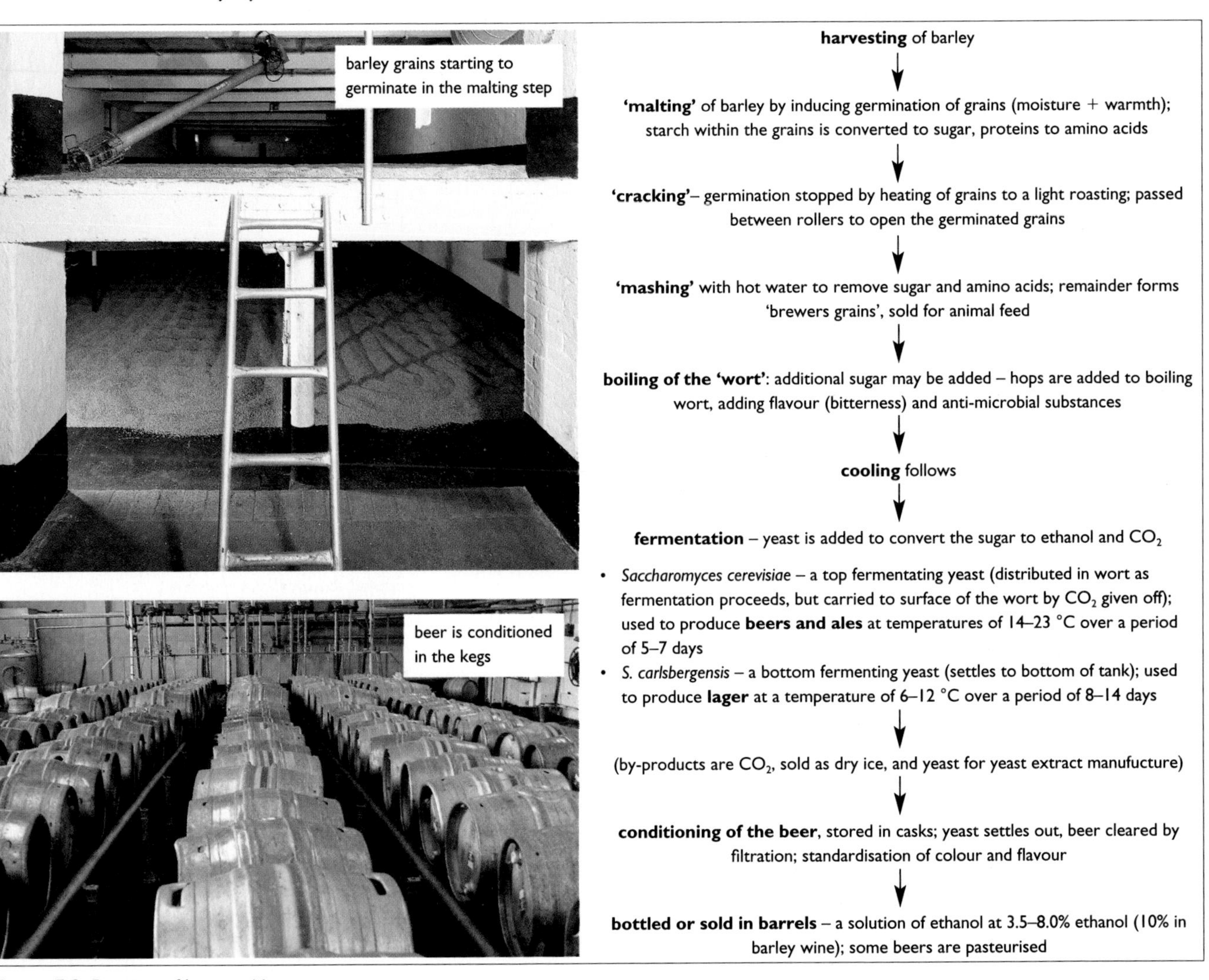

Figure 7.2 Brewing of beer and lager.

The significance of alcoholic drinks for human societies

Archaeological records show that brewing was an advanced skill in several human societies more than 6000 years ago. Today, brewing wine, beer and lager are huge industries. There are important consequences of this.

- Before the development of clean, piped water (in relatively recent times in developed countries), beers and wines were safer to drink than water. Pathogenic organisms are unlikely to survive in beers due to the acidity, the antimicrobial compounds in hops, and the concentration of ethanol present. It was wiser to consume beer rather than water (and give beer to children, too) in times when drinks made from boiled water (e.g. tea and coffee) were too expensive for most people.
- All alcohols are poisonous substances – ethanol is the least poisonous of this family of organic chemicals. Now that people tend to live longer and be affluent enough to consume more alcohol, we are aware of the harm that excessive alcohol consumption has on health (due to brain, heart and liver damage); on public behaviour ('lager louts'); and in road accidents (drinking and driving kills).

1 In what ways is a dilute solution of ethanol potentially toxic to humans?

Fermented foods

Microorganisms are also involved in food production from raw materials of animal origin (e.g. milk) and plant origin (e.g. soya bean). Many different types of fermented foods are known; together they contribute about one-third of human food intake worldwide. The following examples introduce the diversity that exists, and some of the ways microorganisms are involved in their production.

Cheese

Fresh milk has a very short shelf life. Cheese, the product of the fermentation of milk, is a way of converting the nutrients of milk to a solid form that preserves them for later use. The necessary bacteria live on the skin of cows, and tend to contaminate raw milk during the milking process. Today, pasteurised milk (page 65) is used for cheese production, and selected starter cultures of particular strains of bacteria help to create the distinctive flavour and texture of different cheeses.

Yoghurt

This form of fermented food has been popular around the world for a long time; in the UK it is a more recent addition to diets.

Figure 7.3 Hand milking.

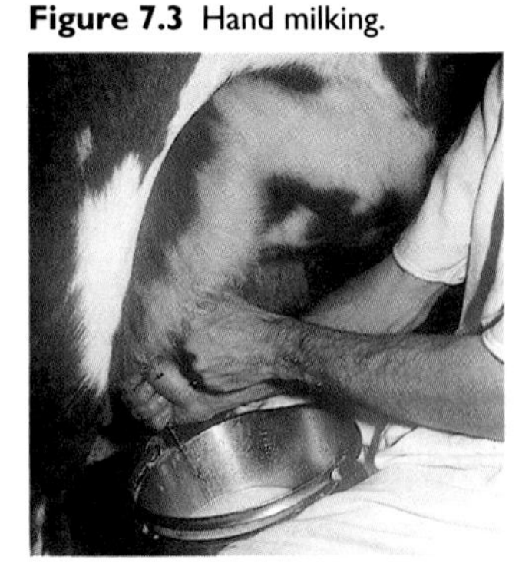

2 Milk may be contaminated with antibiotics. How may this occur, and what problem does this impose in yoghurt production?

Figure 7.4 Manufacture of fermented foods from milk.

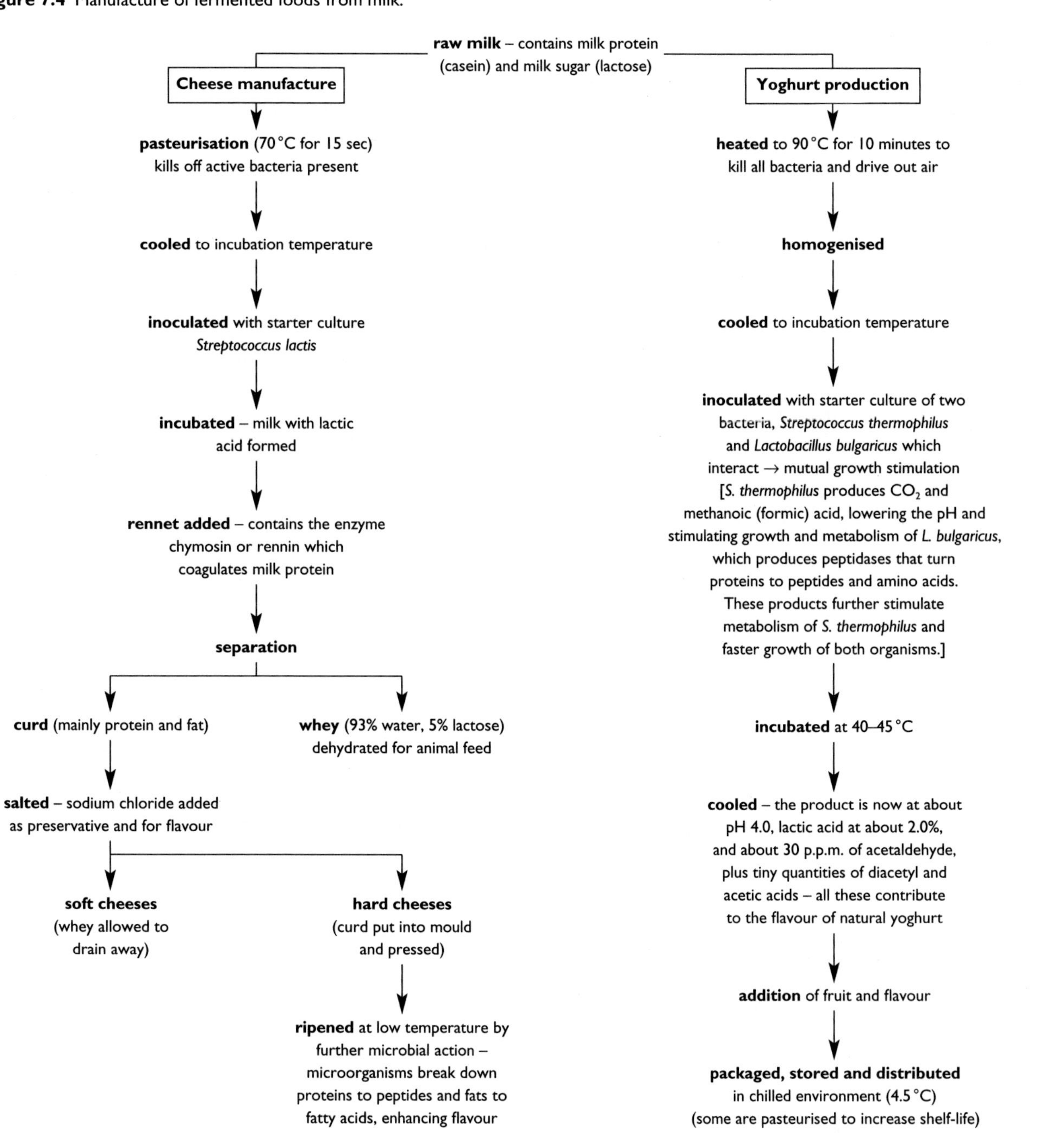

Sauerkraut

An example of a vegetable preserved by fermentation is sauerkraut, which is derived from fresh cabbage. The process involves a succession of microorganisms, each producing lactic acid from sugars in the cabbage tissues at increasingly lower pH. The acidity drives away other bacteria that might otherwise cause decay.

cabbage plants prepared, core leaves finely shredded

↓

salt added (2.5% by weight), packed into press under anaerobic conditions – the following fermentation takes over 1 month at about 18 °C

↓

fermentation of cabbage – a succession of bacteria are involved:

phase I *Leuconostoc mesenteroides* initiates fermentation, forming lactic acid and reducing the pH to 4.0

phase II after 6–8 days, *L. brevis* and *L. plantarum* increase

phase III after 16–18 days, *L. plantarum* has reduced all sugars to lactic acid

↓

finally **sauerkraut** has a pH of 3.8 and contains about 2.0% lactic acid; other microorganisms are inhibited

Figure 7.5 Sauerkraut production.

Food from soya bean

Soya bean (*Glycine max*) is one of the oldest cultivated crops, originating in south-east Asia. There, various fermented foods have been derived from it. Today it is grown very much more widely, and the beans are used mostly for oil, animal feed and soya flour preparation.

Soy sauce is a food flavour and colour derived from soya bean by fermentation.

cooking of soya beans (with rice) – sterilising and softening phase

↓

inoculated with mould fungus, *Aspergillus oryzae*

↓

incubated at 25–30 °C for 2–3 days; aerobic breakdown of starch to sugar, proteins to amino acids, and of fibrous tissues by enzymes from fungus

↓

addition of salt to raise salt concentration to 16–19%

↓

anaerobic fermentation in deep tank for 6–8 months by lactic acid bacteria and yeasts; sugars are converted to lactic acid and alcohol

↓

pressed to extract soya sauce, followed by **pasteurisation** at 70–80 °C to prevent further fermentation

↓

bottled

Figure 7.6 Production of soy sauce.

Food from microorganisms

The culturing of microorganisms for the manufacture of protein-rich foods is now an important industry. The product is called **single-cell protein**.

Initially this approach was applied to the manufacture of animal feeds, for there were early doubts about whether such products would ever be acceptable in human diets. In a pilot plant built close to an oil and gas terminal in County Durham, methanol (a waste product of natural gas) was used, and a product called 'Pruteen' resulted. However, the project failed for economic reasons – one or more alternative sources of biomass (e.g. soya bean, cereal grains, fish meal, oilseeds) provided substantially cheaper sources of animal feed. Meanwhile, public doubts about human consumption of such products were dispelled – anxieties about meat as a source of prions (page 86) may have played a part.

Mycoprotein – Quorn

Quorn is produced from the filamentous fungus *Fusarium graminearum*, grown in an industrial fermenter (Figure 7.8). The liquid culture medium contains sugars, typically produced from hydrolysed starch or molasses, but enriched with ammonia as a combined nitrogen source. The resulting biomass is high in dietary fibre, but low in lipids (but with a high ratio of unsaturated to saturated fatty acids), and low in sodium ions. There is about 45% (dry weight) protein present, with an amino acid profile that compares favourably with that of beef. Protein is generally the more expensive component of the diet, so it is the protein component of mycoprotein that is most valuable.

Figure 7.7 Quorn products have proved popular as well as nutritious.

Figure 7.8 Production of Quorn.

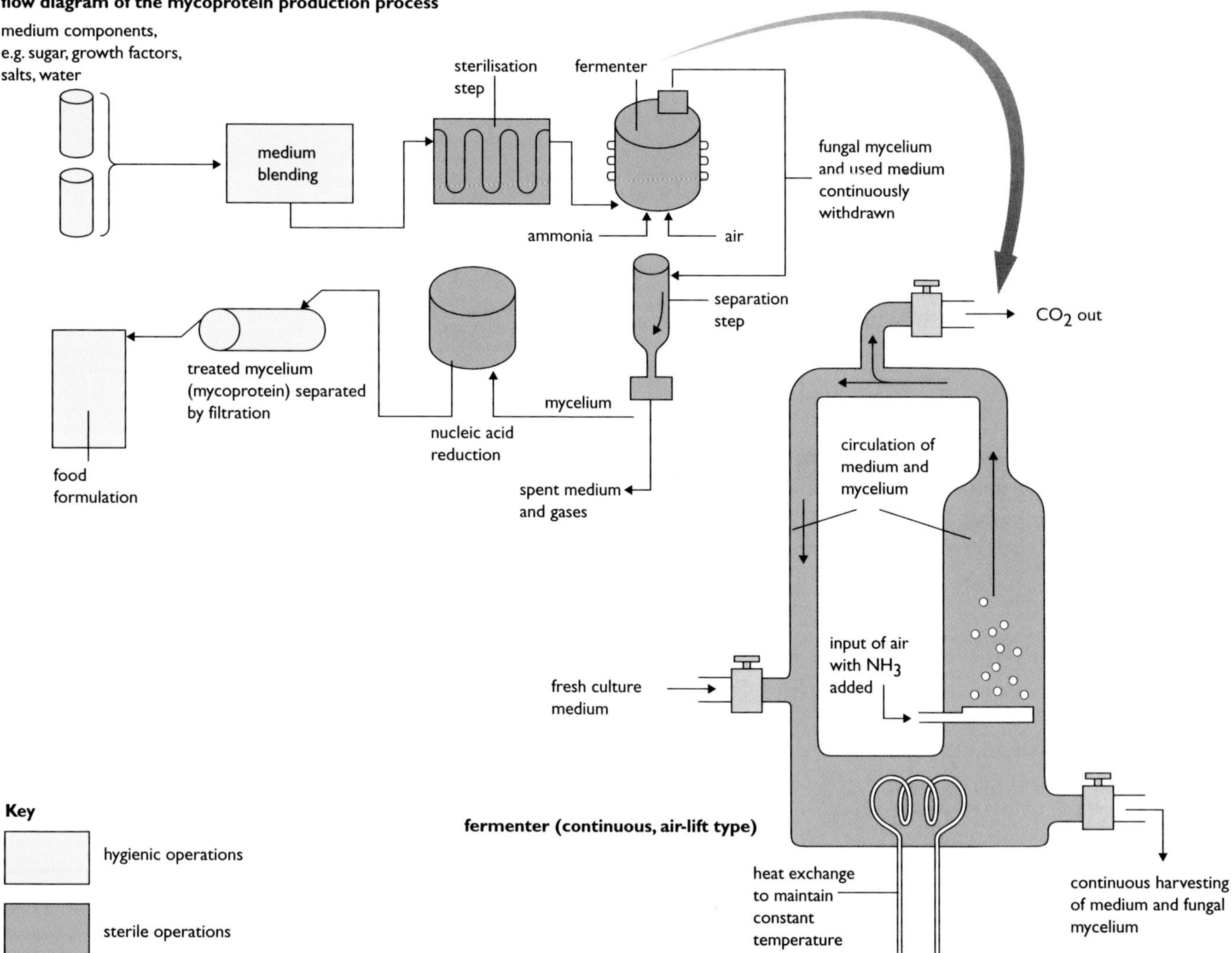

Fermenter technology

Fermenters are vessels used for the **growth of microorganisms in liquid culture**. There are two types of fermenter vessel in use.

- **Open fermenter tanks** are non-sterile systems, normally open to the atmosphere. These include brewing vats in alcoholic fermentation (Figure 7.2, page 45), and the open, aerobic fermentation tanks with activated sludge of the sewage works (Figure 6.10, page 34). The rumen of cows and sheep is a naturally occurring, non-sterile fermenter 'tank' (page 32).
- **Closed fermenter systems** are enclosed, sterilised vessels housing a single culture of one (or occasionally more) microorganisms of a particular strain. The air-lift fermenter (Figure 7.8) is an example, as is the fermenter for industrial enzyme production (Figure 7.10, page 50), and that for antibiotic production (Figure 7.15, page 53).

There are two ways that **fermentation operations** are carried out in fermenters.

- **Batch culture**, in which an initial fixed volume of culture medium and microorganisms is processed in the fermenter until the products have accumulated.
- **Continuous culture**, in which the fermenter is run for an extended period, with nutrients added and products removed at a steady rate.

In both batch and continuous culture processes, conditions in the fermenter vessels are monitored, and adjustments may be made to maintain favourable growing conditions.

3 Classify the use of fermenters shown on this page, and on pages 32, 34, 42, 44, 45, 46, 48, 51 and 53, as examples of **a)** batch culture and **b)** continuous culture of microorganisms.

Single-cell protein from *Spirulina*

Spirulina is a cyanobacterium (page 16), adapted to growth in highly alkaline (up to pH 11) pond waters. In these environments few other microorganisms survive. As a photosynthetic bacterium it requires carbon dioxide, light and a supply of essential ions, in addition to water. *Spirulina* forms dense, floating mats (O_2 produced is trapped in gas vacuoles and makes the short, spiral-shaped filaments buoyant). Traditionally, in central Africa and South America these mats are drained from pond water and dried in the sun. The resulting blue-green 'biscuits' are rich in protein, containing up to 70%. They are eaten by humans or fed to livestock, and have been used in this way for many hundreds of years – *Spirulina* is an ancient example of single-cell protein. The cell contents are accessible because the cell walls (not cellulose) are relatively easily digested. Because it is a prokaryote, the nucleic acid content is low (compared with a eukaryotic cell). These features increase its food value.

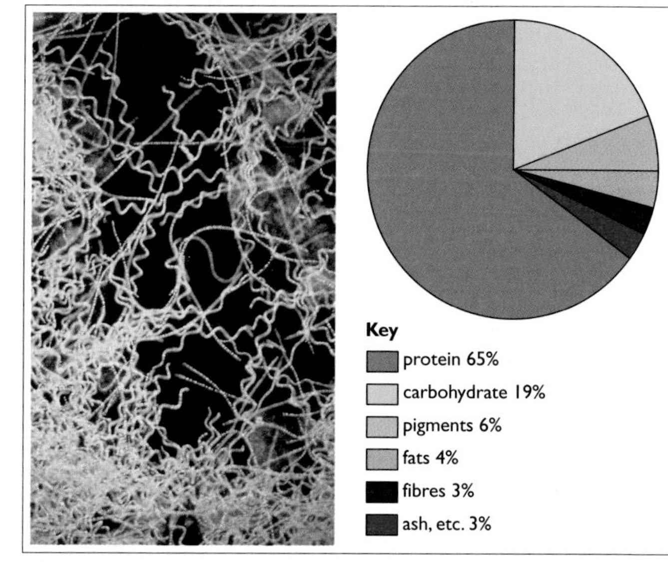

Figure 7.9 SEM of *Spirulina*, and the food value of *Sprirulina* biscuits (× 100).

Table 7.1 Single-cell protein as a food source – an evaluation

Advantages	Disadvantages
Microorganisms grow rapidly, compared with the growth rate of traditional crops/herds. Typically, microorganisms double their mass in about 50 minutes, compared with weeks (e.g. for chickens) or months (e.g. for young cattle).	In eukaryotic microorganisms (fungi, algae and protoctista) the ratio of nucleus to cytoplasm may lead to toxic levels of nucleic acid in food products. These have to be corrected (at a cost). However, prokaryotic microorganisms are comparatively low in nucleic acids.
Prokaryotes are more easily modified genetically than animals or plants (page 59).	The initial product from microorganisms grown as food is colourless, odourless, tasteless and without texture. The product requires further treatment to produce a palatable food.
The protein content of microorganisms is generally at a higher concentration than in eukaryotes.	
Microorganisms can be cultured in fermenter vessels by continuous culture.	
Microorganisms can be grown on a wide range of raw materials, including waste materials.	

Industrial enzymes

Catalysts are substances used to speed up reactions. They are frequently used in processes in the chemical industry. For example, iron is the catalyst in the Haber process (the industrial production of ammonia from nitrogen and hydrogen – Figure 6.20, page 41). Without iron there would be no significant production of ammonia.

In many modern industrial developments, **enzymes** (biological catalysts) are used. In brewing, the enzymes of barley are exploited to turn stored starch into sugar prior to fermentation (Figure 7.2, page 45); in the tanning of leather, the proteases in the faeces of carnivorous animals are used to soften hides; in cheese manufacture, bovine chymosin is produced by the genetically modified bacterium *Kluyveromyces lactis*. The use of enzymes as industrial catalysts is widespread because they are:

- highly specific, catalysing changes in one particular compound or one type of bond
- efficient – a tiny quantity of enzyme catalyses the production of a large quantity of product
- effective at normal temperatures and pressures, so a limited input of energy (e.g. as heat and high pressure) is required.

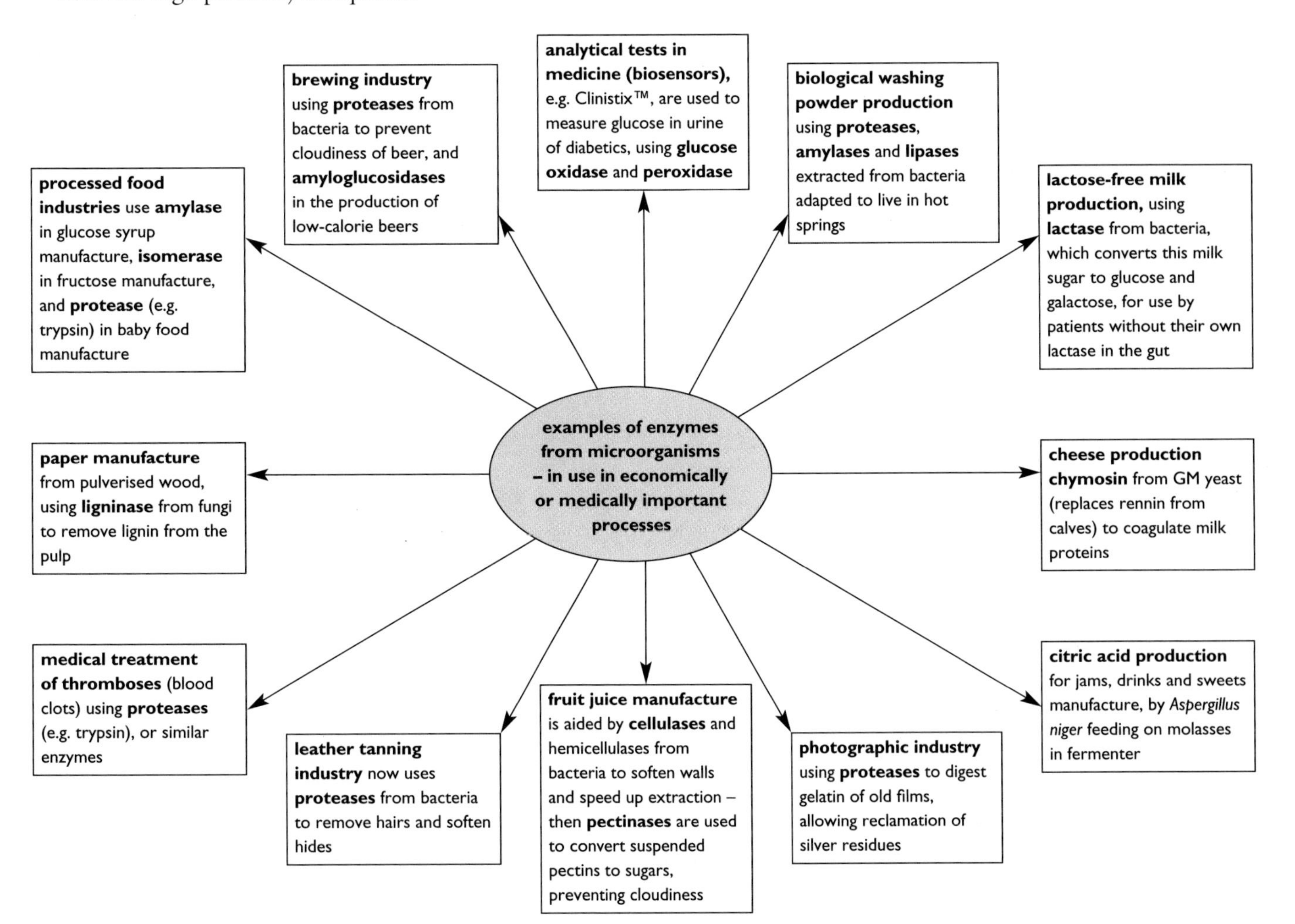

Figure 7.10 Uses of enzymes from microbiological sources.

Industrial production of enzymes from microorganisms

Most industrial enzymes are obtained from microorganisms (mainly bacteria and fungi) because these organisms:

- may be grown economically in fermenters (page 49) throughout the year, rather than being limited to a short growing season
- tend to grow quickly and produce large quantities of enzymes relative to cell mass
- are sometimes found under extreme environmental conditions (page 17), and have enzymes adapted to function under abnormal conditions (e.g. extremes of pH or temperature)
- may be modified genetically with comparative ease (e.g. to produce human insulin; page 56).

4 In the industrial production of enzymes, water is removed from the spent medium (Figure 7.11), often by reverse osmosis. From your knowledge of osmosis, design a pilot plant to demonstrate this.

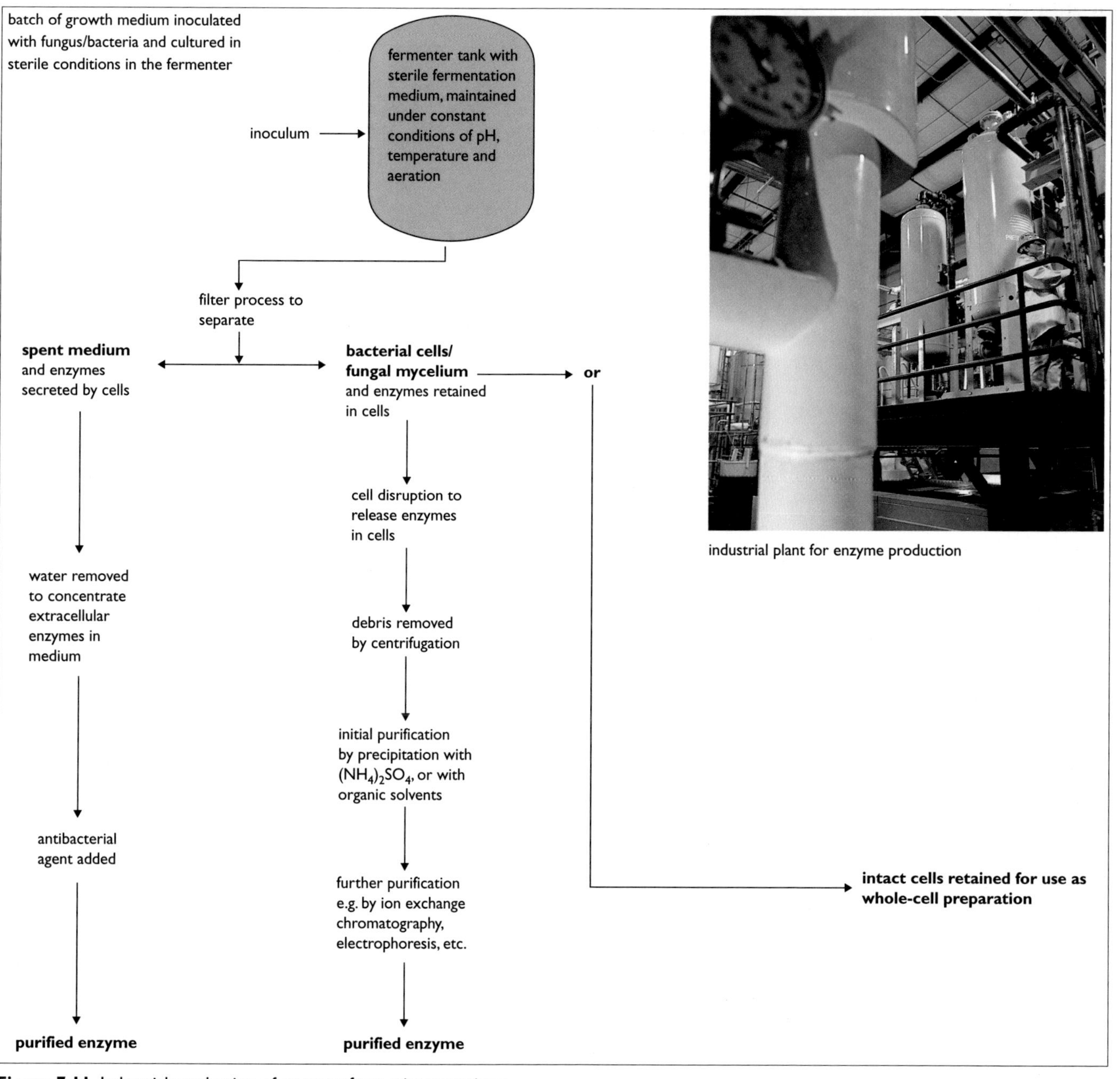

Figure 7.11 Industrial production of enzymes from microorganisms.

Using enzymes in industry and commerce

Enzymes may be used as whole-cell preparations or as cell-free extracts. They may be added to reaction mixtures, or they may be immobilised and the reactants passed over them. The advantages of these alternatives are summarised in Figure 7.12.

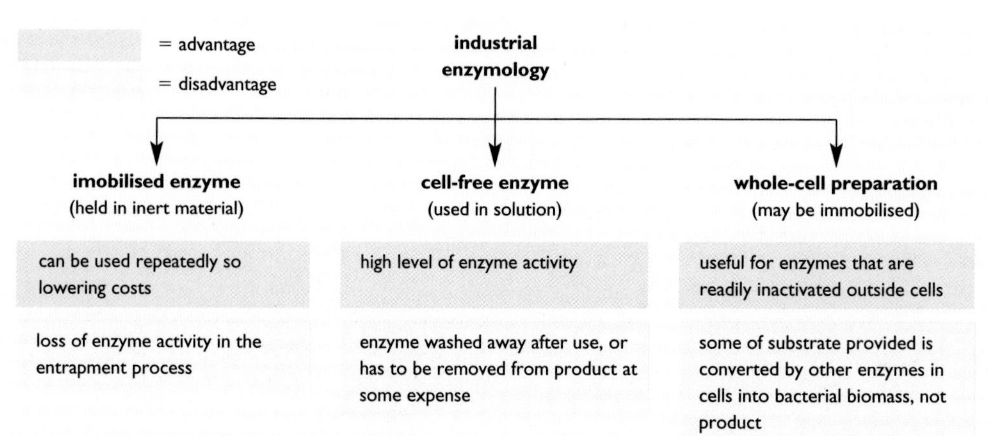

Figure 7.12 How enzymes are used.

Antibiotics from microorganisms

Antibiotics are naturally occurring chemical substances, obtained mainly from certain fungi and bacteria commonly found in the soil. When antibiotics are present in low concentrations, they inhibit the growth of other micro-organisms or cause their outright death, as demonstrated in Figure 7.13.

Figure 7.13 Investigating sensitivity to antibiotics.

to the bacterial lawn of a known species was added a mast ring* with each 'arm' impregnated with a different antibiotic (colour coded). Then the plate was closed and incubated. From the result (below) there is evidence that growth of this bacterium is more sensitive to certain antibiotics (e.g. CM, A) than to others (e.g. S, I).

*obtained from Philip Harris Scientific

different antibiotics are contained in the arms of the mast ring, so that sensitivity to many antibiotics may be tested simultaneously

Obtaining new antibiotics

Antibiotic resistance

Many different antibiotics have been discovered, and new ones are continually being searched for. Some new discoveries are effective – and are not toxic to patients. However, problems arise with antibiotics with time. Sooner or later some pathogenic bacteria in a population will develop genes for resistance. Resistant genes develop either by mutations, or as a result of gene transfer between bacteria by conjugation (page 5). Consequently, resistant bacteria survive exposure to the drug. Once competition with other (non-resistant) bacteria is removed by the antibiotic, the resistant pathogenic bacteria flourish. In the longer term, the pharmaceutical industry faces the challenge of producing new antibiotics more quickly than bacteria can develop resistance to them.

Screening for antibiotic activity

New antibiotics may be found by design or by screening (Figure 7.14). In the screening process, possible antibiotic-producing organisms are obtained from nature and plated out to form colonies. Indicator microorganisms susceptible to antibiotics are grown alongside. If zones of growth inhibition develop around particular colonies, the bacteria in these colonies are subcultured and any antibiotic they secrete is investigated. New antibiotics must be toxic to pathogens, but not to humans.

5 **Antibiotics are widely used as prophylactics in animal husbandry. What does this mean, why does this happen, and what possible dangers arise from this use of antibiotics?**

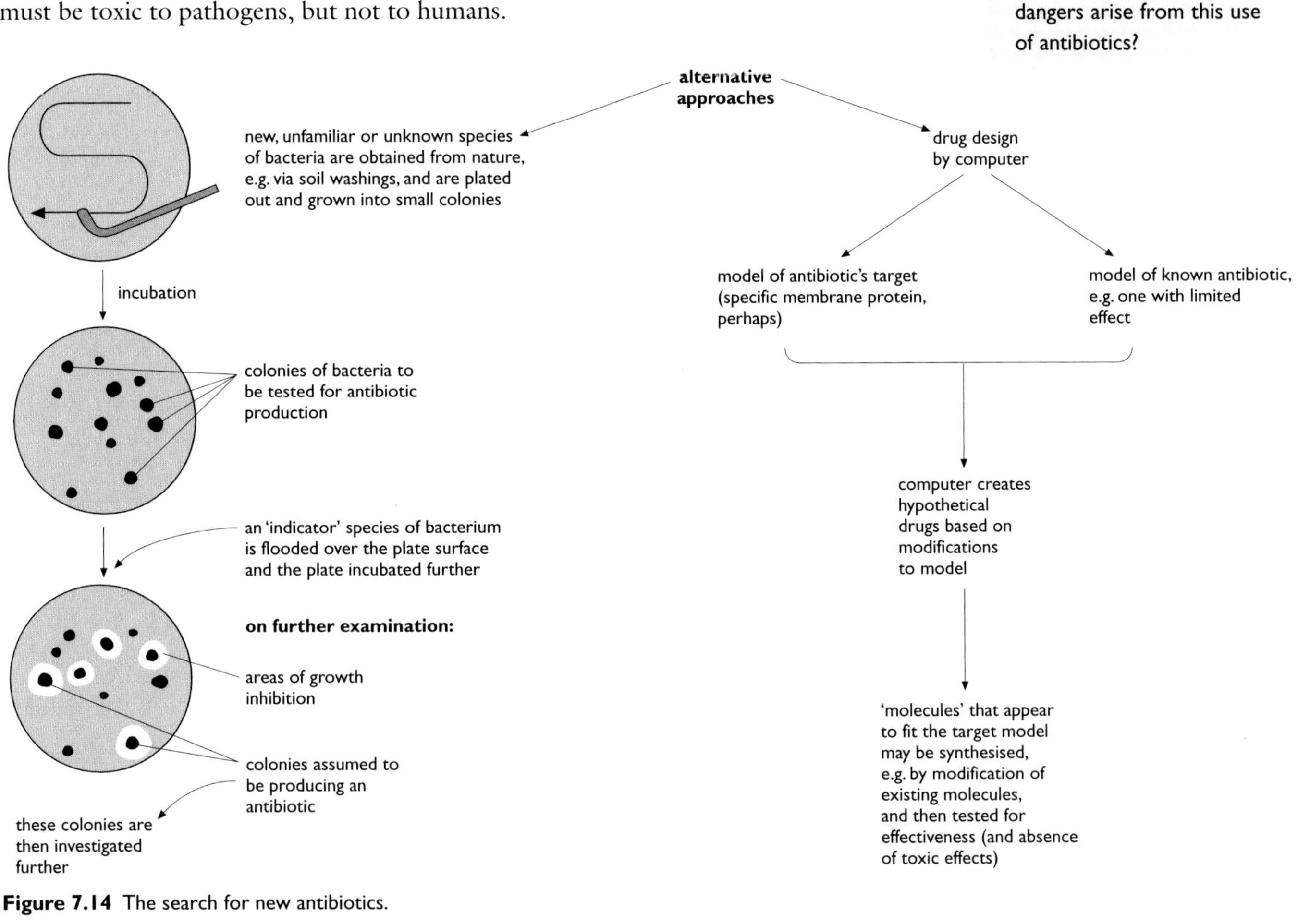

Figure 7.14 The search for new antibiotics.

From laboratory culture to industrial production – the scale-up process.

Once a new potential source of antibiotic has been discovered, it is grown up in the laboratory. Subsequently the microorganism is cultured in a sequence of fermenters, in order to discover the nutrient requirements and environmental conditions for maximum yield, and to ensure the same conditions can be maintained economically in a commercial fermenter.

Figure 7.15 Industrial production of antibiotics.

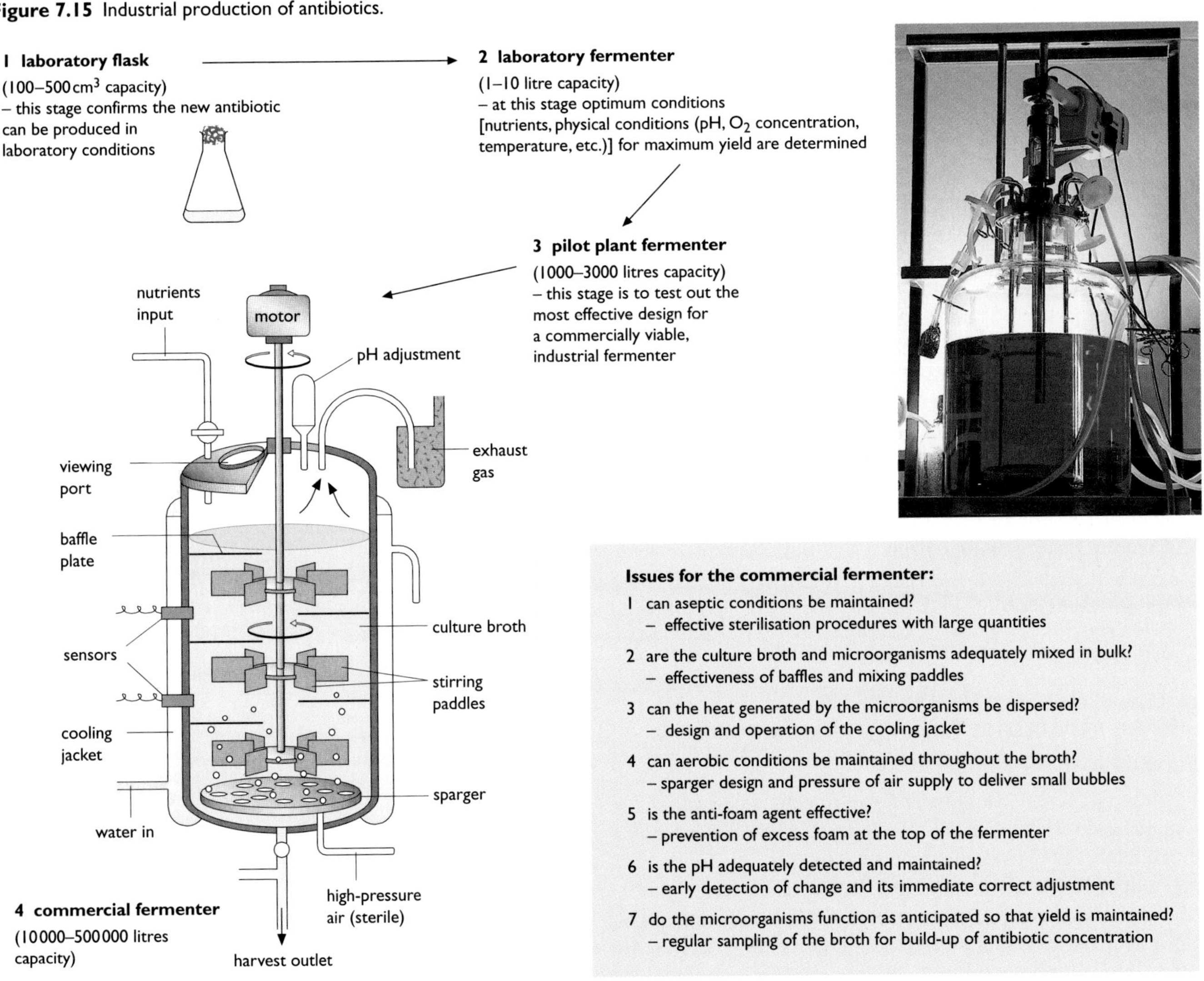

Downstream processing

After harvesting, the antibiotic must be recovered from the spent broth, then purified prior to packaging. Downstream processing may involve the antibiotic molecule being chemically modified and improved using enzyme technology.

Figure 7.16 Downstream processing.

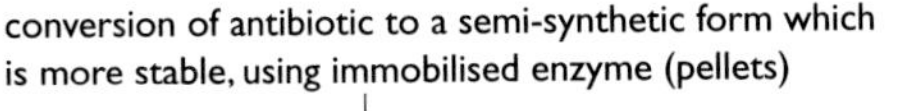

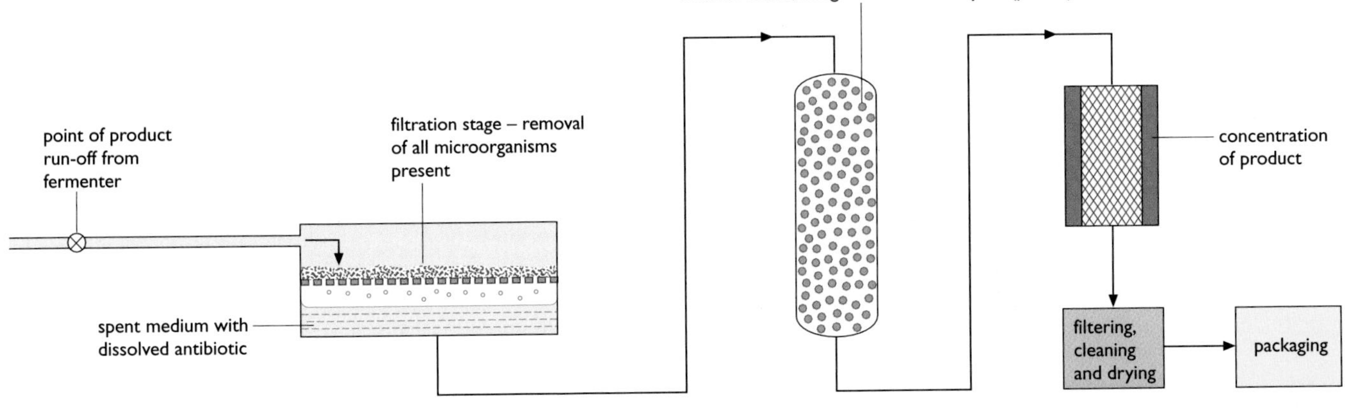

8 Genetic engineering of microorganisms

Key terms

Gene:

- the basic unit of inheritance
- a specific region of a chromosome that is capable of determining the development of a specific characteristic of an organism
- a length of nucleic acid in a chromosome, typically about a thousand bases long, which codes for a length of messenger RNA (directly) and a polypeptide (indirectly) – polypeptides typically function as enzymes in cell metabolism.

Genome:

- the haploid complement of genes of a cell or organism.

A gene or genes from one organism can be transferred to the set of genes of another organism. This process is called **genetic engineering**, and the organism is said to have been **genetically modified** (GM). Many of the GM organisms produced so far are bacteria, but eukaryotes may also be engineered, including yeast. This relatively new technology has important applications in biotechnology, pharmaceuticals manufacture, medicine, agriculture and horticulture. Genetic engineering may raise environmental problems, and certainly raises ethical issues (page 61).

1 What are the key steps by which a gene dictates the type of protein a cell manufactures?

One uncontroversial application is the transfer of the human genes for insulin production to a strain of the bacterium *Escherichia coli* (Figure 8.1). Insulin consists of two short proteins (polypeptides) linked together by sulphide bridges (–S–S–). This hormone enables body cells to regulate blood sugar levels, and a supply of human insulin is needed to help in the treatment of diabetes. Cultures of *E. coli* have been engineered to manufacture and secrete insulin when cultured in a bulk fermenter with appropriate nutrients.

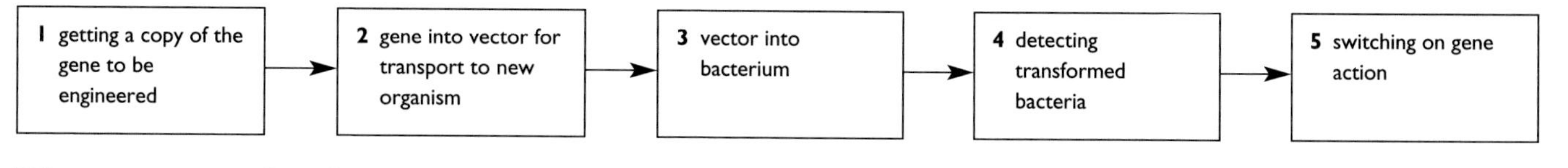

Figure 8.1 The steps to genetic engineering of *E. coli* for insulin production.

Three strategies for copying a gene

Starting with messenger RNA

A human pancreas contains patches of cells (islets of Langerhans) where insulin is produced. The relevant genes in the nuclei of these cells are transcribed to produce messenger RNA. This messenger RNA passes out to the ribosomes in the cytoplasm. Here the base sequence of the RNA is translated into a linear sequence of amino acids of the insulin proteins.

In the laboratory, messenger RNA for insulin can be isolated from a sample of human pancreas tissue. Using this RNA and the enzyme **reverse transcriptase** (itself obtained from a type of virus called a retrovirus, page 20), a copy of the gene can be synthesised, as shown in Figure 8.2.

human pancreas – here insulin is synthesised in β cells of islets of Langerhans – so **mRNA for insulin production** may be extracted from this tissue

purified mRNA coding for insulin

AUGGAACACUGGCACCGUUGCUGU

reverse transcriptase enzyme added – this synthesises a complementary strand of DNA (using a pool of nucleotides) by base pairing with the sequence of bases of the mRNA

mRNA being used as a template for DNA synthesis

mRNA strand is then discarded

AUGGAACACUGGCACCGUUGCUGU
TGTGACCGTGGCAACGACA

cDNA strand – DNA complementary to base sequence of mRNA

cDNA being used as a template for DNA synthesis

DNA polymerase enzyme added – this synthesises a second DNA strand, complementary to the base sequence of cDNA

TACCTTGTGACCGTGGCAACGACA
TGGCACCGTTGCTGT

the two DNA strands are the **gene for insulin**

TACCTTGTGACCGTGGCAACGACA
ATGGAACACTGGCACCGTTGCTGT

Figure 8.2 Using reverse transcriptase to build the gene for insulin.

Working backwards from a protein

Where the primary structure of a protein (its linear sequence of amino acids) is known, the genetic code for the protein can be worked out. With this information a copy of the gene may be synthesised, using recently developed automated machines for amino acid sequencing and for synthesis of DNA (Figure 8.3).

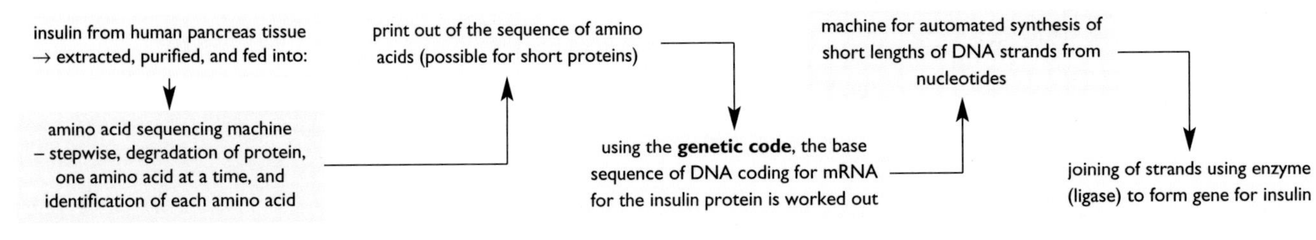

Figure 8.3 Synthesising a gene using knowledge of the protein structure it codes for.

Obtaining a gene from the chromosome it occurs in

The DNA of human chromosomes may be 'cut' into short fragments by an enzyme called a **restriction enzyme** (itself obtained from bacteria, where it combats 'infection' by nucleic acid from invading viruses). Fragments have then to be sorted according to size, and the location of the fragment containing the insulin genes found using a DNA probe (Figure 8.4).

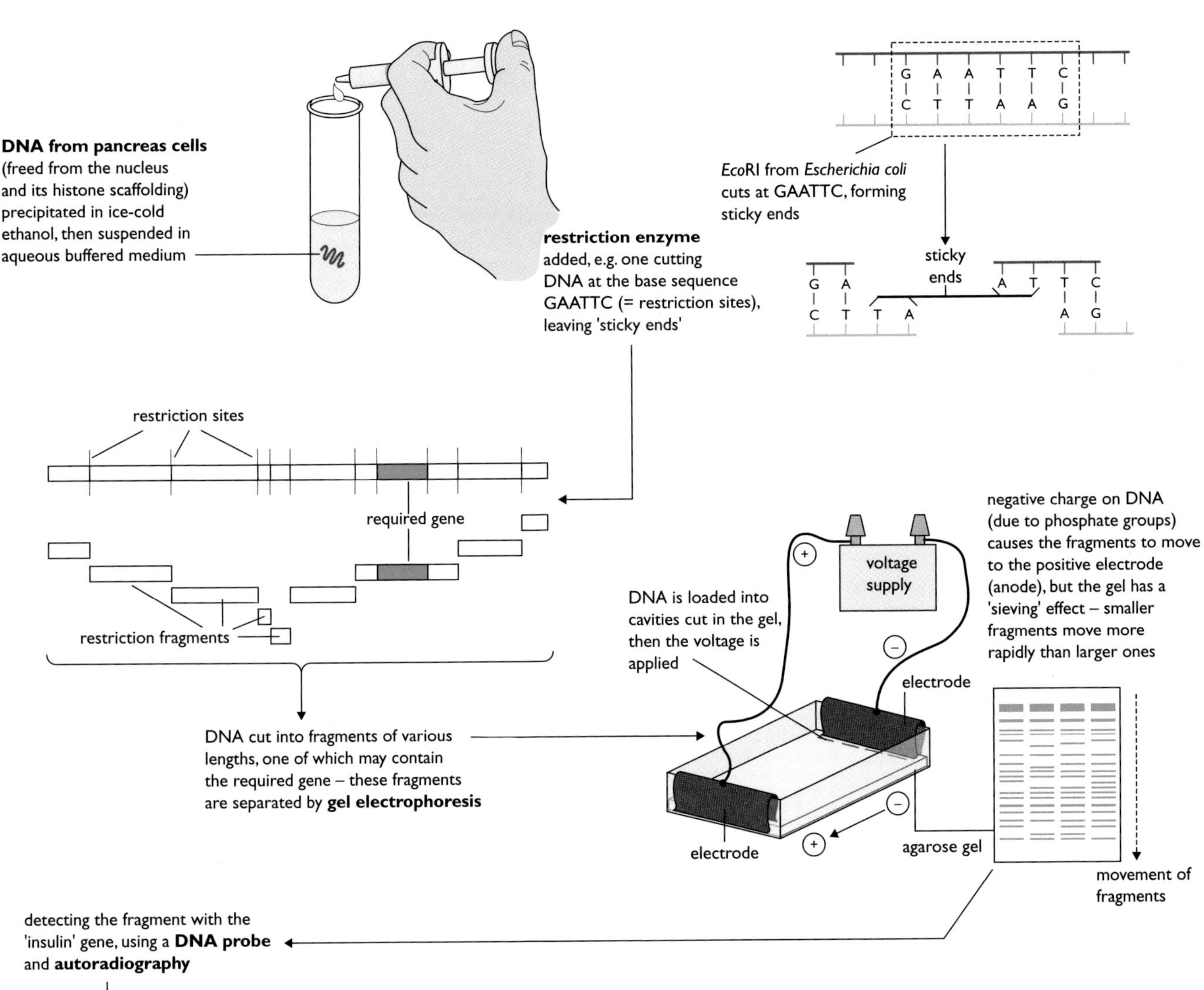

Figure 8.4 Obtaining the insulin gene by cutting it from the DNA of chromosomes.

Insulin gene into a vector

In biology, a vector is a transport agent involved in movement between organisms. In genetic engineering, the **plasmids** typically found in bacterial cytoplasm (page 3) are commonly chosen as vectors for genes (bacteriophages may be used as alternative vectors).

The DNA of the plasmid is cut open using a restriction enzyme, and the same restriction enzyme is used to cut the chromosome into gene fragments. This creates complementary sticky ends (Figure 8.4, page 55), so the gene can then be spliced into the plasmid.

Figure 8.5 A plasmid as vector for the insulin gene.

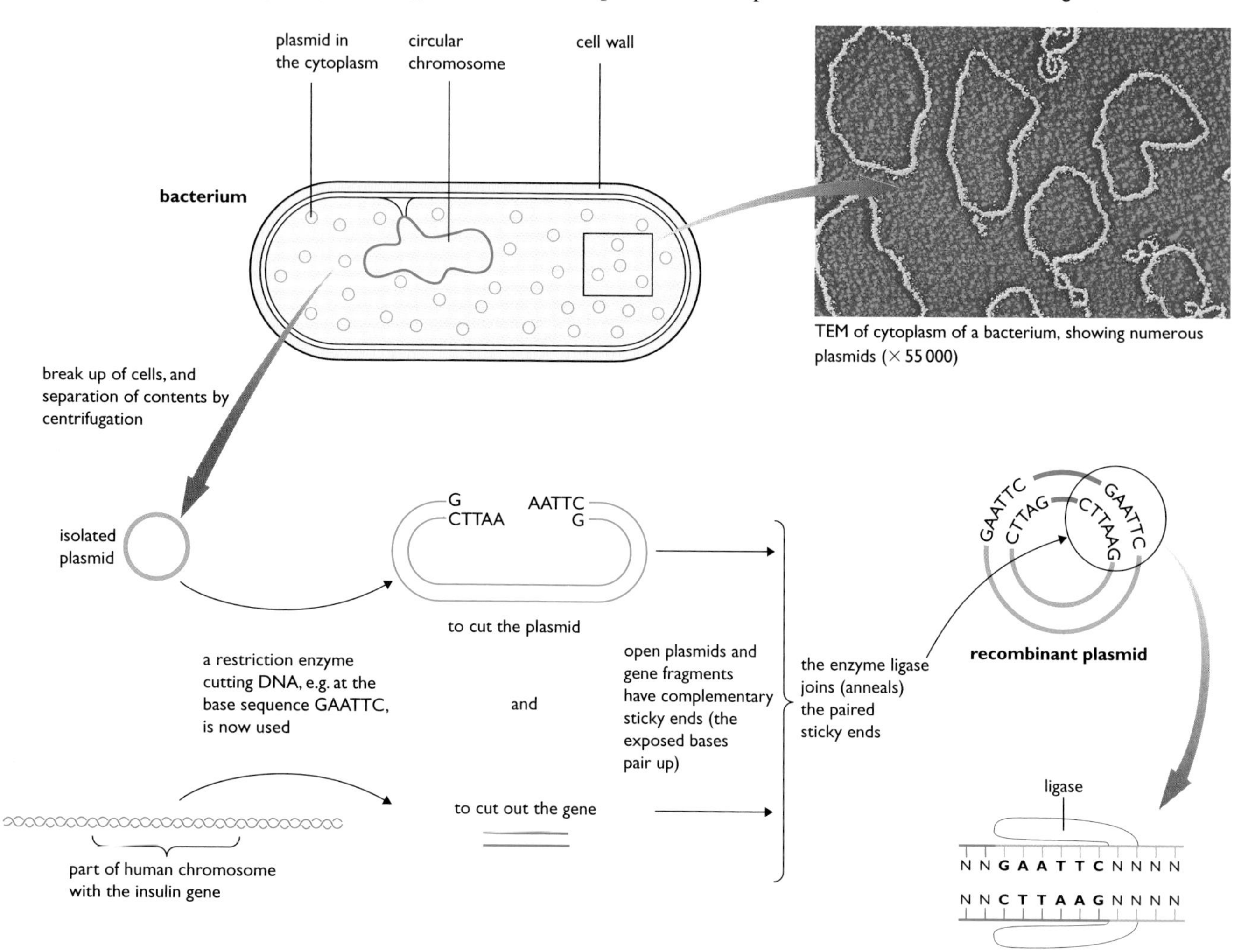

Vector into bacterium

In the next step, recombinant plasmids have to be returned to bacterial cells. This is a challenge, as the wall is a barrier to entry. The ways this may be done are shown in Figure 8.6.

Figure 8.6 Return of recombinant plasmids to the bacterium.

recombinant plasmids

entry to intact bacterial cells by

'zapping' – treatment with brief electric shock

or

treatment with ice-cold $CaCl_2$ solution, followed by brief incubation at 40+ °C

temporary creation of tiny pores in walls of host bacteria, allowing entry of recombinant plasmids

bacterial cell

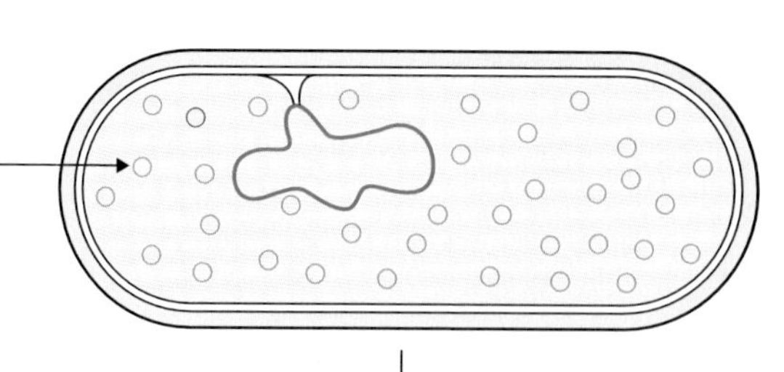

transgenic bacteria

Finding the engineered bacteria

Very few bacterial cells that have been treated as in Figures 8.5 and 8.6 will have successfully taken up recombinant plasmids, and thus be able to synthesise insulin. Yet it is only these GM bacteria that should be grown up in a fermenter, if a significant amount of product (insulin) is to be produced. How they are selected, using a plasmid with **antibiotic resistance genes**, is summarised in Figure 8.7.

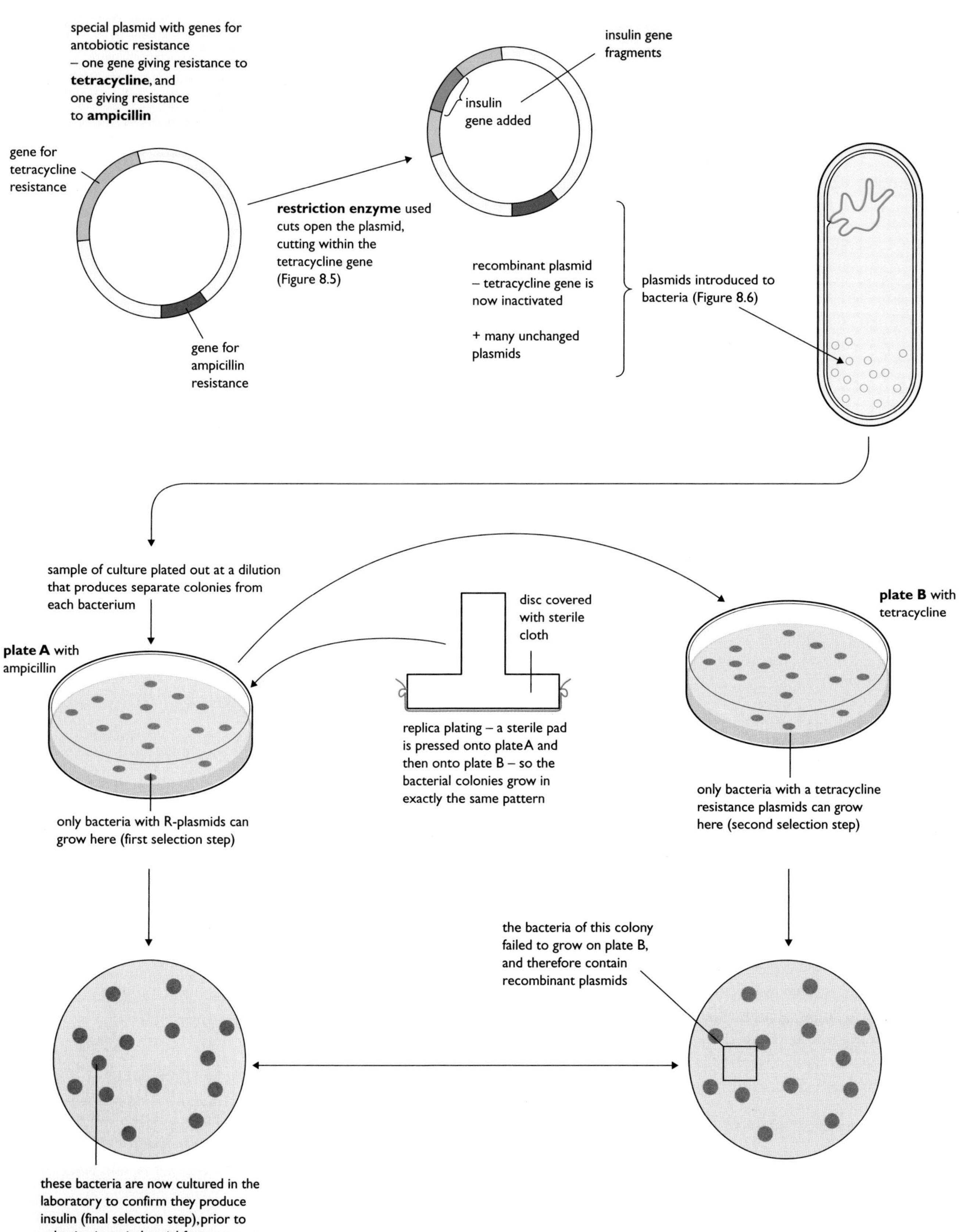

Figure 8.7 R-plasmids for selecting genetically modified bacteria.

Switching on gene action

In the living cell, some genes are expressed throughout the life of the cell because the protein coded for is required constantly. Other genes have to be activated, or 'switched on', first. In bacteria, one switch mechanism is the **lactose operon**. The genes of this switch mechanism may be inserted alongside the insulin gene. The result will be that when lactose is present in the medium, the recombinant bacteria will be activated to produce insulin.

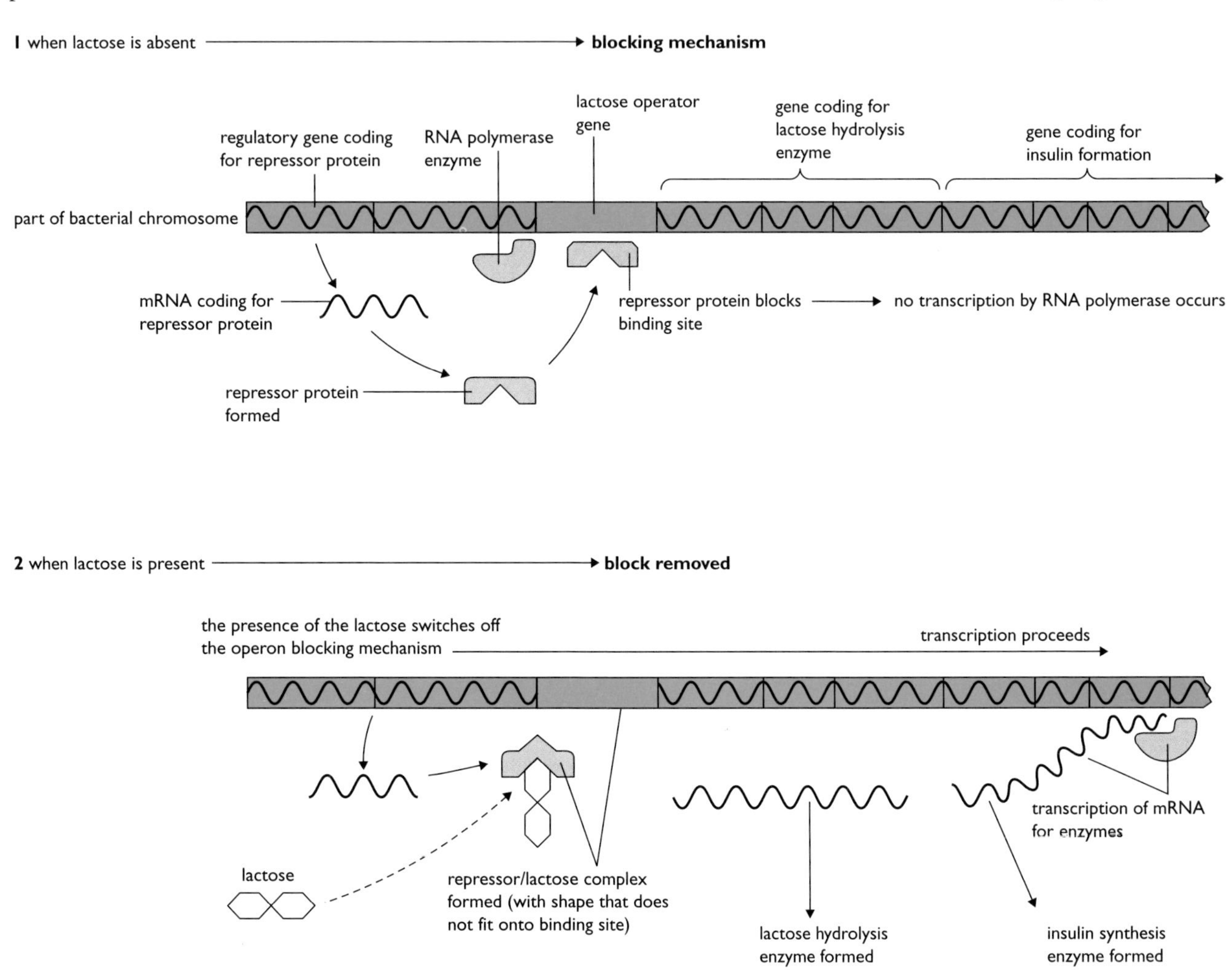

Figure 8.8 A mechanism for switching on gene action.

Other applications of genetic engineering in prokaryotes

Human growth hormone

Another protein synthesised by the pharmaceutical industry using genetically modified bacterial cells is human growth hormone (**hGH**), also known as **somatotrophin**. This protein is one of a group of hormones normally synthesised in the pituitary gland and circulated around the body in the blood stream. hGH stimulates the growth of body cells, particularly of the skeleton and skeletal muscles. It works by enhancing cell protein synthesis, and the uptake of amino acids by the plasma membrane from the plasma and tissue fluid. Also, it enhances conversion of glycogen to glucose in the liver, so tending to cause the blood sugar level to rise. Failure to secrete sufficient hGH in children and young people in their growth years leads to a condition called **pituitary dwarfism**. This can be overcome by a series of injections of hGH. A similar hormone extracted from other mammals is not effective in humans, so the supply of hGH from GM bacteria is important. hGH is produced from cultures of *E. coli* that have been engineered to carry the gene for this hormone. The biotechnological techniques involved are very similar to those used in insulin production (pages 54–57). The hGH gene may be activated by the lactose operon (above).

Figure 8.9 Human patient suffering from deficiency of growth hormone. This 11-year-old required treatment with hGH hormone to overcome the effects of childhood deficiency that resulted in the condition of dwarfism. A programme of regular injections was administered by the doctor and the effect monitored.

Bovine somatotrophin

Bovine somatotrophin (**BST**) is a growth hormone of similar chemical structure to hGH. It is produced in the pituitary glands of cows at a steady rate, and so is naturally present in the blood stream at low concentration. The gene for BST has been obtained from cows and engineered into a bacterium, as described above, and this hormone is also produced on an industrial scale. It is marketed for injection into lactating cows – additional BST in the blood stream raises the milk yield per cow. The issues generated by BST use are illustrated in Figure 8.10.

improved milk yield issues

- injection of BST into diary herds in USA since 1993 has raised yield per cow by 10–15%
- in Europe, excess milk is already produced (→ milk 'lake' in the countries of EU); production is capped on dairy farms by a quota system, and this is steadily tightened to reduce total production
- biotechnology firms argue that use of BST to raise yields per cow increases efficiency (i.e. conversion of expensive concentrates into milk) and therefore improves profitability; however little milk is produced

herd and human health issues

- milk yield per cow has almost doubled in the past 40 years; cows may suffer physiological stress if made to increase yields still further
- herds treated with BST have increased incidence of mastitis (bacterial infection of udder, treated with penicillin)
- the bovine growth hormone, IGF1, occurs at significantly higher concentrations in milk of BST-treated herds; if this protein is not digested in the stomach of milk drinkers, it may trigger growth of gut-lining cells at up to twice the normal rate

Figure 8.10 Issues in the use of recombinant BST for enhanced milk production.

Prokaryotes are easier to engineer than eukaryotic cells

Engineering the genes of prokaryotes is easier than those of eukaryotes because:

- prokaryote life cycles are generally much shorter, and they grow quickly
- there is a single chromosome present, rather than pairs of chromosomes with two (or more) alleles to each gene, as in eukaryotes
- tiny rings of nucleic acid, called plasmids, commonly occur in the cytoplasm of prokaryotes – plasmids are excellent vectors for transferring genes between cells (yeasts, although eukaryotes are also relatively easy to engineer as their cytoplasm also contains plasmids).

Genetic engineering of eukaryotes using yeasts

Yeasts are eukaryotic cells, so they possess a nucleus with paired chromosomes (Figure 4.6, page 26). Unusually, though, their cytoplasm contains plasmids very much like those of prokaryotic cells. Also, yeasts can be grown in liquid culture or on the surface of agar plates, just as bacteria are. So yeasts are the most frequently used eukaryotes in genetic engineering. They accept their own plasmids, or those from prokaryotes that have been engineered, as vectors. Re-entry of plasmids is induced by treatment with lithium salts.

The pharmaceutical industry has produced recombinant yeasts in the processes of manufacturing:

- a vaccine against hepatitis
- hirudin, an anti-coagulant produced by leeches to aid their feeding mechanism, now used as a drug in the treatment of blood-clotting disorders
- an enzyme – superoxide dismutase – used to prevent tissue damage by free radicals in the post-operative care of heart patients.

2 Transfer of the genes controlling root nodule formation from leguminous plants to cultivated members of the grass family has not yet proved possible. In what ways would this development, should it come about, be advantageous?

GM plants using *Agrobacterium tumefaciens*

The otherwise harmless, soil-inhabiting bacterium *Agrobacterium tumefaciens* can invade the tissues of dicotyledonous plants (but not of monocotyledonous plants such as grasses and cereals). The infection occurs at the junction of stem and root, where it causes the growth of a tumour (called a plant gall). The gene for tumour formation occurs in a plasmid, the **Ti plasmid**, in the bacterium. Useful genes (e.g. for resistance to a viral infection) can be added to the Ti plasmid (using **restriction enzyme** and **ligase**, page 55), and the recombinant plasmid returned to the bacterium. When a crop plant is infected by that bacterium, recombinant plant gall tissue forms. Next, the gall tissue is cultured into independent plants – these are GM plants, carrying a recombinant gene, e.g. for virus resistance (Figure 8.11).

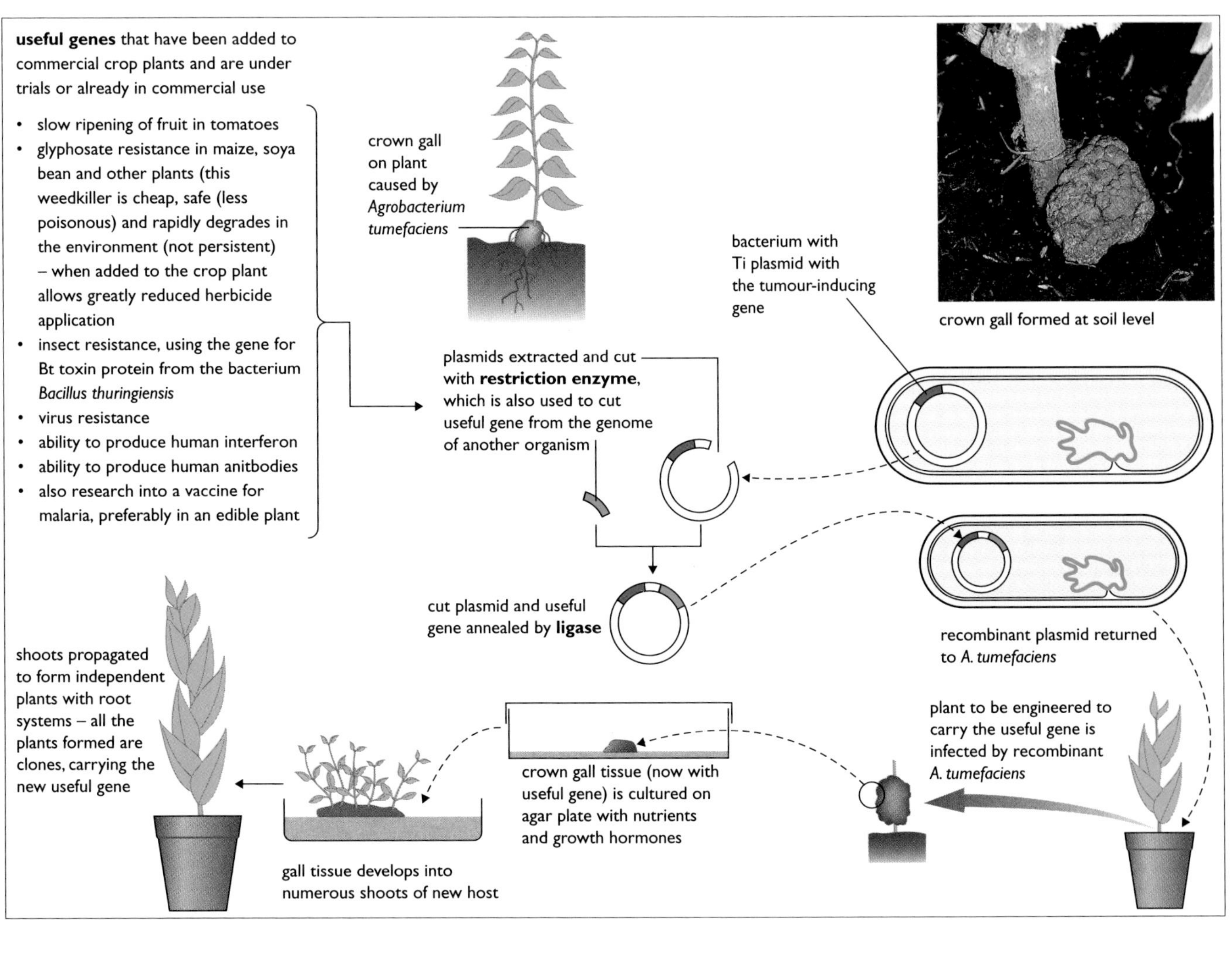

Figure 8.11 Forming transgenic crop plants with Ti plasmids from *Agrobacterium*.

Despite the potential benefits of growing many of the GM crop plants developed so far, there has been strong opposition to this development in the UK. There are anxieties about whether genes will move into other organisms around the GM crop. On the other hand, at the international level there is recognition that GM crops offer significant benefits (Figure 8.12).

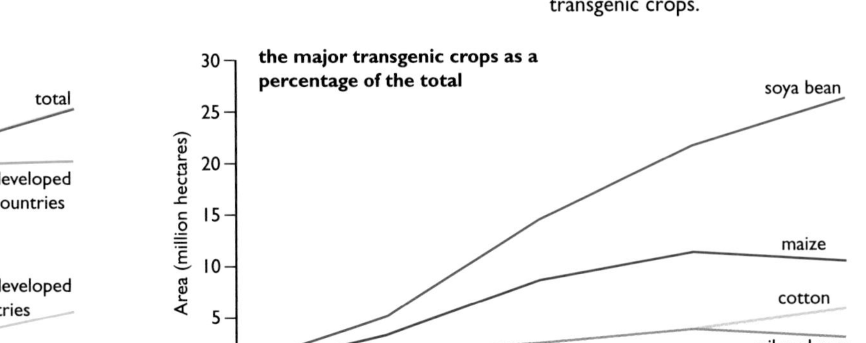

Figure 8.12 Global area of transgenic crops.

Vaccines from plants using a virus as vector

Vaccines for animals and humans (page 77) have traditionally been produced by growing the disease-causing organism in culture, and then producing dead or attenuated (weakened) forms to inject into healthy people, in order to induce immunity. The vaccine contains one or more proteins of the outer surface of the pathogen that are recognised by the human immune system on entry (antigens). Protective antibodies are then produced against the antigen, and the ability to produce them in future is set up (page 75, memory cells).

An alternative approach currently being researched involves the use of a plant and a specific virus that infects it. By genetic engineering, an antigen from the animal pathogen can be added to the plant virus protein coat. If this GM plant virus is used as a vaccine, it will trigger the formation of antibodies against the pathogen's antigen, conferring future immunity against invasion by the actual living pathogen. The plant virus chosen must be chemically robust, and harmless to all organisms except its plant host. This technique is summarised in Figure 8.13.

3 What potential problems may arise from the exposure of microbes, genetically modified for laboratory and industrial processes, to populations of similar microorganisms in the outside environment?

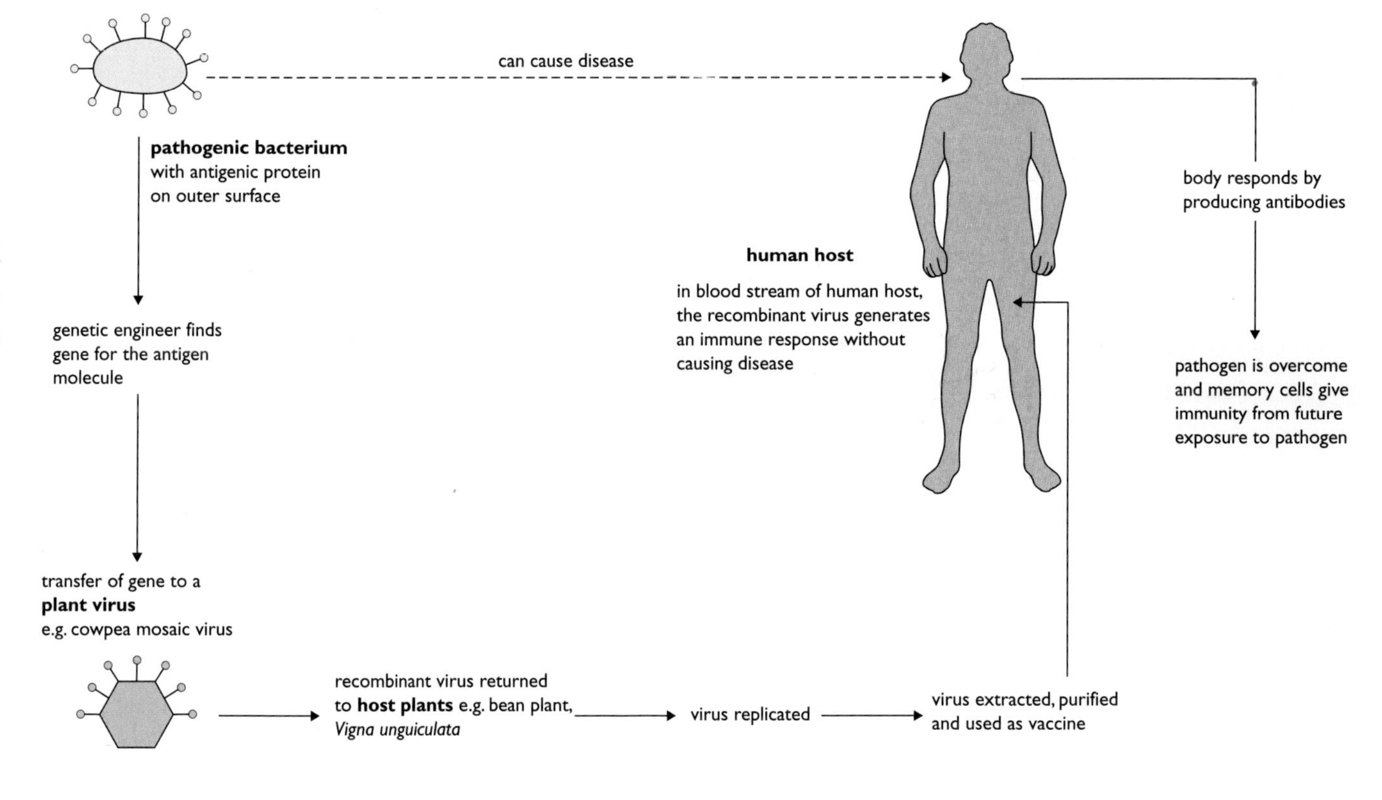

Figure 8.13 Steps to vaccine production using plant viruses.

Environmental and ethical issues in genetic engineering

There are great benefits from the genetic engineering work illustrated here, but there are also real and potential dangers. A decision to go ahead with a particular project may raise ethical or environmental issues, some of which are highlighted below. Can you think of others?

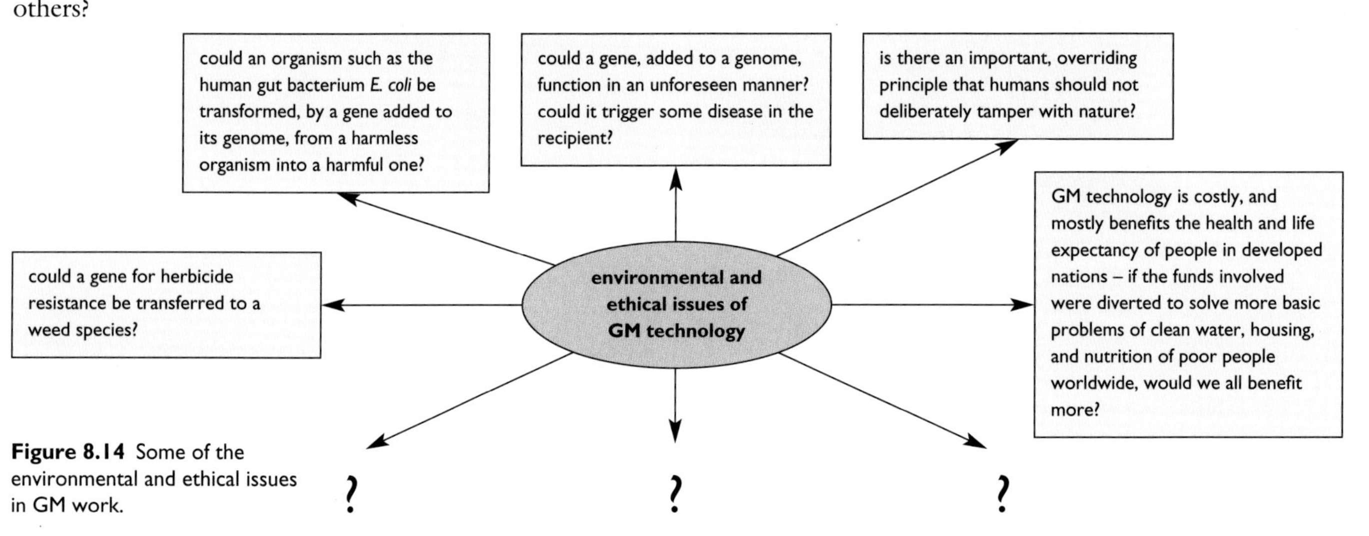

Figure 8.14 Some of the environmental and ethical issues in GM work.

9 Microorganisms, biodeterioration and preservation

The saprotrophic nutrition of bacteria and fungi causes the breakdown of dead organic matter, and allows the recycling of nutrients within the environment (Chapter 6, page 30). When that organic matter is waste (dead organisms, leaf litter, faeces, etc.), we describe this process as **biodegradation**, and we recognise it as an indispensable environmental process.

But saprotrophic organisms are just as likely to attack valuable organic resources. If conditions are suitable, foods, fabrics, fuels and many building materials are all vulnerable to decay caused by microorganisms, and damage of this type is frequently seen. The microbiological decay of economically useful materials is known as **biodeterioration**. The prevalence of biodeterioration makes preservation a huge industry.

walking shoes – the leather is damp and the surrounding air is humid

dry-rot fungus of wet roof timbers, in a non-ventilated roof space

Concorde in flight – in supersonic flight the aeroplane 'skin' is warm, and fuel stored in the wings becomes 'food' for microorganisms which come to block the fuel supply pipes; fuel additives are required to prevent this growth

cheese with surface mould that has grown in moist storage conditions (food preseveration is illustrated on page 64)

Figure 9.1 Examples of biodeterioration.

Plastics – a case of low biodegradability as a problem?

The term plastic refers to any solid material which becomes mobile when heated and can thus be cast into moulds, extruded to form tubes, or rolled out into sheets and layers. Today's plastics are various synthetic polymers, manufactured from fossil fuels (oil or coal), including polythene, polypropylene, polyvinyl chloride, perspex and polystyrene. Plastics occur in most manufactured goods, especially in packaging where they have largely taken over from paper products. Consequently, plastics are a major component of rubbish. This is a problem because many plastics biodegrade in the environment only very slowly (Figure 9.2).

Figure 9.2 Plastics as an environmental problem. These discarded items were deposited on the beach by wave and tide action.

A biological source of commercial plastic

Alternative sources of plastic are being sought because fossil fuels will eventually run out. A bacterium, *Alcaligenes eutrophus*, has the ability to produce granules of a plastic called polyhydroxybutyrate (PHB) and store them in the cytoplasm. The bacterium produces most PHB when supplied with excess carbohydrate nutrients, but is deficient in nitrate and phosphate ions. This natural plastic is held as a food reserve in the cytoplasm, but can be extracted under industrial conditions, and forms a plastic similar to polypropylene. The addition of varying amounts of proprionic acid to the medium when *Alcaligenes* is grown in an industrial fermenter allows the strength and flexibility of the product to be adjusted. Unfortunately, production of this plastic is still quite an expensive process, compared with traditional plastic made from fossil fuels. However, it has been used to manufacture bottles, films and fibres. PHB and its derivatives are strong materials, and have the added advantage of quite speedy biodegradability when discarded in contact with soil microorganisms.

An alternative source of PHB is under investigation, from species of flowering plants, such as oilseed rape, that store oils as food reserves. A weed plant, thale cress (*Arabidopsis thaliana*), has been successfully genetically engineered to produce and store granules of plastic in its chloroplasts, in the light. In this case the genes for PHB synthesis were obtained from the bacterium *Alcaligenes*. However, plastics from fossil fuels are still cheaper to produce.

1 What would be the environmental gain if plastics were manufactured using oils from crops such as oilseed rape?

Figure 9.3 A biological source of plastics.

seeding of thale cress, *Arabidopsis thaliana*, genetically engineered to produce PHB in the chloroplasts and store it there

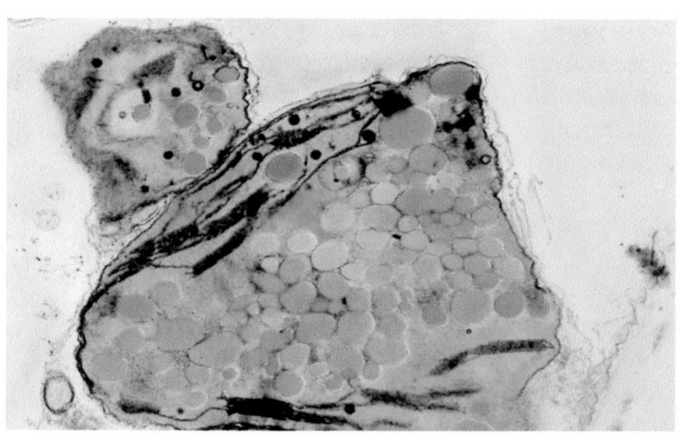

coloured TEM of chloroplast of *Arabidopsis*, with stored granules of PHB (coloured yellow)

Making traditional plastics more biodegradable?

Some packaging plastics that are otherwise only slowly broken down by soil microorganisms may be made more speedily biodegradable by the addition of tiny starch granules in the plastic film. The presence of this starch aids faster disintegration, but the particles of synthetic plastic themselves remain slow to decay.

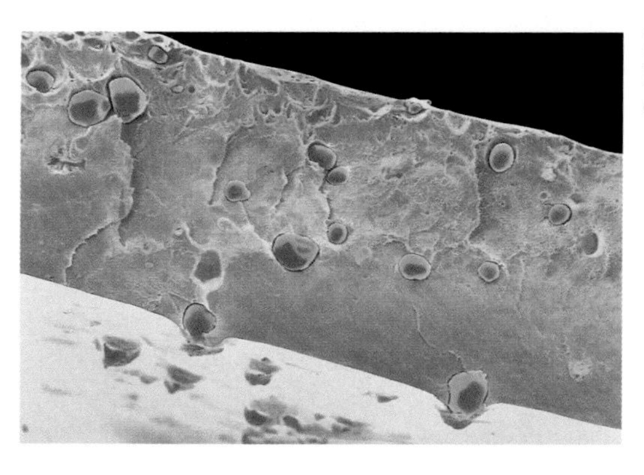

Figure 9.4 Making plastic degradable.

starch granules have been introduced into this plastic film during manufacture, increasing the speed of disintegration as it ages

Food preservation by processing

Microorganisms are ubiquitous, so their presence in food is inevitable. If food items are not consumed as soon as they have been prepared, decay by microorganisms must be prevented in some way. Figure 9.5 summarises possible approaches to food processing for preservation. For any particular type of food, the method of preservation chosen must maintain the original character (flavour and appearance), as well as its nutritional value, as far as possible.

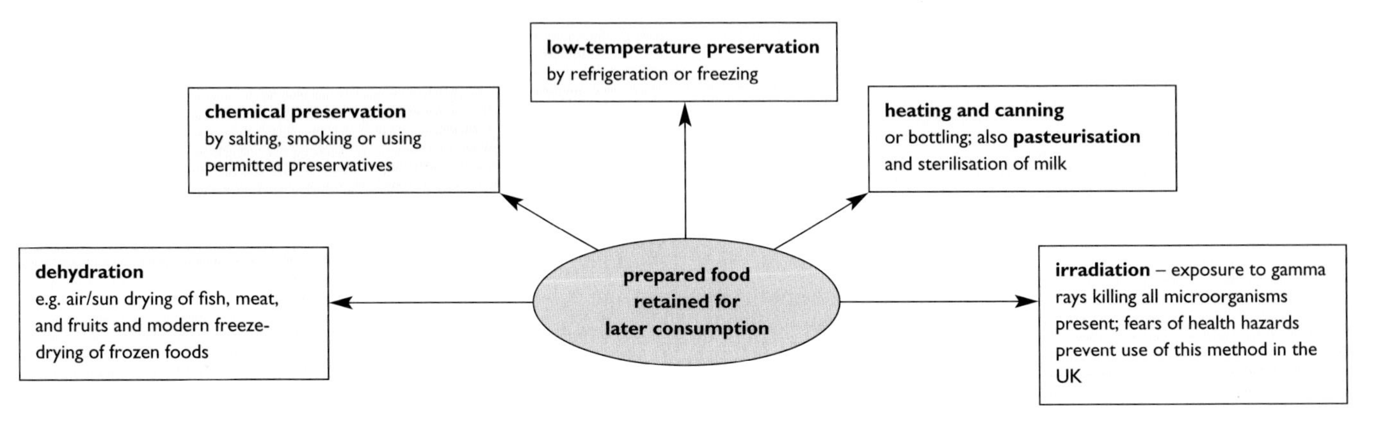

Figure 9.5 Techniques of food preservation.

Chemical preservatives

Chemicals such as sodium chloride, sugars, vinegar, alcohol and wood smoke discourage or prevent the growth of microorganisms, and they have been used in food preservation for many hundreds of years. Other chemicals, toxic to microorganisms but also toxic to humans to various degrees, have been used. For example, milk stays fresh much longer if treated with formalin (a preservative for zoological specimens), and this was widely used at one time. However, formalin is very toxic to humans and its use is now banned. Today, the use of chemicals as preservatives in foods is regulated. Permitted preservatives are identified by numbers (Table 9.1). Where use of the preservative is agreed throughout the European Community, the letter E precedes the number.

Refrigeration and freezing

Low temperatures slow the rate of metabolism of microorganisms contaminating food, and so aid preservation. The domestic fridge holds its contents at about 5 °C. This is helpful for short periods, but microorganisms can still grow and reproduce. Enzyme action in the stored food also continues slowly, leading to a loss in quality. So for long-term storage food must be frozen and stored at low temperature. For example, domestic freezers store food at –18 °C.

The rate at which food is frozen is also significant – for good quality, food should be frozen quickly, reducing the temperature within the food to about 0–4 °C in no less than 30 minutes. Chilling food more slowly than this allows large water crystals to form, and these disrupt the cells and tissues. The damage this causes becomes evident on thawing. The loss of nutritional value in freezing and low-temperature storage is very small. The disadvantage of this method lies in the cost: electricity is needed continuously, and relatively expensive equipment must be maintained.

Table 9.1 Examples of permitted chemical preservatives, and their typical uses

Preservative	Uses
E2000 sorbic acid	Soft drinks, fruit yoghurt, processed cheese
E201 sodium sorbate E202 potassium sorbate E203 calcium sorbate	Frozen pizza, flour, confectionery
E210 benzoic acid E211 sodium benzoate E212 potassium benzoate	Beer, jam, salad cream, soft drinks, fruit pulp, fruit-based pie fillings, marinated herring and mackerel
E220 sulphur dioxide E221 sodium sulphite E222 sodium hydrogen sulphite	Dried fruit, dehydrated vegetables, fruit juices and syrups, sausages, fruit-based dairy products, cider, beer and wine, also used to prevent browning of raw peeled potatoes and to condition biscuit dough
E230 diphenyl E231 2-hydroxydiphenyl E232 sodium 2-phenylphenate	Surface treatment of citrus fruits
E233 thiabendazole	Surface treatment of bananas
E234 nisin (an antibiotic)	Cheese, clotted cream
E239 hexamine	Marinated herring and mackerel
E249 potassium nitrite E250 sodium nitrite E251 potassium nitrate	Bacon, ham, cured meats, corned beef and some cheeses
E280 propionic acid E281 sodium propionate E282 calcium propionate	Bread and flour, confectionery

Source: Hassen, M. (1984) *E is for Additives: The Complete E Number Guide.* London: Thorsons.

Dehydration

Preservation of food by dehydration is an ancient tradition in many communities, typically involving fish, meat and fruit (Figure 9.6). The process, reducing the water content of food to the level where microorganisms cannot grow, is highly practical where there is sufficient sunlight. An added advantage is that the product occupies very little storage space. Some loss of food value (e.g. of vitamin A) occurs.

A modern development is freeze-drying, in which frozen foods are dried under high vacuum. This method is expensive as it involves special equipment and the use of electricity. However, the product is dehydrated without loss of colour, and the vitamin content of food is not reduced.

Figure 9.6 A fish catch is laid out on the sandy shore to dry in the sun.

Heating

In **canning**, food is sealed in a container and then heated to a temperature that kills all microorganisms and their spores, which might otherwise grow during storage. For vegetables and meat this is about 115 °C, but for fruit 100 °C is sufficient. Today, higher temperatures may be used, which allows the duration of heat treatment to be reduced. Some nutrient loss occurs during the heating process, e.g. of vitamin C and thiamine (from meats).

A few viable microorganisms may remain, but conditions must be extreme enough to kill the heat-resistant spores of the bacterium *Clostridium botulinum*. This organism is an anaerobic bacterium, and produces a toxin which is the most virulent poison known. However, even if heat treatment has been insufficient, this species of *Clostridium* can only survive if the pH is above 4.5. Commercial canners are well aware of the dangers of botulism, and see to it that products above pH 4.5 have the necessary rigorous heat treatment.

Canned foods should be stored in cool, dry conditions to prevent the can rusting externally, and to prevent growth of any thermophilic bacteria that have (unusually) survived heat treatments.

2 How does the addition of quantities of sugar or salt, or the drying of foods, prevent decay when foods are exposed to the air?

Heat treatment of **milk** is necessary as it is an ideal medium for the growth of microorganisms, and is a difficult product to keep free from contamination from the air. Occasionally an unhealthy cow may be the source of harmful bacteria, including *Mycobacterium tuberculosis* (causing TB); *Brucella abortus* (causing brucellosis); or *Streptococcus pyogenes* (causing sore throats and scarlet fever). **Pasteurisation** of milk (72 °C for 15 seconds, then rapid cooling to below 10 °C) kills over 99% of the bacteria present. Pasteurised milk keeps longer, too, because bacteria that turn lactose (milk sugar) into lactic acid (which causes milk to turn sour) are also killed off. Alternatively, milk can be rendered bacteriologically sterile by ultra-high temperature (UHT) treatment (homogenised, then heated to 132 °C for 3 seconds, then cooled).

Figure 9.7 Plant for the pasteurisation of milk.

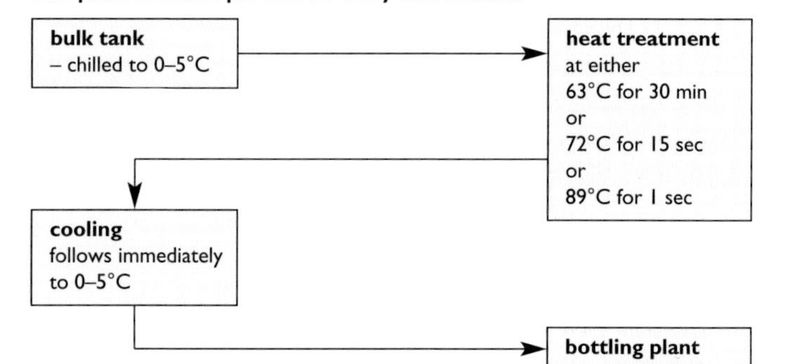

10 Microorganisms and disease

Disease is defined as an unhealthy condition of the body. Many diseases are due to parasitic organisms invading the body. Such organisms are referred to as **pathogens**, and the diseases they cause are called **communicable diseases** because the pathogen may be passed from host to host in various ways (Figure 10.1).

Microorganisms cause the majority of communicable diseases, but only relatively few of the vast numbers of microorganisms are pathogens (most are free-living saprotrophs). Despite this, the pathogenic microorganisms generally appear to have earned all microorganisms a bad name, at least with the general public. People tend to equate bacteria with disease, despite the many ways our quality of life and environment are maintained by them.

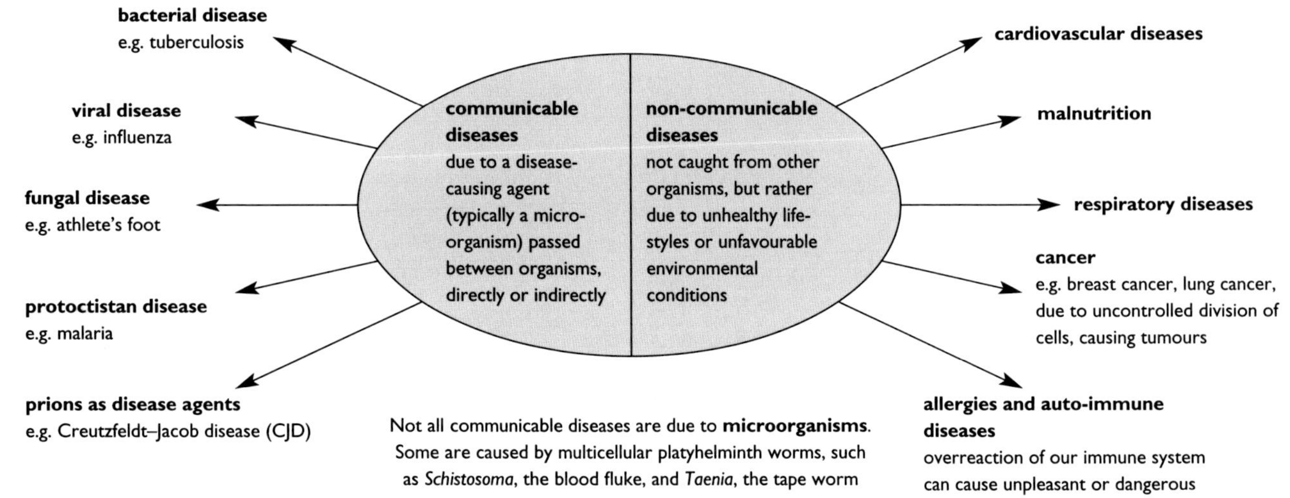

Figure 10.1 Types of human disease.

Knowing which is the pathogen

When a human is ill with a disease, it may be difficult to tell which, if any, of the organisms present is the pathogen. Our bodies have a huge flora of microorganisms. We have to differentiate the harmless from the pathogenic microorganisms if a disease is to be understood and then treated successfully. This problem was tackled by the microbiologist Robert Koch (1843–1910), and his rules are known as **Koch's postulates** (Figure 10.2).

the huge numbers of microorganisms inhabiting the human body consist of permanent residents, plus some occasional visitors (including occasional unpleasant pathogens).

in all, about 200 species (mostly bacteria) make up this microflora.

80 of these species are commonly found in the mouth.

typically:

- 10 million bacteria occur on an average cm^2 of external skin
- 10 billion bacteria occur on an average cm^2 of gut epithelium
- 100 billion bacteria are excreted daily, in the faeces, by an adult

Which bacterium found on an ill patient has caused the disease?

The rules by which this question is resolved are:

- the same bacterium must be found in all cases of the disease
- the pathogen must be isolated and grown in pure culture from the infected host
- bacteria of the pure culture must be shown to cause the disease in a healthy host
- the bacterium must be isolated from the experimentally infected host

Figure 10.2 Koch's postulates.

Prevention is better than cure

Preventive medicine is a discipline that rarely catches the headlines, compared to work on curing the very ill by devising new, exciting techniques. Nevertheless, preventive medicine is frequently more important to the majority of people's health. Here the emphasis is on the conditions and lifestyle that help us to avoid contracting diseases in the first place, including:

- eating good food in the right proportions
- access to clean water
- safe disposal of sewage
- avoidance of atmospheric pollution
- provision of uncrowded living conditions, and a secure position in society.

1 Given the prevalence of heart disease and cancer in developed countries, how may preventive medicine effectively contribute to combating them?

Bacterial diseases – how bacteria cause disease

The few pathogenic bacteria that enter our bodies tend to do so through internal surfaces, typically those of the respiratory, alimentary and urinogenital systems. The external skin is such an effective barrier that it is normally only crossed at a break, such as an open wound or insect bite. The steps to infection are summarised in Figure 10.3.

Figure 10.3 How pathogenic bacteria may cause disease.

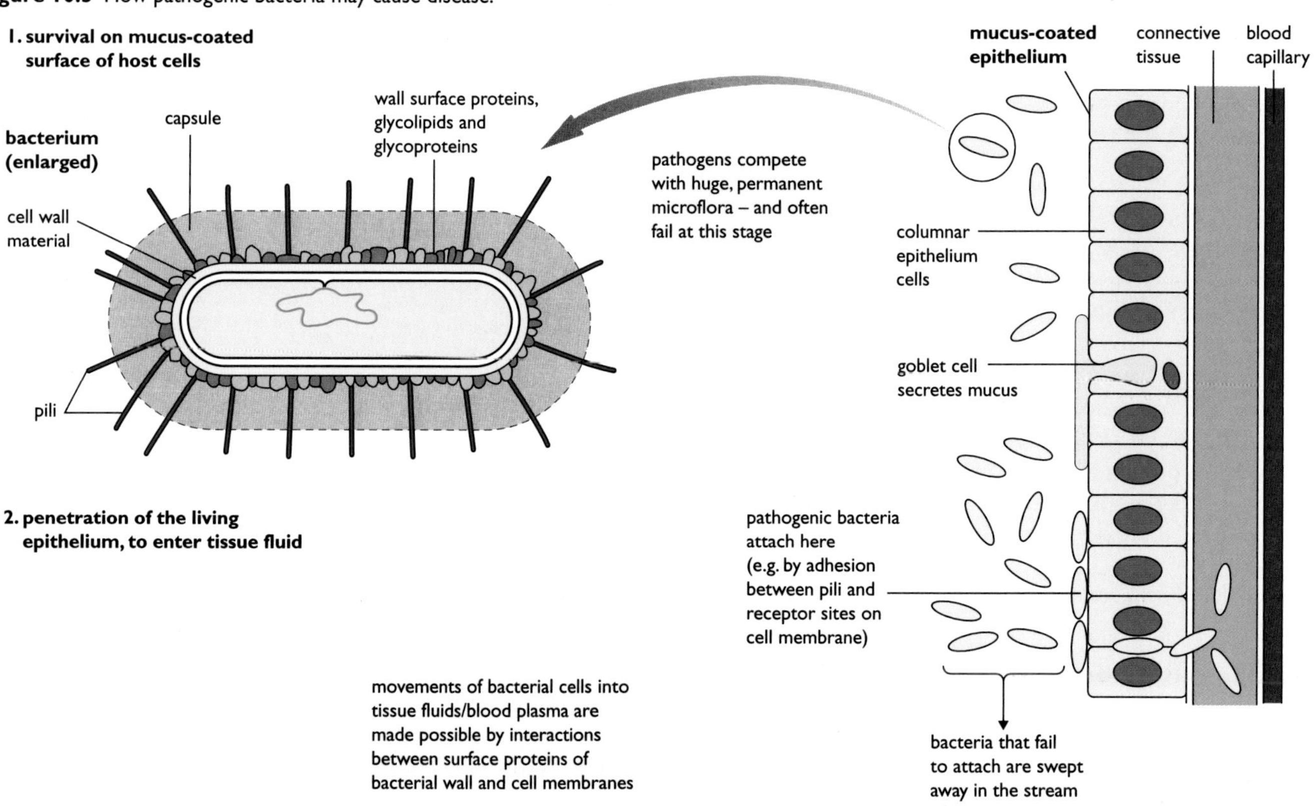

3. survival and growth in the host, by overcoming the chemical defences

arrival of foreign cells in tissue fluid is detected and triggers a cascade of **complement proteins** (Figure 10.19, page 74); and specific **antibodies** of the immune system secreted by plasma cells (Figure 10.21, page 75)

only pathogenic bacteria that **inactivate these defence mechanisms of the host** (using the proteins, glycoproteins and glycolipids of the walls) survive and cause an infection

4. damage to the host

pathogens then move through the lymph vessels and blood system
– often **lodging themselves in particular tissues**, e.g. bacteria causing meningitis, which lodge in brain membranes called meninges

symptoms of disease are caused by:
1. poisons (toxins – specific protein molecules) originating in or on the pathogen

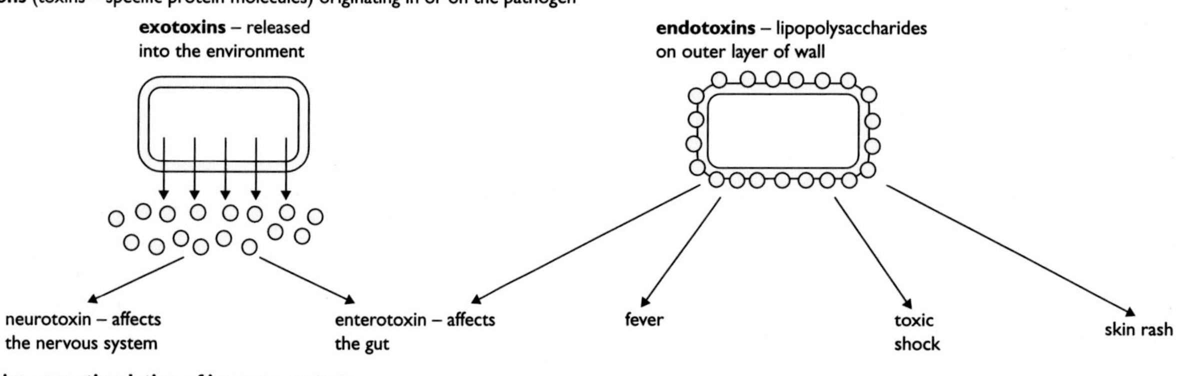

2. massive over-stimulation of immune system
e.g. accumulation of phagocytes in lungs of TB patient (page 72), releasing enzymes that destroy lung tissue

Food poisoning (enteritis) caused by bacteria

Enteritis is an **acute** disease (arising suddenly, but of short duration) that results from the ingestion of food containing certain microbes (usually bacteria) or microbial toxins. The organisms involved may reproduce themselves in food (e.g. when prepared food is not refrigerated), or in the host. The bacteria or toxins affect the linings of the small and large intestine. Typically, the patient develops abdominal pain and diarrhoea, with or without vomiting and fever.

The rise and rise of food poisoning

The number of reported cases of food poisoning has been rising at an accelerating rate recently (Figure 10.4). Possible reasons include:

- fewer meals are prepared and cooked at home, and more ready meals are bought in
- more fast foods are eaten whilst on the move, with little opportunity to wash hands
- demand for cheaper food has led to intensive farming of livestock, followed by slaughter at centralised abattoirs with larger throughputs of carcasses
- changes (mutations) in the microorganisms themselves
- greater public awareness of food-poisoning symptoms, improved diagnosis by doctors, and better detection at hospital pathology labs may mean more cases are now reported and recorded.

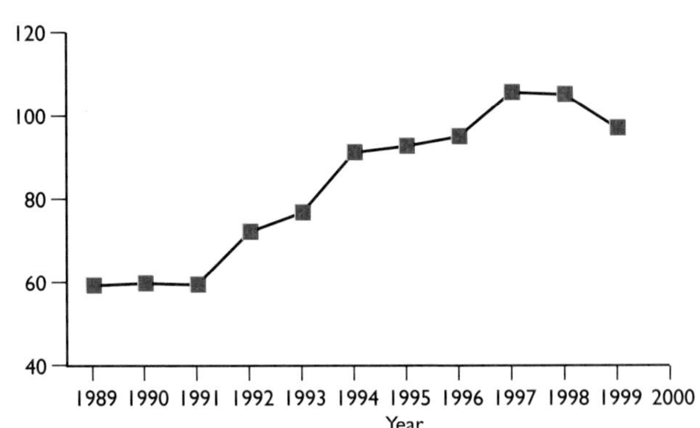

Figure 10.4 Number of food poisoning cases reported in the UK, 1989–99.

Two types of food-borne diseases

- **Food poisoning** – which results from the ingestion of food with pre-formed microbial **toxins**. Here the organisms responsible for the toxin do not have to grow in the host, and may even no longer be alive when the contaminated food is eaten, e.g. *Staphylococcus aureus.*
- **Food infection** – which results from the ingestion of food with **living pathogens** in sufficient numbers to cause a disease in the host, e.g. *Salmonella enteritidis.*

Staphylococcus aureus*, a food poisoning organism*

Staphylococcus aureus is a Gram-positive coccus (page 15), which produces enterotoxins. Handling cooked foods (e.g. cream-filled pastries, chicken and other meat products, puddings, creamy salad dressings) with contaminated hands may introduce the bacteria. If food is kept refrigerated after preparation, the bacteria fail to multiply. But if food is left in a warm place (in the kitchen, or out of doors at a picnic), *Staphylococcus* will grow and secrete toxins. When contaminated food is eaten, nausea, vomiting and diarrhoea occur within 1–6 hours. Re-cooking of contaminated food does not remove these toxins – they are heat-stable.

Figure 10.5 How *Staphylococcus aureus* causes food poisoning.

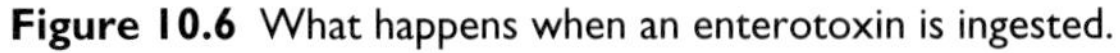

Figure 10.6 What happens when an enterotoxin is ingested.

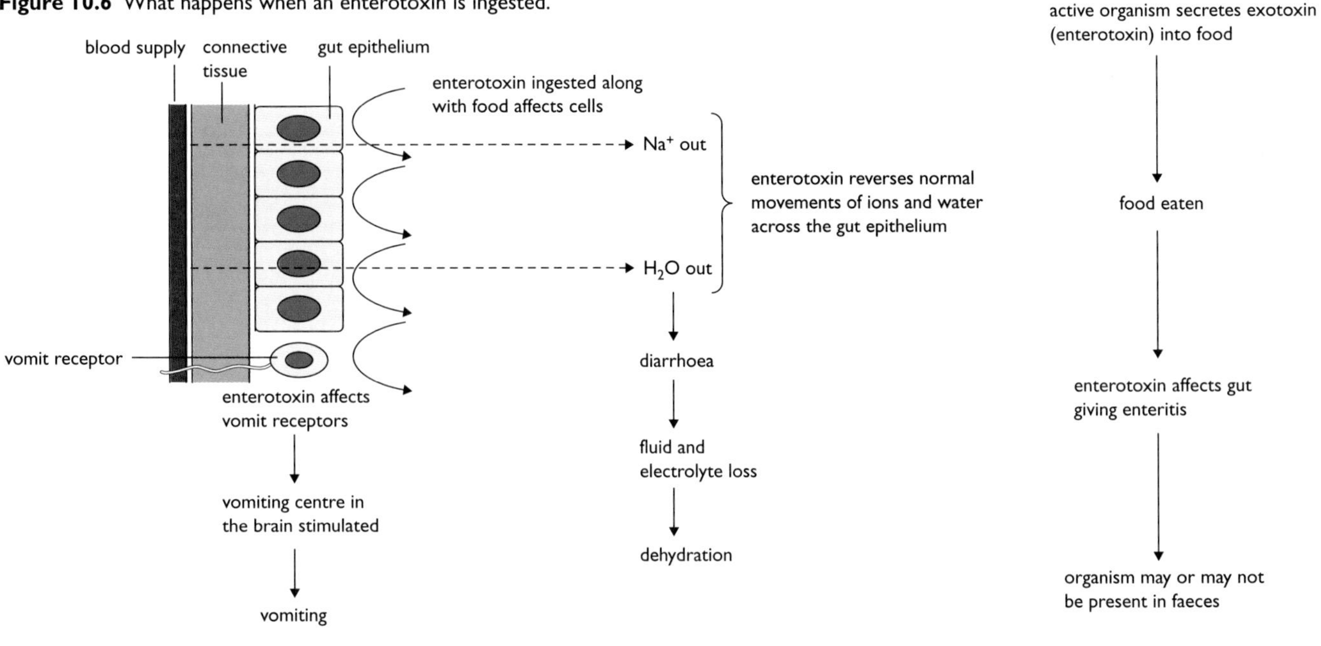

Salmonella, a food infection organism

Salmonella food poisoning is caused by the presence of a large number of Gram-negative bacilli, either *Salmonella enteritidis* or *Salmonella typhimurium*, in the intestines. Identification of these pathogens is by their surface chemistry, detected by reaction with various antisera. There are over 2000 different serotypes, and not all cause disease. The bacteria originate from the intestines of birds and other animals. Humans may receive the bacterium from contaminated foods such as beef products, poultry, eggs and egg products (particularly any made with uncooked eggs). Properly cooked foods are safe if eaten immediately or stored below 4 °C. Cooked food that is handled by an infected person and then left for long periods without refrigeration, particularly in warm conditions, is a common source of the illness.

An illness occurs several days after eating contaminated food, and only after the pathogen has multiplied and then attacked the intestinal mucosa (Figure 10.8). The symptoms of salmonellosis are the sudden onset of headache, chills, vomiting and diarrhoea, followed by fever that lasts a few days. The illness can be fatal to the very old, the very young, and those who are already suffering from other diseases.

Figure 10.7 How *Salmonella* causes a food infection.

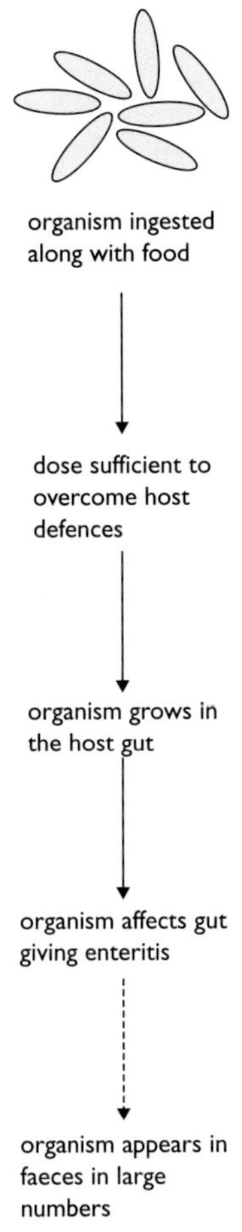

organism ingested along with food

dose sufficient to overcome host defences

organism grows in the host gut

organism affects gut giving enteritis

organism appears in faeces in large numbers

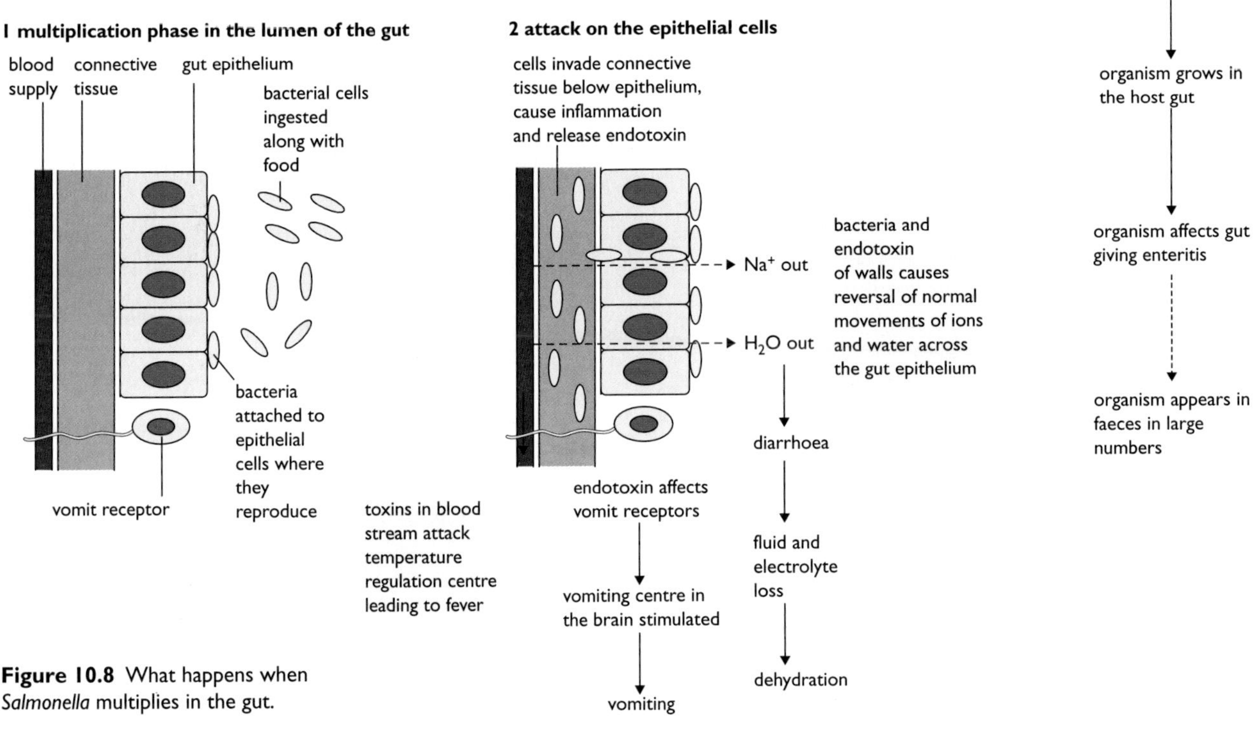

Figure 10.8 What happens when *Salmonella* multiplies in the gut.

How to prevent food-borne disease

Food poisoning can be prevented if the hazards are identified and controlled (Figure 10.9).

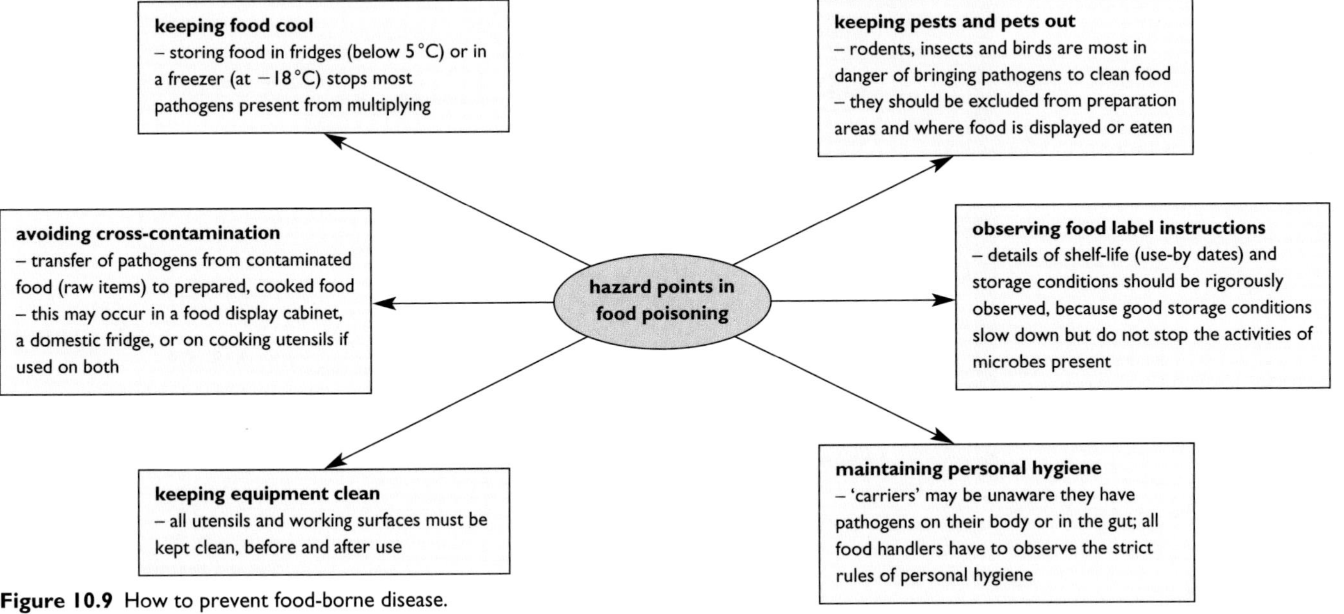

Figure 10.9 How to prevent food-borne disease.

Cholera

Cholera is caused by the Gram-negative, curved rod bacterium *Vibrio cholerae* (Figure 10.10). This pathogen is acquired almost exclusively from drinking water that is heavily contaminated by the faeces of patients (or of carriers of the pathogen). However, it is possible the bacterium may also be introduced on food contaminated by flies that have previously fed on faeces, or via eating raw shellfish taken from waters polluted with untreated sewage rich in *V. cholerae.*

Cholera was once endemic in Europe, but has now disappeared following the introduction of effective sanitation and the treatment of drinking water. Today, cholera is endemic in India, Pakistan, Bangladesh, and Central and South America.

A large number of the cholera bacteria must normally be ingested for the disease to develop; fewer than 100 000 000–1 000 000 000 (10^8–10^9) organisms would be ineffective (unless anti-indigestion tablets that neutralise stomach acid have been taken beforehand). Once the bacteria have survived the stomach acid and reached the intestine, the incubation period is from 24–72 hours.

In the intestine, the pathogen increases in number and attaches itself to the epithelium membrane. The release of enterotoxin follows. The effect of the toxin is to trigger a loss of ions from the epithelium. Outflow of water follows. The patient loses a massive amount of body fluid – 15–20 litres may drain from the body as watery diarrhoea. At the same time the patient has severe abdominal muscle cramps, vomiting and fever. Death may easily result from dehydration, as the severely reduced levels of body fluids cause the blood circulation to collapse.

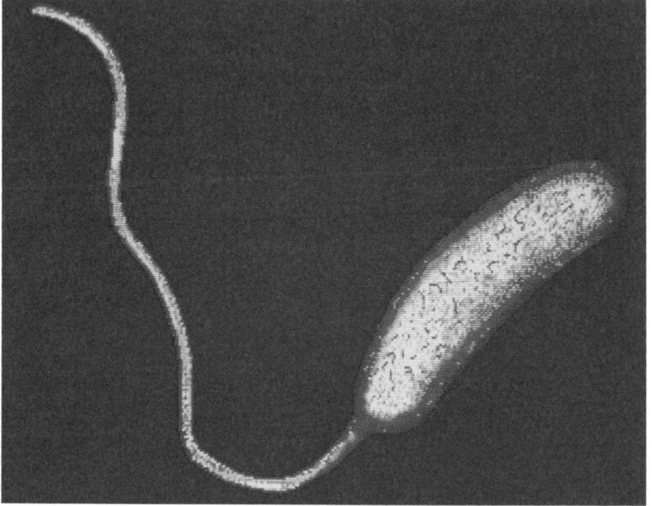

Figure 10.10 TEM of *Vibrio cholerae*, a Gram-negative bacterium causing cholera in humans (× 16 500).

How the cholera toxin works

The cholera enterotoxin consists of a two-protein complex. The B part is the binding protein which attaches the toxin complex to a particular binding site, a glycolipid, on the plasma membranes of intestine epithelial cells. The A part is an enzyme that activates the enzyme systems of the plasma membrane of the epithelium cell to which it is attached, causing secretion of chloride ions into the gut lumen and inhibiting any uptake of sodium ions. Hypersecretion of chloride ions results, and is followed by water loss (Figure 10.11).

Figure 10.11 Action of the cholera enterotoxin.

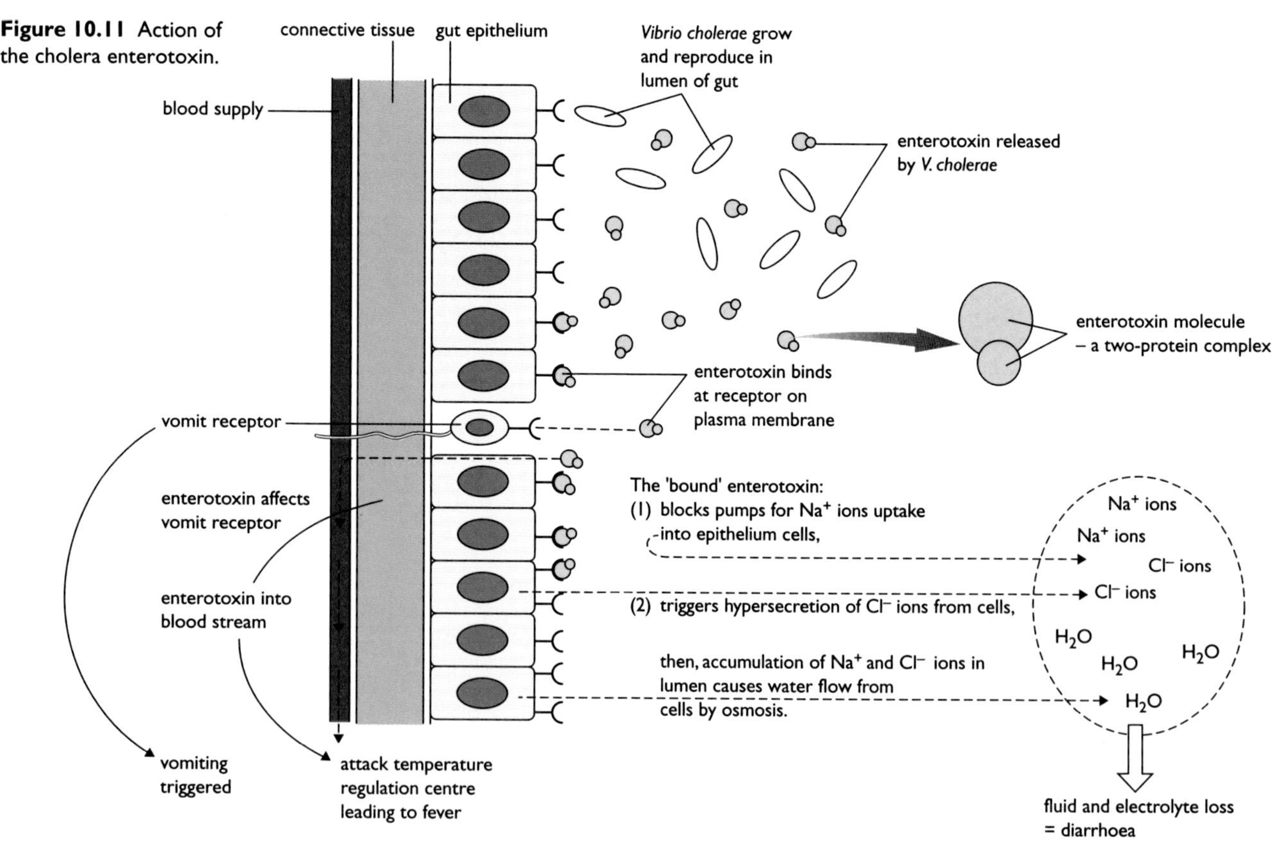

The treatment of cholera

About 50% of untreated cholera cases are fatal; properly treated, generally less than 1% are likely to die from this very unpleasant disease. A cholera patient requires immediate oral administration of a dilute solution of electrolytes in order to make good the fluid and ions lost from the body (Figure 10.12). This might seem simple to people in the developed world, but in those places where cholera is endemic, conditions may be very different. Many who contract the disease are weakened by shortage of food. Drinking water, of dubious purity, may have to be carried some distance. Boiling the water to make it safe to use in the rehydration fluid requires scarce fuel.

An additional treatment, taking antibiotics such as streptomycin or tetracycline, will assist in ridding the gut of the pathogen. However, these drugs are of little use unless the fluid and electrolyte replacement has succeeded in restoring normal body fluid composition.

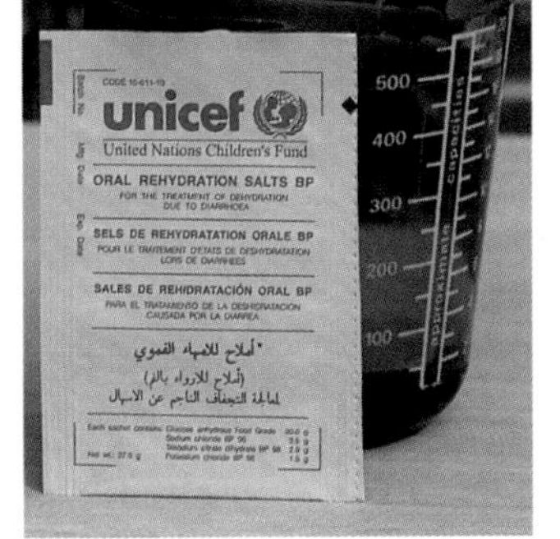

Figure 10.12 Medical pack of electrolytes for the treatment of cholera.

Prevention of cholera

In endemic areas the spread of infections between households may be prevented by boiling drinking water. The faecal contamination of food by flies must also be prevented. In the longer term, the disease can be virtually eliminated by satisfactory processing of sewage and the purification of drinking water (page 34), including a chlorination phase.

Extension: John Snow (1813–58), the first epidemiologist

John Snow, a young medical apprentice in Sunderland, worked with victims of a cholera epidemic in 1831. He developed a belief that cholera was due to some water-borne organism, a novel idea then. (Pasteur introduced his germ theory of disease in the 1860s, and the cholera bacterium was not described by Koch until 1884; page 1).

In the 1849 outbreak, Snow was working in the Broad Street area of London. He noticed that the official records of deaths showed that most victims lived around, and had made use of, the Broad Street pump for their drinking water. He assumed this was contaminated by raw sewage, and he had the pump's handle removed. The numbers of cholera cases dropped immediately. Today, the John Snow pub stands on the site of the Broad Street pump.

In the 1854 outbreak, Snow was working in south London, in an area where two companies (the Lambeth Company and the Southwick and Vauxhall Company) supplied water. Snow's careful research showed that most victims of cholera in this outbreak took their supplies from the second company. The Vauxhall Company drew water from the Thames in the locality of the outbreak, near where raw sewage (including excrement from cholera patients) was discharged into the river. The other company, whose clients had a better health record, drew water from the Thames at a point upstream of the city (Figure 10.13).

2 In cholera, toxin protein triggers the exit of chloride ions from epithelium cells into the lumen of the gut. How does this ion movement differ from the movement of water that follows ion transport?

you can read more about Dr John Snow and his study of the epidemiology of cholera on www.ph.ucla.edu/epi/snow.html

the **Southwick and Vauxhall Water Company** withdrew water for drinking from the Thames below the point where raw sewage from its customers' homes was discharged into the river

most homes here were supplied by the **Southwick and Vauxhall Water Company**

the **Lambeth Water Company** took water from the Thames just outside London (then)

many homes here were supplied by the **Lambeth Water Company**

Soon after the 1854 cholera outbreak a new waterworks was opened at Hampton, above Teddington Lock. Here the Thames is not tidal, so sewage from London was not carried to these works. This water was piped to south London, and people's health was further improved.

Kew
Mortlake
Putney
Richmond
Teddington
Hampton
Kingston

a new pipeline was completed (1855)

Southwark, Vauxhall and Lambeth – areas of south London, where John Snow practised medicine.

John Snow noted (1854) the incidence of cholera in houses and homes:
supplied by **Southwick and Vauxhall Water Company** = 315 deaths/10000
supplied by **Lambeth Water Company** = 37 deaths/10000
of the rest of London = 59 deaths/10000

Figure 10.13 Victorian London, and the epidemiology of cholera.

Tuberculosis

Tuberculosis (TB), a major, worldwide public health problem of long standing, is caused by a rod-shaped bacterium, *Mycobacterium tuberculosis.* The bacteria were first identified by Robert Koch in 1882. Viewed under the microscope after staining by the Ziehl–Neelsen technique (basic dye fuchsin and phenol) the cells show up bright red. They are known to bacteriologists as **acid-fast bacilli** because an acid–alcohol rinse after the staining step does not remove the red coloration. This is unique to bacteria of the genus *Mycobacterium*, and is due to dye staining the quantities of wax and other lipids in the walls of cells of this genus. This is how the pathogen is identified in infected patients.

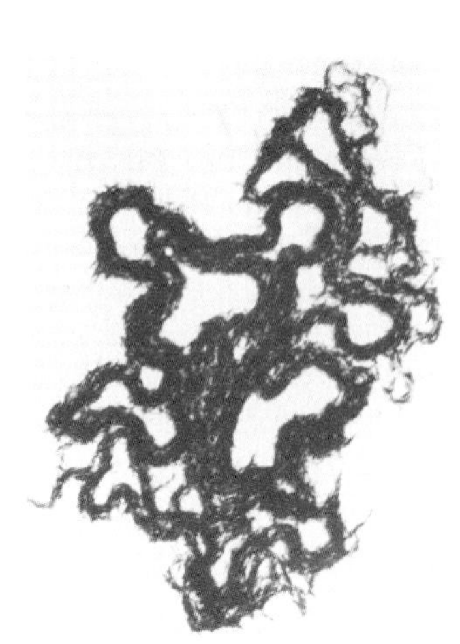

Figure 10.14 Photomicrograph of a colony of *Mycobacterium tuberculosis.* These rod-shaped bacteria grow in aggregates called 'cords'.

How the disease spreads

People with pulmonary TB cough persistently. The droplets they spread in the air are infected with live *Mycobacterium* (Figure 10.15). Tuberculosis is chiefly spread by droplet infection in this way. Because of their lipid-rich cell walls, the bacilli are protected from drying out, so the pathogen may survive for many months in the air and in the dust of homes. This is another source of infection. Overcrowded and ill-ventilated living conditions are especially favourable for transmission of the infection. However, it still requires quite prolonged contacts with a viable source before people will succumb, for the bacterium is not strongly infectious.

A bovine form of TB occurs in cattle, and the bacillus can enter the milk. Unpasteurised milk from infected cows is another potential source of infection. This type of milk was consumed in some rural communities, especially by children. Today, all milk in Britain is supplied by tuberculin-tested cows that have been certified free of *Mycobacterium.* Milk is pasteurised too. In the UK, TB is not now contracted by this route.

Figure 10.15 Droplet infection.

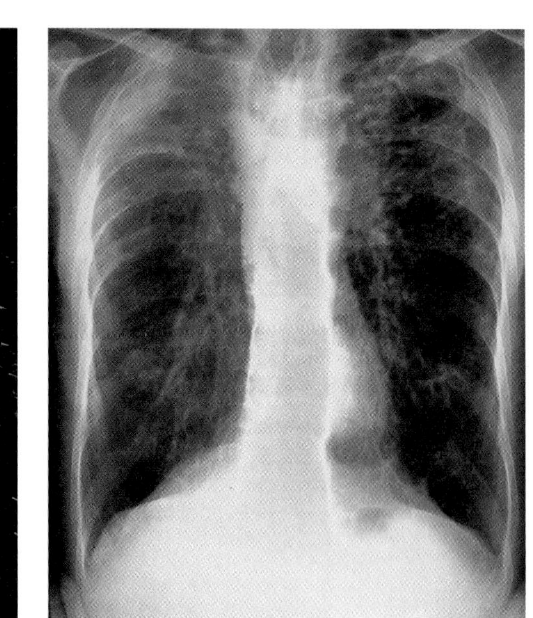

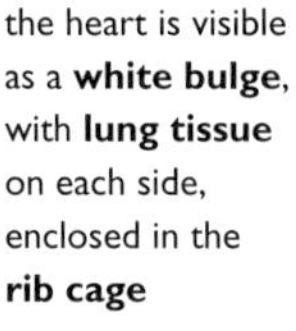

Figure 10.16 Chest X-ray showing TB in the lungs.

How TB develops

Once inside the lungs, the bacteria are engulfed by macrophages in the alveoli and bronchioles. If the recipient is in good health, these white cells (with the help of the T lymphocytes which migrate in from the lymph nodes) kill the pathogen. Alternatively, and particularly with strains of *Mycobacterium* that are more virulent, the pathogens may remain alive within the macrophage, although localised and effectively controlled by the immune system. But if recipients are malnourished or in inferior health, with a weakened immune system, a chronic infection may develop, typically within the lungs. Cavities appear as bacteria destroy the lung tissues. Blood vessels are broken down and fluid collects. The patient coughs blood in the sputum. The structural damage to the lungs can be seen by X-ray examination (Figure 10.16).

The pathogen, carried in the blood stream, can lead to TB in almost any part of the body, including the meninges of the brain or bone tissue, the lymph glands, the liver, the central nervous system, the kidneys or the genital organs. Generally, patients show loss of appetite, loss of weight, excessive sweating, and decline in physical activity.

Treatment of TB

Today TB is treatable, but it is still a killer disease if not diagnosed early in the infection. From 1995 to 2000, deaths due to TB in England alone increased from 5426 to almost 6000 per year.

The disease is contagious, so on confirmation of a case of active tuberculosis, all contacts of the patient are traced and screened by a community public health team. Infectious patients are isolated and treated with specific anti-TB drugs until they cease to cough up viable bacilli. Then, anti-TB drugs continue to be administered to the patient, now back in the community, until it is clear that the infection has been eradicated from the body.

We have seen that microorganisms develop resistance to antibiotics and other drugs used against them, with time. In the case of TB, patients are now treated with several drugs simultaneously because of the emergence of **multiple-drug-resistant** TB (**MDR**-TB). This is particularly prevalent in the case of AIDS patients. MDR-TB is especially common in the USA and south-east Asia.

3 How may strains of TB-causing organisms develop resistance to the drugs used to treat the disease?

Prevention of TB

All secondary school children should receive immunisation with the vaccine known as BCG. This consists of a tiny quantity of an attenuated (weakened) form of *Mycobacterium* derived from cattle. It is injected under the skin of the upper arm (Figure 10.17). The body develops immunity but no infection. Immunity typically lasts for at least 15 years – longer if the individual is re-exposed to TB bacilli, for example by accidental encounter with it. However, immunity wanes with time, and disappears entirely from the elderly.

Before vaccination takes place, people are tested for **existing immunity**, as these people will be hypersensitive to the vaccine. In the tuberculin test, a protein from *Mycobacterium* tuberculosis is injected just below the skin with an instrument with six needles arranged in a circle (the **Haef test**; Figure 10.17). A reaction at the site of injection appears as redness and hardening within a few days. If this occurs it suggests active TB in the patient (in children) – or existing immunity from previous immunisation or exposure to mild infection (in older people), and is investigated further. Only those people with no reaction may receive the vaccine.

Figure 10.17 The Haef test and a TB vaccination.

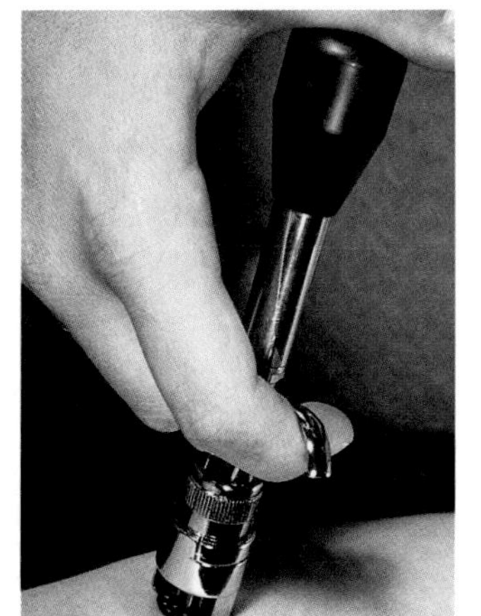

1 instrument with six needles by which the skin is punctured and a small amount of protein, extracted from TB bacilli, is introduced below the skin (**Haef test**)

2 after 2 or 3 days, this test has proved positive, i.e. the patient has TB or has already acquired immunity

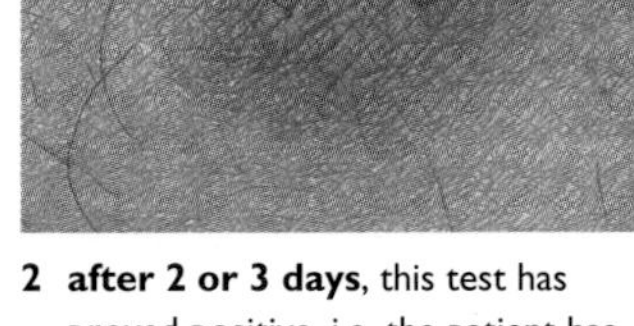

3 in a TB vaccination, the needle introduces the BCG vaccine just below the skin – a blister forms and eventually subsides, leaving the patient protected against TB

Changing pattern of the disease, 1900–2000

There is evidence that TB was present in some of the earliest human communities, and it has persisted as a major threat to health where vulnerable people live in crowded conditions. In Koch's time, one in seven of all deaths among Europeans were due to TB.

Today, in the UK, this disease is relatively rare, but about 20% of the world's population are infected. The reduction in TB in the UK throughout most of the past century was due to steadily improving living conditions, including housing and diet (Figure 10.18). Today there is evidence of a resurgence of the disease in the UK, probably due to the globalisation of travel, as well as to the emergence of MDR-TB.

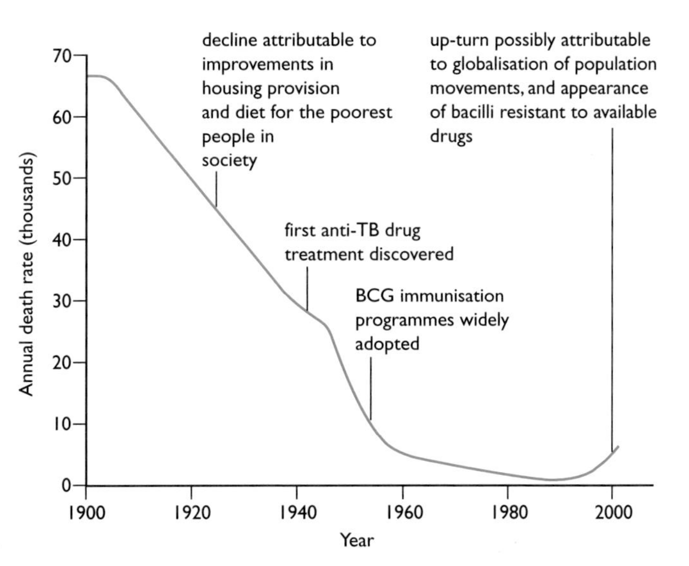

Figure 10.18 Decline in deaths from TB in England and Wales, 1900–2000.

Our defences against microbes that may invade the body

Pathogens are excluded from our bodies as far as possible – for example, the external skin is a highly effective barrier. If the skin is damaged, the body responds with a defensive **inflammation response**, in which the volume of blood in the damaged area is increased, and white cells and plasma accumulate outside the enlarged capillaries. If the blood capillaries are actually broken, then the **blood-clotting mechanism** creates a protective scab, prior to full tissue repair being carried out. Our internal surfaces (e.g. of the respiratory, digestive and urinogenital systems) are lined by moist epithelia **protected by mucus**, and are more vulnerable to attack (Figure 10.3, page 67). So, by one route or another, pathogenic microbes do sometimes gain entry to the body. How are the body's defences maintained?

4 A blood parasite such as *Plasmodium* (page 83) is present in blood plasma only briefly – mostly it occurs in blood or liver cells. What advantage does this give the parasite?

Complement proteins at work

Invading microbes in the body are identified as non-self (foreign matter). Complement proteins, made in the liver and circulated in the blood stream, work with phagocytic white cells to dispose of them (Figure 10.19).

Antibodies

At the same time, the **immune system** of white cells (lymphocytes) recognises foreign matter (antigens), and works to destroy them by producing specific antibodies (Figure 10.20). Millions of different antibodies are produced in response to the diversity of antigens that exist. The lymphocytes of the immune system are of two types, known as **B-** and **T-lymphocytes**. Both are stored in the lymph glands and circulated in the blood stream, so they are present wherever antigens appear. These lymphocytes have distinctive roles in the production, activation and regulation of antibodies (Figure 10.21).

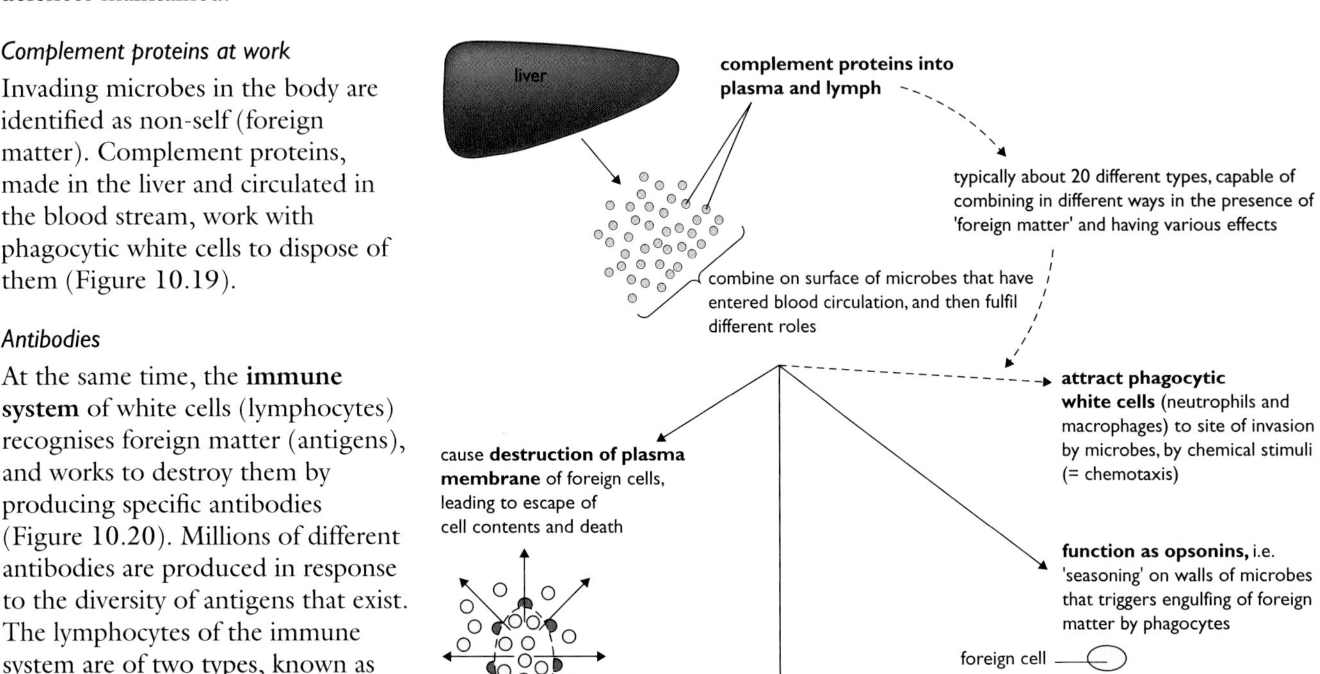

Figure 10.19 Complement proteins at work.

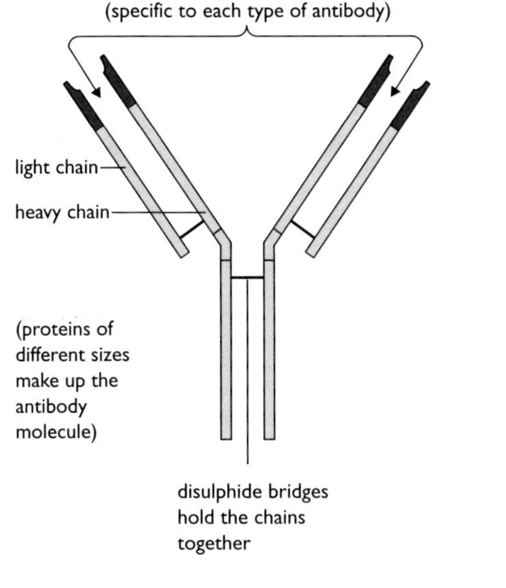

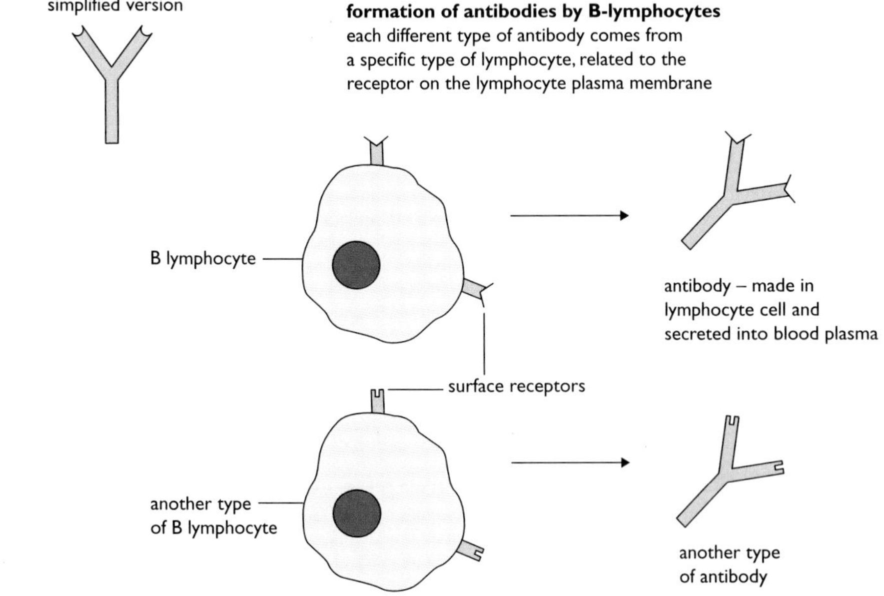

Figure 10.20 The structure and formation of antibodies.

Figure 10.21 Formation and actions of the lymphocytes of the immune system.

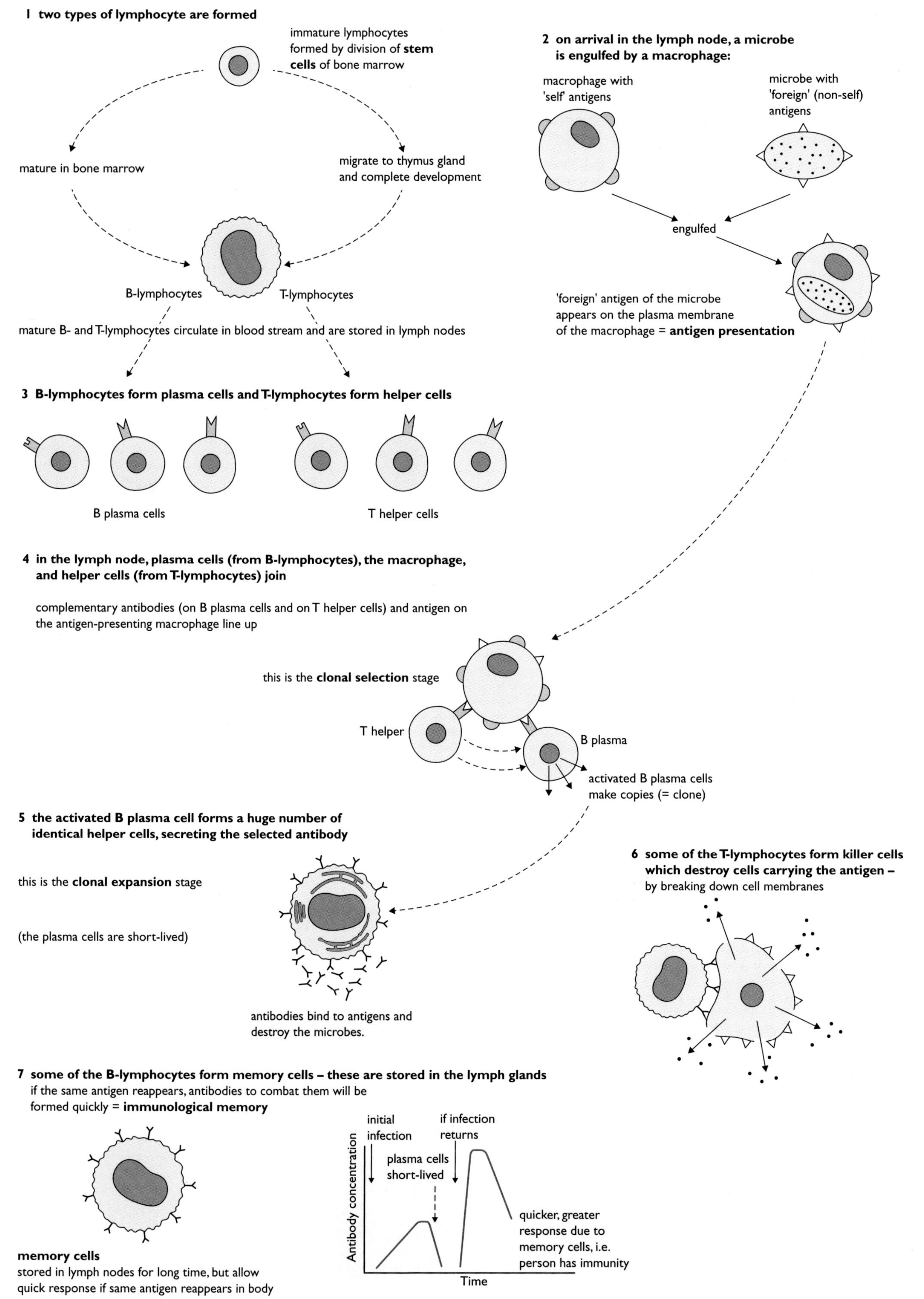

Immunity and immunisation

People who had recovered from the plague or smallpox rarely contracted these diseases again. Thus the possibility of a patient **acquiring immunity** ('immunological memory'; Figure 10.21, page 75) came to be appreciated as a possibility long before the mechanism was understood.

The first modern attempt at immunisation (artificial immunity) was made by **Edward Jenner (1749–1823)**, a country doctor from Gloucestershire. At the time, many who had smallpox died of it, but not those who had earlier contracted cowpox (workers who handled cows typically did so at some stage). Jenner extracted fluid from a cowpox pustule on an infected milkmaid, and injected it into the arm of an 8-year-old boy. The child got a mild infection, but remained healthy when exposed to smallpox. This technique was named **vaccination**, after vaccinia (cowpox).

The French scientist **Louis Pasteur (1822–95)** discovered that cultures of chicken cholera bacterium, allowed to age for 2–3 months, produced only a mild infection of cholera when inoculated into chickens. The old cultures had become less pathogenic (**attenuated**). Fresh, virulent strains of the chicken cholera bacterium failed to infect chickens previously exposed to the mild form of the disease. Pasteur, one of the greatest experimentalists in the field of microbes, had notable successes in immunisation. But he recognised the contribution of Jenner to the discovery of immunity by using the name '**vaccine**' for injections of the attenuated organisms that he developed to prevent chicken cholera disease, and (later) anthrax (due to a bacterium), and rabies (due to a virus).

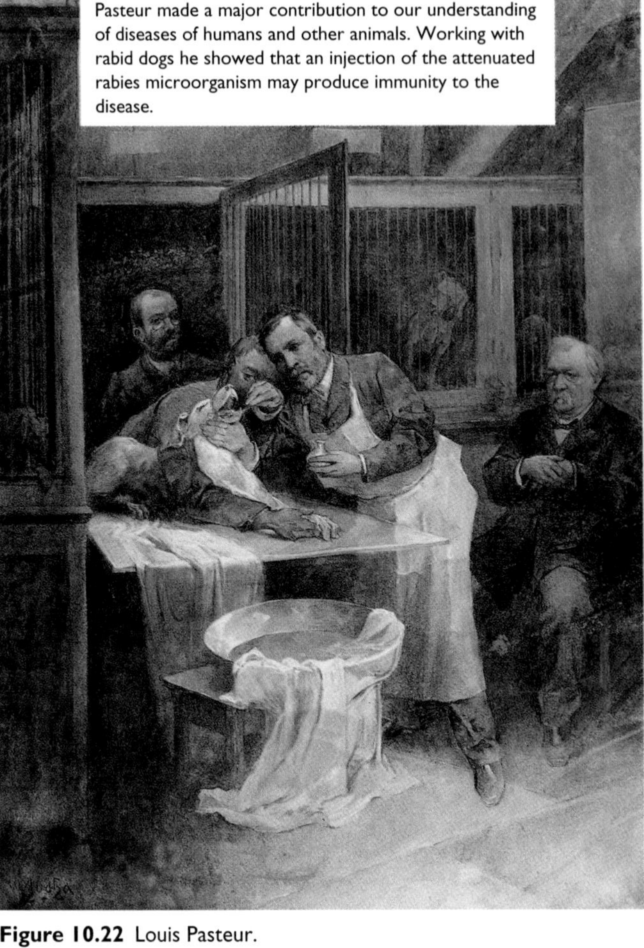

Pasteur made a major contribution to our understanding of diseases of humans and other animals. Working with rabid dogs he showed that an injection of the attenuated rabies microorganism may produce immunity to the disease.

Figure 10.22 Louis Pasteur.

Table 10.1 Types of immunity

	Passive immunity: no exposure to antigen, so immune system not activated; no memory cell formation, so immune response not maintained	**Active immunity**: exposure to antigen occurs, so immune system is activated; memory cells formed, so immune response can be maintained
Natural immunity: antibodies are naturally produced	Fetus receives the mother's antibodies across the placenta; and the baby also, in the mother's milk	As a result of an infection, contracted by chance
Artificial immunity: antibodies present as a result of human actions	Injection of human antibodies, e.g. to protect against toxins from exposure to rabies or a snake bite	Immunisation with prepared vaccine

Smallpox – how it was eradicated

Smallpox, once endemic throughout the world, killed or disfigured all those who contracted it. This disease has now been eradicated (the last case of smallpox occurred in Somalia in 1977). The development of a vaccine played an important part in this achievement, which was supervised by the World Health Organization (WHO). How was smallpox eradicated, when so many other diseases continue? The reasons include:

- the disease was easily identified, with clinical features not easily overlooked
- it had a short period of infectivity – about 3–4 weeks
- transmission was by direct contact only
- on diagnosis, patients were isolated and all their contacts traced and immunised
- patient who recovered did not harbour the virus (carriers did not exist for smallpox), and no animals acted as 'reservoirs' of infection.

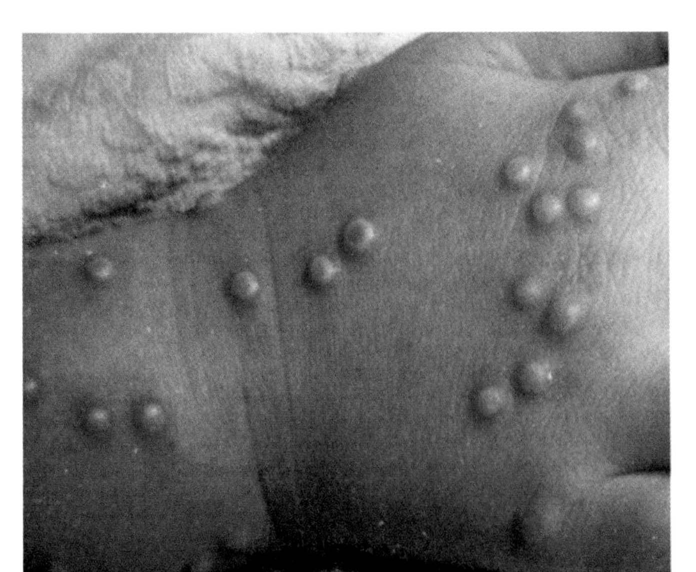

Figure 10.23 Smallpox skin vesicles.

Immunisation practices today

The types of vaccine currently available are illustrated in Table 10.2. The recommended schedule of vaccinations for children and young people can be accessed via the website of the Department of Health – www.doh.gov.uk.

Table 10.2 Examples of the types of vaccine available

Vaccine type	Treatment for:
Killed bacteria	Whooping cough Typhoid fever Cholera
Attenuated (weakened) strain: bacterium virus virus	Tuberculosis (BCG) Measles Rubella
Inactivated virus	Influenza
Purified polysaccharide from wall	Meningitis Bacterial pneumonia
Toxoid	Diphtheria Tetanus
Recombinant DNA produced by genetic engineering	Foot and mouth of sheep, cattle etc. Hepatitis A Hepatitis B

Current issues in vaccination

WHO and poliomyelitis

The polio virus is stable and remains infectious for long periods in food and water, by which it may be transmitted. Inside some human hosts the virus reaches the central nervous system and destroys motor neurones, causing permanent paralysis. Since the development of vaccines, the incidence of polio has decreased markedly (Figure 10.24). The WHO is working on the worldwide eradication of polio, and this is anticipated in the next few years.

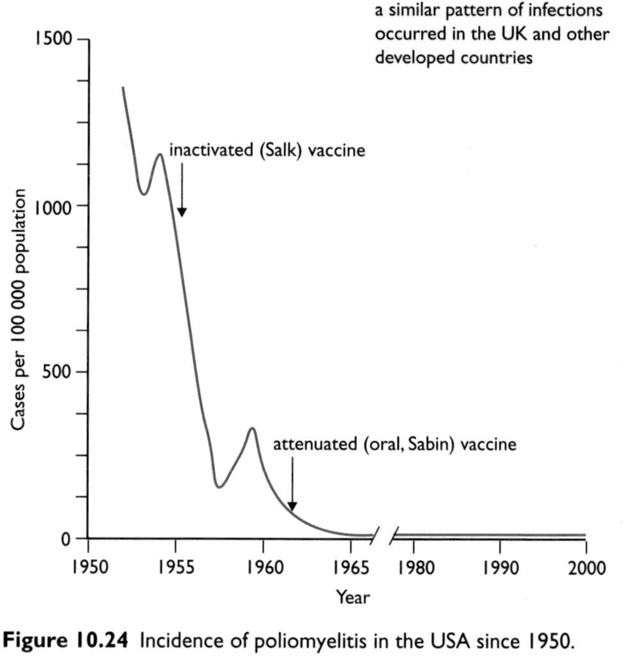

Figure 10.24 Incidence of poliomyelitis in the USA since 1950.

When a disease is conquered, people may grow careless

When the incidence of a life-threatening or seriously disabling diseases falls due to an immunisation programme, rare side-effects of the vaccine, appearing in a very small proportion of those vaccinated, can become unacceptable to the general public. Whooping cough vaccination suffered such adverse publicity in the 1970s, and this was reflected in a subsequent increase in the disease (Figure 10.25). In fact, whooping cough was the most lethal of childhood infections in the 19th and early 20th centuries, a fact we overlook at our peril. Similar current anxieties about the MMR (measles–mumps–rubella) vaccine may lead to increases in these serious childhood diseases.

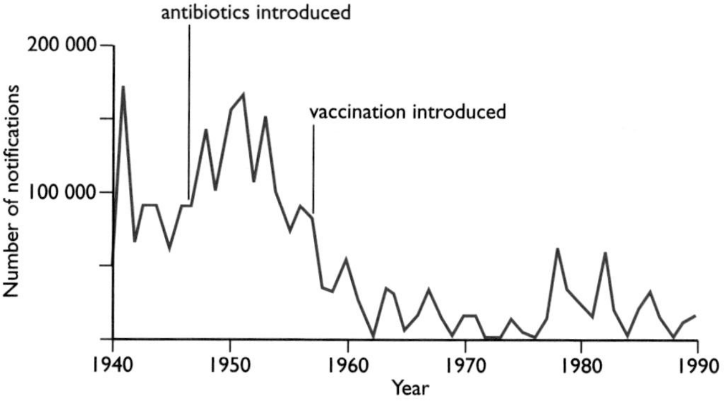

Figure 10.25 Number of whooping cough cases notified in England and Wales, 1940–90.

Microbes that are never eradicated from a healthy person's body

After an infection, the microbe may persist in the healthy patient. For example, the chicken pox virus can lie dormant in peripheral nerves, but become re-activated years later, causing shingles. The typhoid bacillus exists in some people in a population who show no clinical symptoms of typhoid (we call them carriers). However, they may shed the bacillus daily in their faeces.

Microbes may have effective evasion strategies

Some parasitic microbes mutate and change some of the external proteins by which they are recognised, helping to keep themselves temporarily 'invisible' to the immune system, for example, the influenza virus and human HIV.

5 What anxieties are expressed by **a)** parents and **b)** public health experts about the MMR vaccine?

Virus diseases

Influenza – the sweating sickness

The influenza virus contains eight short strands of RNA (a segmented genome), ranging from about 800 to 2400 nucleotides long. The protein coat (the capsid, page 18) around the nucleic acid is itself surrounded by an external layer of lipid. This combined capsule is a flexible coat, so flu viruses are of no set shape. However, constant features of the viral capsid are protein knobs (known as HG and NA) projecting through the lipid layer. HG and NA function as antigens in the human host.

Figure 10.26 Structure of the influenza virus.

Transmission and replication of the virus, and the symptoms of flu

The virus enters the human host by droplets inhaled via the mouth and nose into the lungs, and parasitises the epithelial cells lining the bronchus and bronchioles. Entry into epithelial cells occurs by endocytosis (Figure 10.27). Replication of the virus then occurs. The RNA of the influenza virus is 'negative strand' RNA (see below). This means it does not serve directly as messenger RNA (mRNA). Rather, in the host cell it has to be transcribed into a complement form, which then functions as mRNA.

New RNA segments are produced in the nucleus and the capsid proteins in the cytoplasm of the host cell. The new virus particles are then assembled, and the lipid capsule is added as the virus leaves the dying host cells (Figure 10.28). Toxins are released at this stage, and destruction of the epithelial cells paves the way for serious secondary bacterial infections of the host lung tissues, typically bacterial pneumonia.

Figure 10.27 Entry of the flu parasite into a host cell.

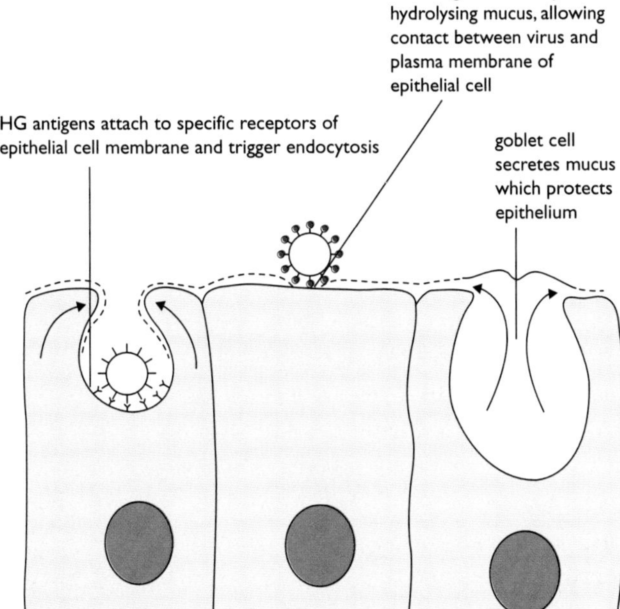

Figure 10.28 Replication of the flu virus.

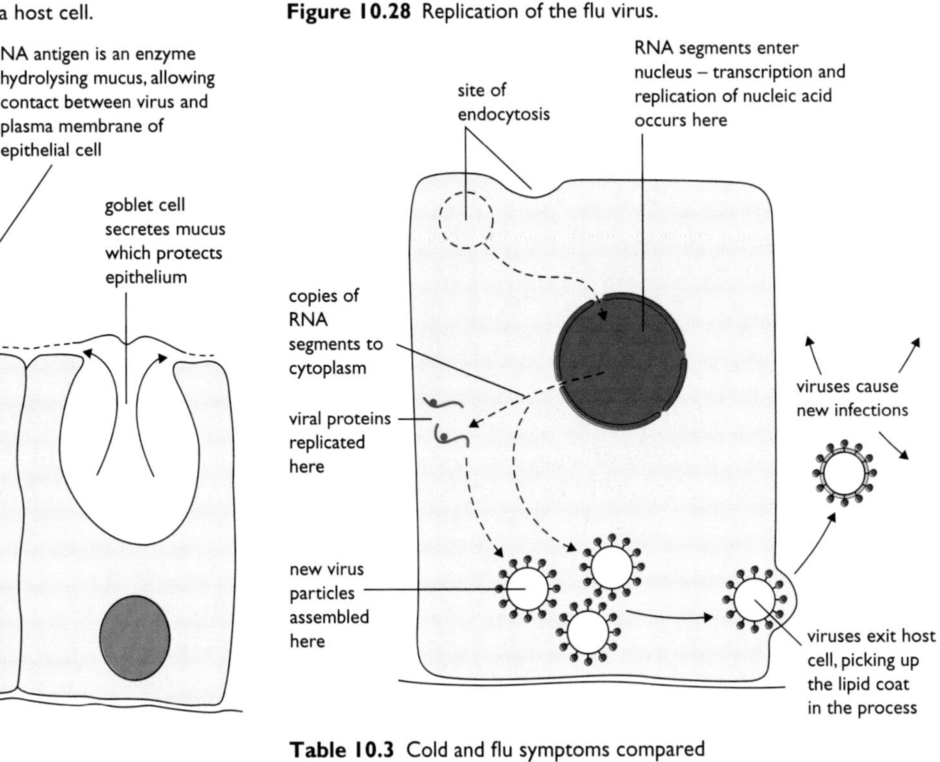

Signs and symptoms of flu

Both the common cold and flu can make the patient feel miserable for a short period, but they are quite different diseases (Table 10.3), caused by viruses of very different types.

Table 10.3 Cold and flu symptoms compared

Cold	Symptoms	Flu
Rare	**Headache**	Common
Rare	**Fever – raised body temperature**	Yes, 39–40 °C and sudden onset
Abundant	**Nasal discharge**	Uncommon/slight
Common	**Sore throat**	Uncommon
No	**Vomiting and/or diarrhoea**	Common
Slight	**General malaise**	Usually severe, lasting many days/weeks

Extension: other 'negative strand' RNA viruses

Apart from influenza, the rabies virus and the Ebola virus also have 'negative strand' RNA that has to be transcribed within the host cell before it acts as mRNA, leading to an infection.

Genetic variation in the flu virus

The flu virus is one of the most rapidly mutating life forms known. This is partly due to the rate at which spontaneous changes occur in individual genes. However, the flu virus may also be transformed by occasional exchanges of one or more of the eight RNA strands of the virus genome, swapped between different viruses. How has this come about?

The flu viruses occur in various strains, and parasitise the lungs of pigs and birds as well as humans. But pigs are where reassortment of RNA segments ('viral sex') may occur. This is because the pig lung cells have receptors for human and bird flu viruses, as well as for pig flu virus. Consequently, all three types of virus may occasionally occur in the same host lung cell. In a multi-infected cell, reassortment of genomes may create new virulent combinations from time to time (Figure 10.29).

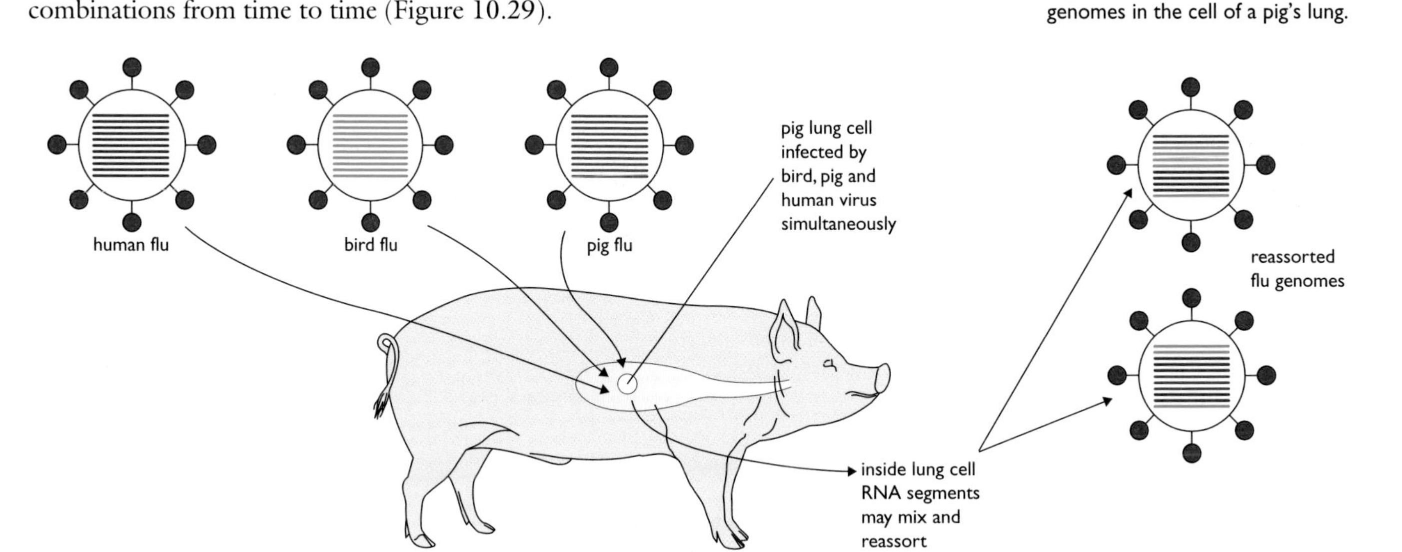

Figure 10.29 Assortment of viral genomes in the cell of a pig's lung.

Influenza epidemics and pandemics

Influenza viruses are classified into antigenic groups according to the HG and NA antigenic proteins of their coat (Figure 10.26). The chemistry of these proteins is controlled by the genome of the virus, which we have seen is quite variable. Alterations of the virus genome lead to new antigenic types, and these often lead to a new strain of flu in humans who have little or no resistance to it. This is because the human's immune system will not have met the new antigens previously, and no memory cells will exist in the lymph nodes. Because of this, flu is a major epidemic disease every year.

6 How does the segmented genome of the flu virus contribute to the genetic variability of this virus?

However, strains arise from time to time with exceptional genes for virulence and transmissibility. A strain with just the right mix of virulence and transmissibility leads to a pandemic in which millions of humans die. For example, a pandemic in 1918–19, immediately after World War I (1914–18), caused the deaths of an estimated 40 million individuals, compared to the 10 million people killed in the war.

Another severe pandemic occurred in 1957, but the pandemic of 1977 was relatively mild. Working from the route-of-spread map for 1957 (Figure 10.30), it appears that some national populations may have had pre-existing resistance to the 1977 flu strain (where only local outbreaks occurred), while other national populations succumbed.

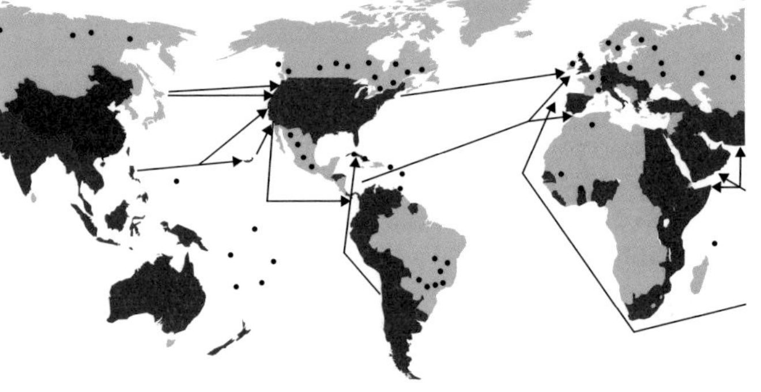

Key
- country of origin
- localized outbreaks
- routes of spread
- countrywide epidemic

Figure 10.30 Spread of the flu pandemic, 1957.

Treatment of influenza

The symptoms of a virus infection can be treated in an attempt to reduce the duration and severity of the infection, but antibiotics are powerless against viruses. However, antibiotics are administered to prevent secondary infections like pneumonia. Prevention of influenza is by means of a vaccine, but this must be made from the prevailing strain of influenza to be effective.

AIDS – acquired immune deficiency syndrome

AIDS is a disease caused by HIV, a retrovirus (page 20). Infection with HIV is possible through contact with the blood or body fluids of infected people, such as may occur during sexual intercourse; sharing of hypodermic needles by intravenous drug users; and during breast-feeding of a newborn baby. Also, blood transfusions and organ transplants will transmit HIV, but donors are now screened for HIV infection in most countries. HIV is not transferred by contact with saliva on a drinking glass, nor by sharing a towel, for example. Nor does the female mosquito transmit HIV when feeding on human blood (page 82).

The spread of HIV and the eventual onset of AIDS in patients are outpacing the current efforts of scientists and doctors to prevent them. The WHO charts the spread of this pandemic (Figure 10.31).

Figure 10.31 Adults and children infected with HIV/AIDS in 2000 (UN/WHO).

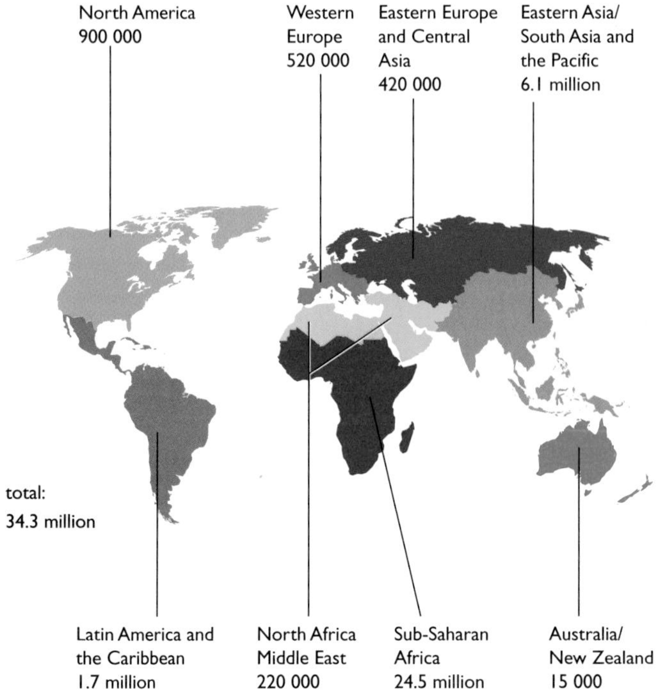

North America 900 000
Western Europe 520 000
Eastern Europe and Central Asia 420 000
Eastern Asia/ South Asia and the Pacific 6.1 million
total: 34.3 million
Latin America and the Caribbean 1.7 million
North Africa Middle East 220 000
Sub-Saharan Africa 24.5 million
Australia/ New Zealand 15 000

Infection of T4 helper cells (CD4 cells) and the eventual development of AIDS

Once in the blood stream, the antigenic proteins of the HIV outer membrane attach the virus to protein receptors of the surface of T-lymphocytes (CD4 helper cells of the immune system), and the core of the virus penetrates to the cytoplasm (Figure 10.32).

Figure 10.32 HIV infection of a white cell

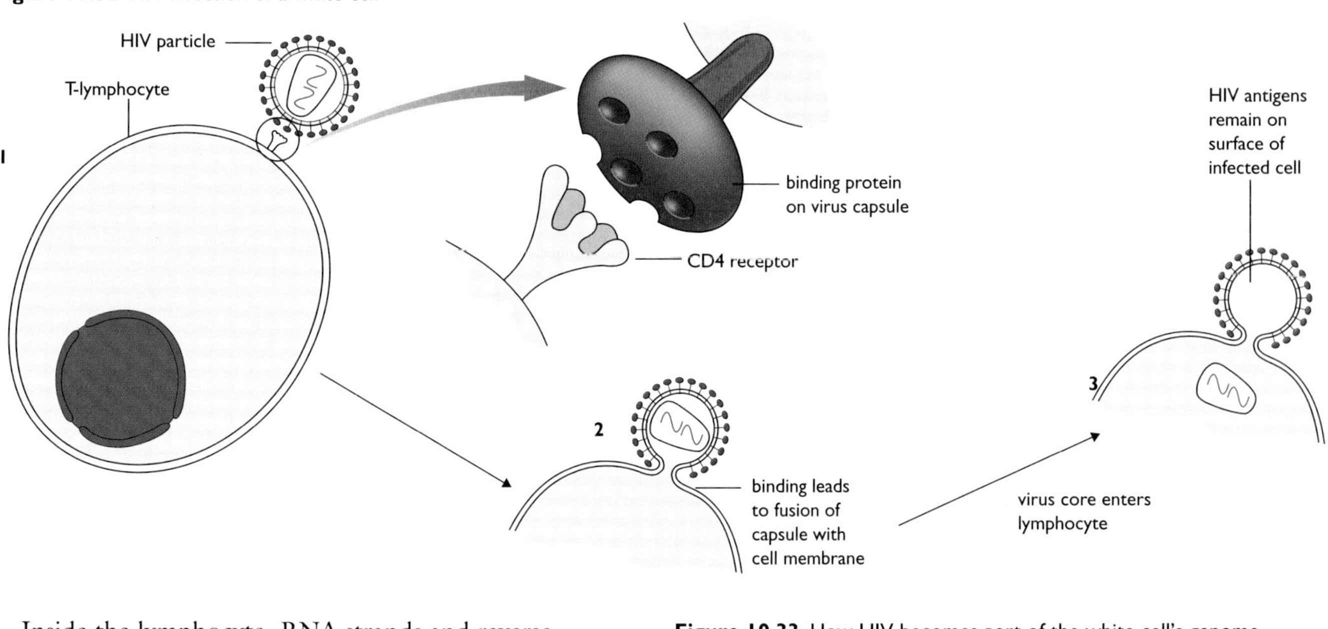

Inside the lymphocyte, RNA strands and reverse transcriptase enzyme are released from the core. Then, using the viral RNA as the template, a DNA copy is formed (action of reverse transcriptase, page 54), and converted to double-stranded DNA (Figure 10.33). This DNA enters the nucleus and attaches to a chromosome, becoming a permanent part of the host cell's genome. It is known as a provirus. After an initial, mild form of AIDS, the provirus remains dormant (latency period).

At some later time the viral DNA replicates, leading to the production of new viruses and the death of lymphocytes (Figure 10.34). It is not clear whether infected T-cells are killed by the virus they harbour, or by the actions of the body's immune system against infected cells. However, AIDS-related symptoms follow. Without an effective immune system, opportunistic infections take hold and the patient's body cannot effectively resist (Figure 10.35).

Figure 10.33 How HIV becomes part of the white cell's genome.

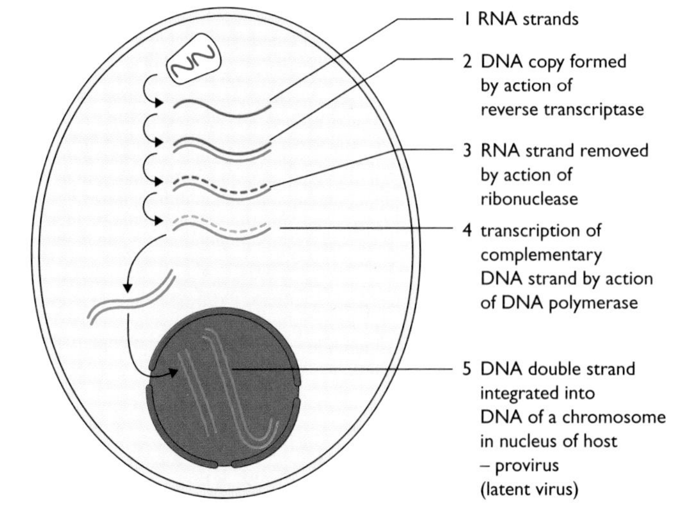

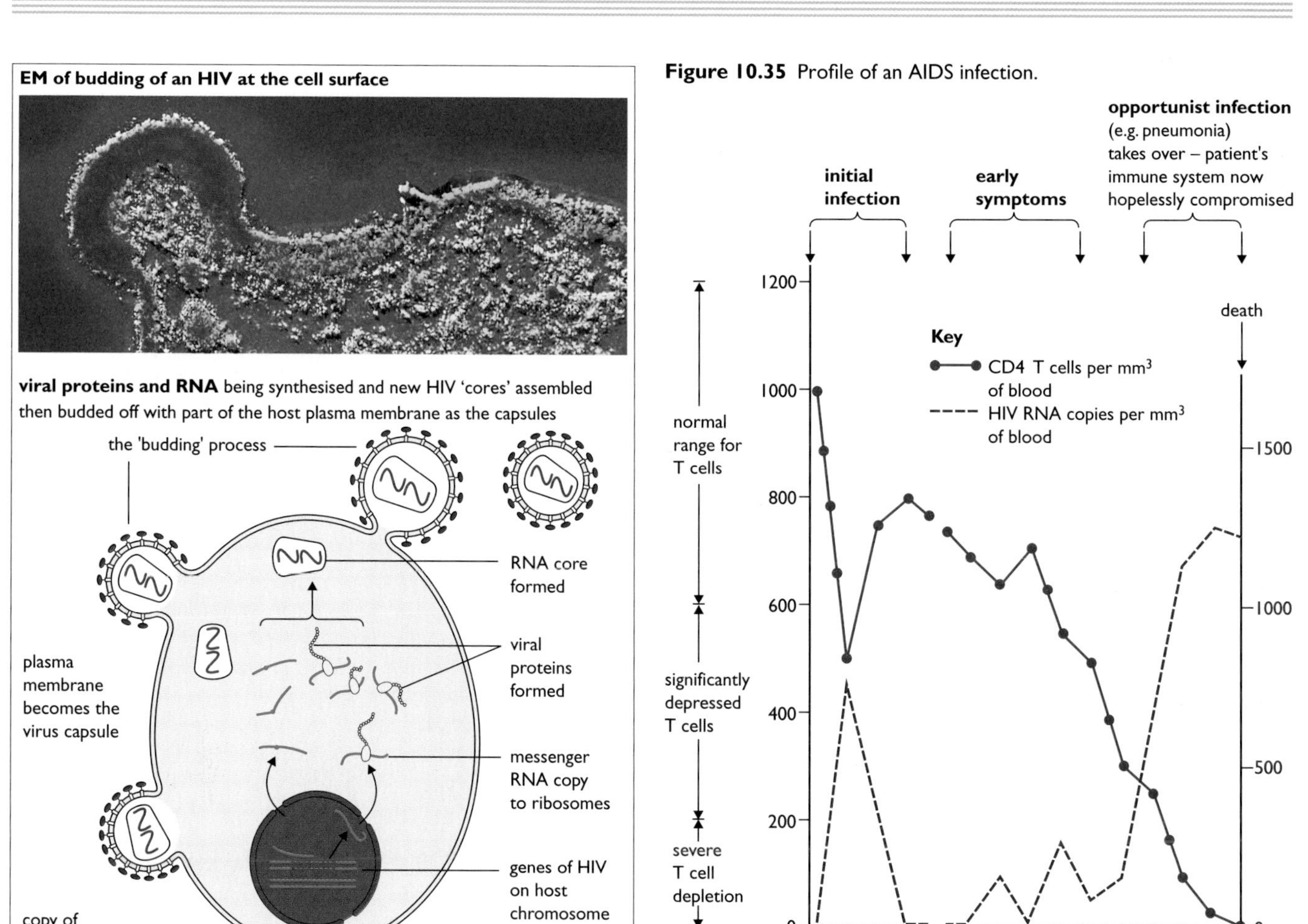

Figure 10.34 Activation of the HIV genome and production of new HIV.

Figure 10.35 Profile of an AIDS infection.

Why AIDS is difficult to treat

Viruses are not usually controlled by antibiotics. AIDS patients are offered **drugs** that slow down the progress of the infection – a combination of drugs reduces the number of HIV-infected cells in a patient, at least temporarily. The three most popular drugs (AZT and two protease inhibitors) interrupt the steps to nucleic acid reverse transcription. A combination of drugs is used to prevent HIV from rapidly developing resistance to any one drug, and to avoid dangerous or unpleasant side effects that some patients experience.

Ideally, a **vaccine** against HIV would be the best solution – one designed to wipe out infected T4 cells and free virus particles in the patient's blood stream. The work of several laboratories is dedicated to this solution. Here the problem is that infected T4 cells in the latent state frequently change the membrane marker proteins that they carry due to the HIV genome within. Effectively, HIV can hide from the body's immune response by frequently changing its identity.

Strategies for preventing the spread of AIDS

Currently, effective measures to prevent the spread of HIV include:

- practising safe sex using condoms to prevent transmission through infected blood or semen; condoms need to be freely available to the sexually active population
- use of sterile needles by intravenous drug users; sterile needles need to be freely available
- effective education programmes so that the vulnerable understand the cause and effects of HIV infection, and the best steps to remain healthy, whether or not they are literate.

Recent studies have shown that the prevalence of AIDS is 2.5 times higher in uncircumcised males, and that circumcised males are less likely to transmit HIV to, or receive HIV from, their partner. This is because the mucous membrane of the inner surface of the foreskin has cells with receptors that HIV can exploit. Consequently, it is possible that infection of future generations with HIV might be reduced if male circumcision was practised more widely. But as health workers point out, if this were to encourage males to participate in unsafe sex with numerous partners, it would defeat the object.

7 What features of the HIV virus and the disease it causes create particular problems for the people and economies of less-developed countries such as Zimbabwe or Zambia?

A protoctistan disease

Malaria, caused by *Plasmodium*, transmitted by mosquitoes

Malaria is the most important of all insect-borne diseases (Figure 10.36). About 80% of the world's malaria cases are found in Africa south of the Sahara, and here some 90% of the fatalities due to the disease occur. It is estimated that about 400 million people are infected, of which 1.5 million (mostly children under 5 years old) die each year.

Malaria is caused by *Plasmodium*, a protozoan which is transmitted from an infected person to another person by blood-sucking mosquitoes of the genus *Anopheles*.

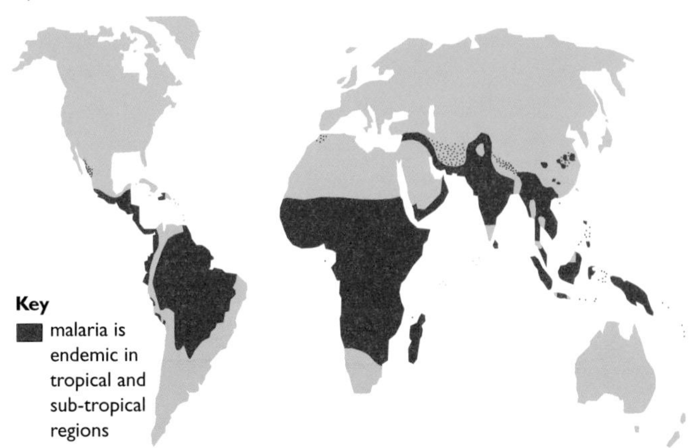

Figure 10.36 World distribution of malaria, 2000.

Transmission of Plasmodium *by the mosquito*

Only the female mosquito is a vector (the male mosquito feeds on plant juices). *Anopheles* is a fluid-feeding heterotroph (not an ectoparasite, as sometimes stated). It detects its human host, lands, and inserts mouthparts (which are formed into a long thin proboscis) into a blood vessel below the skin surface. A meal of blood is taken quickly; there is a danger that an active, alert human will swat the intruder. Mosquitoes tend to feed at night on sleeping victims, but those that alight on patients already ill with malaria are likely to be able to feed unhindered (Figure 10.37). For a blood-sucking insect, the mammal's blood-clotting mechanism presents a problem that has to be overcome. As the mosquito's proboscis penetrates the vein, a secretion from its salivary glands passes down the hypopharynx and inhibits clotting. At this point *Plasmodium* may enter its human host (if the *Anopheles* carries an infective stage). Meanwhile, the *Anopheles* loads up with a blood meal. This is essentially 80% water, and excess water has to be lost from the body as a priority (Figure 10.37).

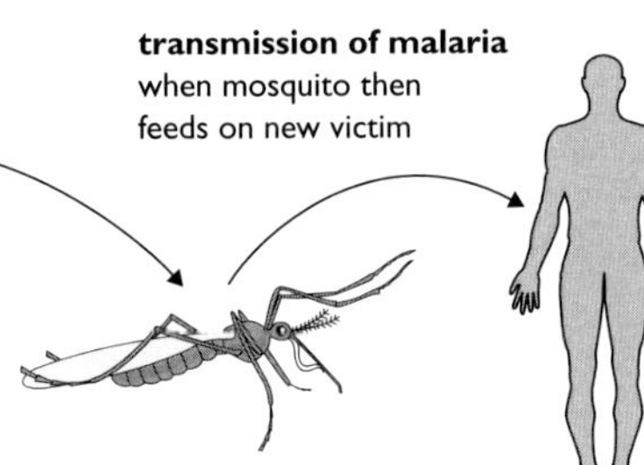

Figure 10.37 *Anopheles* feeding and the transmission of malaria.

Control of malaria

Control measures currently applied include the following.

- Interruption of the mosquito's life cycle by attacking the larval stage – the larva is an air-breathing, aquatic animal. Swamps are drained and open water is sprayed with oil, which forms a surface film blocking off the larva's air tube, and with insecticide.
- Use of insecticides to kill adult mosquitoes on and around the buildings humans occupy.
- Use of drugs to kill the stages of *Plasmodium* found in the blood and liver.
- Use of drugs to kill *Plasmodium* as soon as it is introduced into a healthy person's blood by an infected female mosquito.
- Protection of vulnerable people, e.g. by sleeping under insecticide-treated mosquito nets.

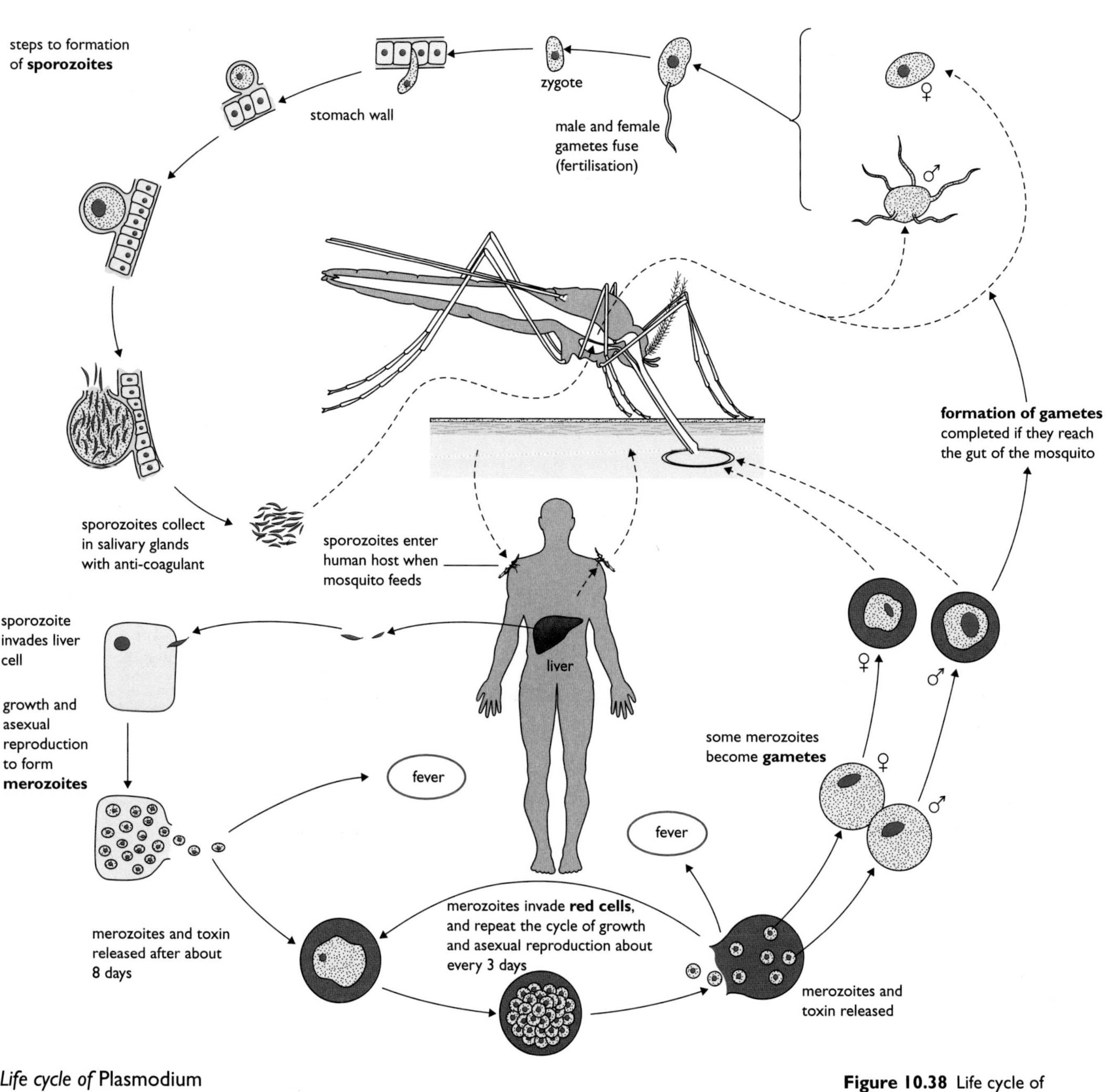

Figure 10.38 Life cycle of *Plasmodium* in vector (*Anopheles*) and primary (human) host.

Life cycle of Plasmodium

The life cycle of the malarial parasite is complex, consisting of a sexual stage beginning in the human host (primary host) and completed in the mosquito (secondary host/vector); an asexual multiplication phase in the mosquito; and a phase of growth and multiplication in the liver cells and red blood cells of the human host (Figure 10.38).

When will malaria be eradicated?

After many soldiers in the wars of the first half of the 20th century became casualties of malaria, money was found for research into its control. Powerful insecticides (DDT, gamma-BHC and dieldrin) became available. By the 1960s, mosquito eradication programmes had success in many countries. However, by the 1970s resistant strains of *Anopheles* had appeared (and also of *Plasmodium* resistant to anti-malarial drugs). The battle against malaria is currently being lost in Africa.

Vaccines against malaria are needed. The WHO urges the development of transmission blocking vaccines (TBV) against malaria, but has reported that the international drugs industry focuses research on vaccines to prevent initial infection. These are needed by tourist travellers and by the military, and would generate significant income for commercial interests. Meanwhile, TBV is most needed in poor countries where malaria is endemic and the lives of millions are permanently at risk. These people do not have the funds to reward the drug firms, however. In the longer term, control of mosquitoes by the careful use of selective insecticides will be necessary, as well as the development of powerful vaccines, before the very well adapted malarial parasite is overcome.

8 The insecticide DDT is harmless to humans at concentrations that are toxic to mosquitoes. This insecticide is a stable molecule that remains active in the environment long after it has been applied. Nevertheless, the use of this compound has been discontinued. Why?

Fungal diseases

Very few fungi are human parasites. The diseases fungi cause, known as **mycoses**, are mostly (but not all) relatively minor infections, but none should be left untreated. By contrast, there are very many fungi that parasitise plants, and these often cause diseases of great economic importance.

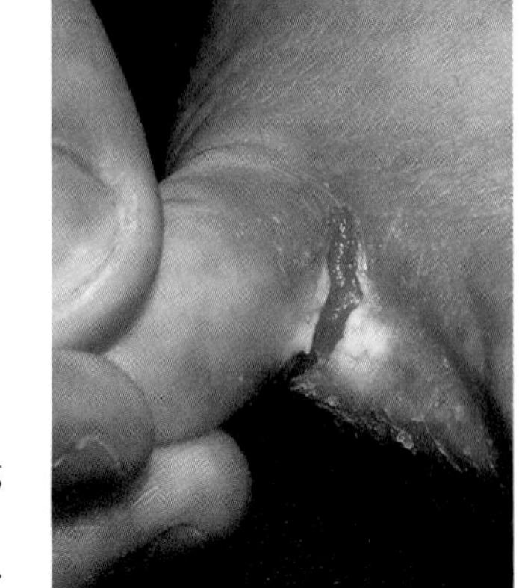

Figure 10.39 Athlete's foot infection.

Fungal diseases of humans

Athlete's foot and ringworm

These and some other diseases are caused by growth of fungi of the genera *Trichophyton*, *Microsporum* and *Epidermophyton* in the superficial layers of the skin. These fungi have the ability to degrade keratin, and may also infect hair and nails. **Athlete's foot** is an infection of the skin around the toes, and is common in people who regularly use communal washing facilities. The symptoms in mild infections are scaly, flaking skin and an itching feeling. The fungus stays alive for long periods in the skin that is shed, so the disease is contagious.

Ringworm typically occurs in farm workers in regular contact with infected farm animals. It is caused by fungi such as *Trichophyton mentagrophytes*. The infection occurs on skin of the body or limbs, and spreads out from a central point of initial infection. It has slightly inflamed leading edge, giving the characteristic ring appearance.

Thrush

Thrush is a fungal disease of the body's mucus membranes. The organism that causes thrush, *Candida albicans*, is a yeast, and a commonly occurring commensal of the microflora of mouth, gut and vagina. Here it normally lives harmlessly as one minor component, in balance with the full range of diverse microorganisms that naturally occur on these body surfaces.

However, *Candida* can change to a pathogenic organism, for example when our immune system is not fully developed or has been compromised. So **oral thrush** is a common infection of AIDS patients, but also may occur in young babies. The latter acquire the infection at birth, and are often slow to throw it off until their immune system is fully functional. Oral thrush takes the form of small white flecks (fungal colonies) that grow at the expense of the mucus membrane of the tongue and soft palate (Figure 10.41).

Figure 10.40 Ringworm infection.

When we take a prescribed course of antibiotics, many members of the natural flora of our bodies are temporarily killed off. However, *Candida* is largely unaffected by antibiotics and will proceed to grow unchecked. This is one of the ways that **vaginal thrush** may be caused. The normal flora of the vagina is of Gram-positive bacteria. These organisms naturally break down glycogen secreted by the epithelial cells. Acids are produced, and the pH of the vagina is low (pH 4.5). This acidity normally inhibits growth of *Candida*. After the menopause, the oestrogen level falls, less glycogen is secreted, and the pH of the vagina rises, favouring growth of *Candida*. Also, during pregnancy there is an excess of sugar secreted there that favours the growth of *Candida*, as does the excess sugar in the urine of diabetic patients. Consequently, courses of antibiotics, menopause, diabetes or pregnancy may all trigger vaginal thrush.

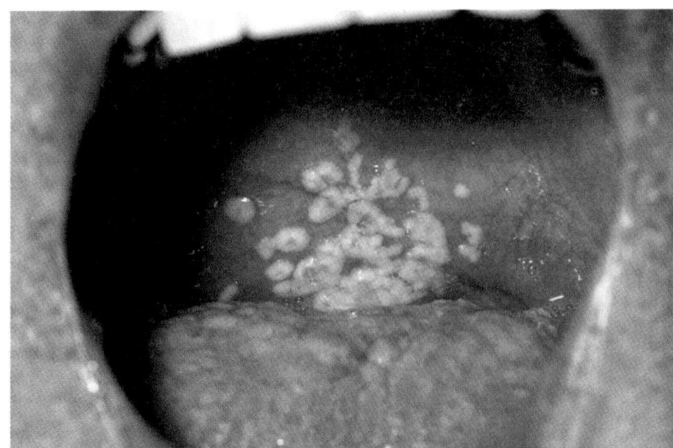

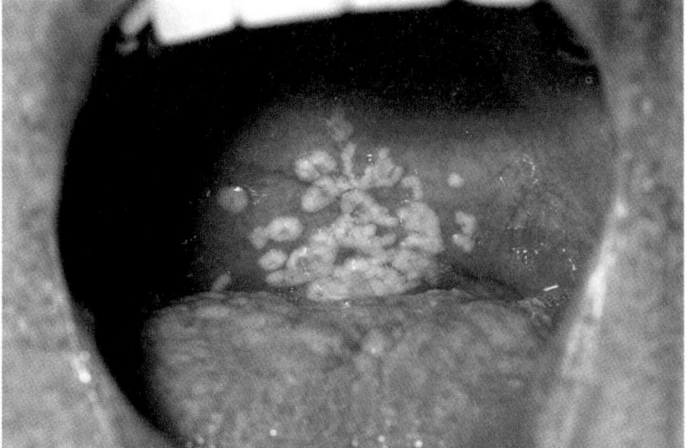

Figure 10.41 *Candida albicans*, and oral thrush.

Fungal diseases of plants

Great economic harm can result from fungal diseases of plants. The boot-lace fungus may destroy established forest trees; the potato blight fungus was responsible for crippling human poverty and a major diaspora of the Irish people in the 19th century.

Boot-lace fungus – Armillaria mellea

This fungus can live saprotrophically on rotting tree stumps. However, if the below-ground strands of hyphae (called rhizomorphs) make contact with the roots of a healthy tree, then it will grow into the living tree and parasitise it, eventually causing its death.

Figure 10.42 The boot-lace fungus.

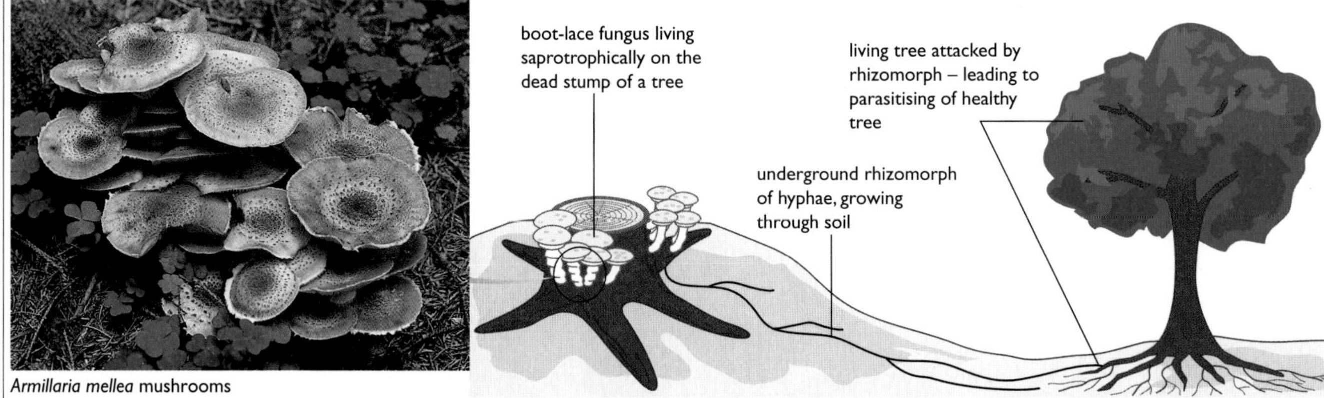

Potato blight – Phytophthora infestans

Phytophthora infestans parasitises the potato plant. This fungus was first recorded in the British Isles in 1845, and was the major cause of the notorious Irish famine. Since then, and until effective fungicides were developed, the disease caused severe damage to potato plants in warmer, wetter areas of the country. *Phytophthora* overwinters in infected tubers, and the mycelium feeds on the new plant that grows up in the spring. Then spores are produced and may infect healthy crop plants, particularly in warm, humid conditions.

Figure 10.43 Life cycle of *Phytophthora infestans*.

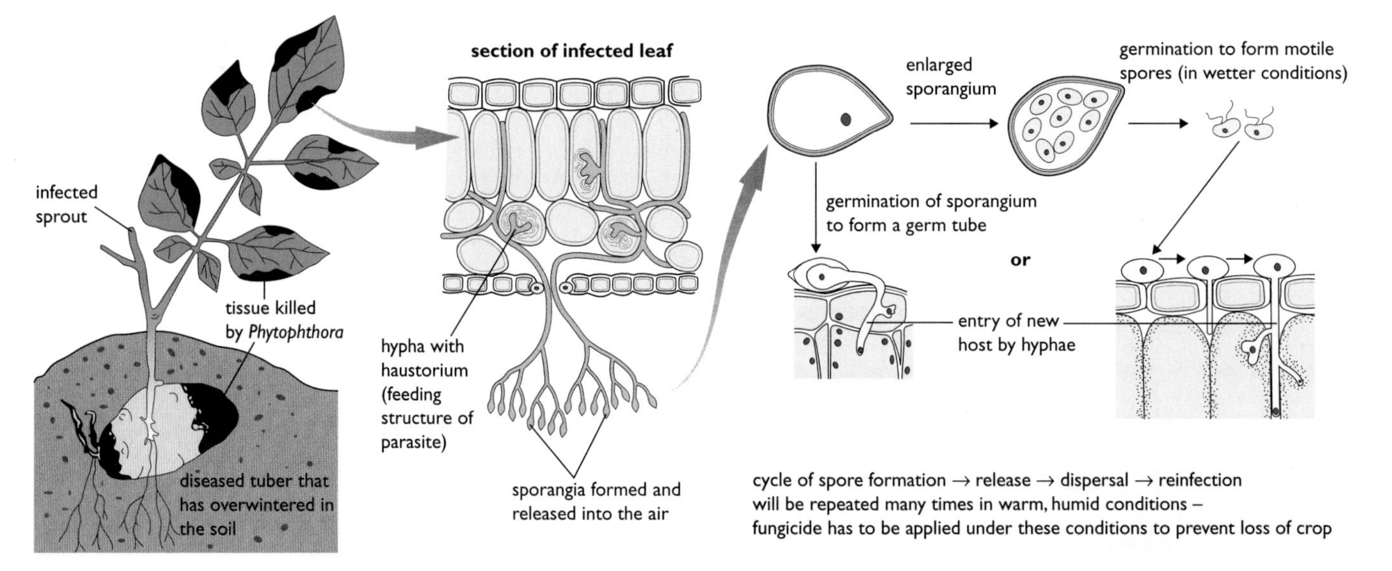

Wild plants and weeds may also be parasitised by fungi

Plants of hedgerows and gardens are occasionally parasitised by fungi. Rusts, smuts, mildews and leaf-spot diseases are found on weeds as well as on crop plants, especially as the former are not routinely treated with fungicides. For example, the wild plant and garden weed dock (*Rumex* sp.) is parasitised by the leaf-spot fungus *Ramularia*. The mycelium feeds on the mesophyll cells of the leaves, and some hyphae grow out of stomata and form and release spores that may be carried by air currents to new hosts.

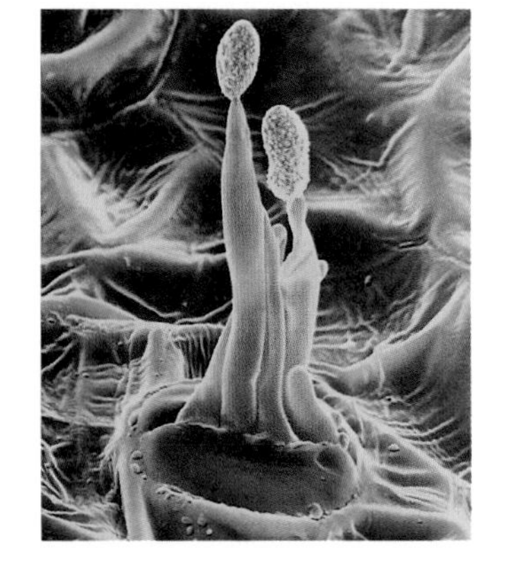

Figure 10.44 SEM of spore-producing hyphae of *Ramularia*, growing out of stomata in the leaf of the host plant (dock) (x 1400).

Prions and the diseases called encephalopathies

Proteins called prions are believed to be the agents that cause diseases known as encephalopathies, in which the brain becomes spongy and forms holes where once there were neurones. Those afflicted – humans (with Creutzfeldt–Jacob disease, CJD); sheep (with scrapie); or cattle (with bovine spongiform encephalopathy, BSE) – lose physical condition and eventually become totally unco-ordinated. The cow in Figure 10.45 had difficulty in standing. In humans the memory is lost, as well as body control, prior to death. The term prion is a contraction, derived from '**pro**teinaceous **in**fectious particle'.

Figure 10.45 A cow with BSE.

What are prion proteins and how may they cause disease?

Prion proteins are natural components of the cells of mammals and also of yeasts. They probably occur more widely. Prions are therefore assumed to have some important role, but this is not yet understood. What has been established is that the normal tertiary structure of prion protein (PrP^c) consists of multiple α-helices, but that this large molecule can unfold and form into a different molecule (PrP^{sc}) where two of the helices become β-pleated sheets (Figure 10.46). The prion is said to have flipped. Once this change in shape has occurred, PrP^{sc} can trigger the same change in shape in normal PrP^c with which it is in contact. Then the mass of PrP^{sc} molecules will coalesce into insoluble fibrils. When this occurs in neurones of the brain, neurone destruction occurs and the encephalopathy condition follows.

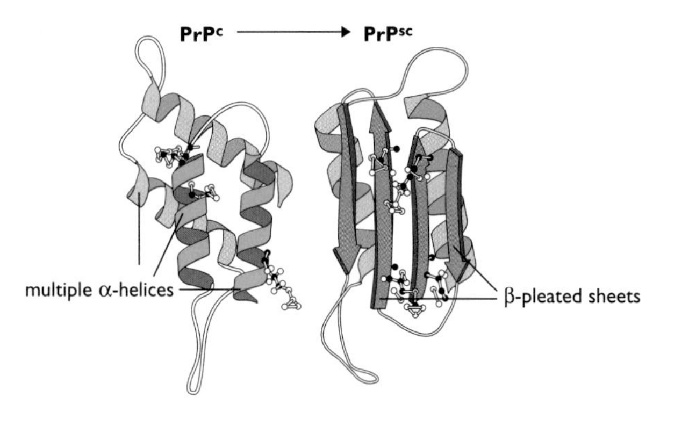

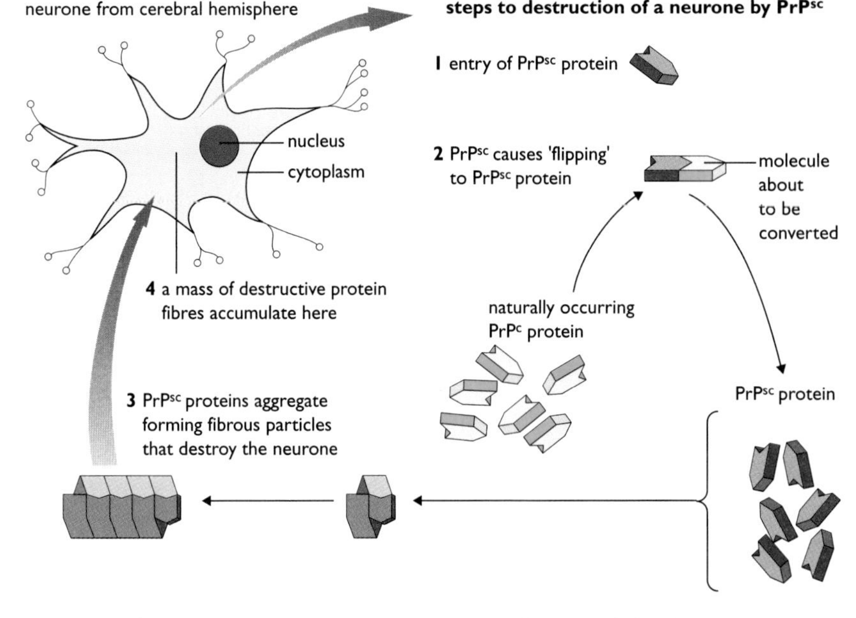

Figure 10.46 Prion protein flipping and the triggering of an encephalopathy.

Encephalopathies are of cannibalistic origin

The first known encephalopathy disease was ***kuru***, observed in people of Papua, New Guinea, whose custom it was to honour their dead by eating them. Men ate muscle tissue, but women and children received brain tissue. Only the latter eventually died of *kuru* – when an ancestor they had honoured had also died of that disease.

Crossing the species barrier

Another prion disease with a long history is **scrapie**. Until recently, encephalopathies had not been known to jump the species barrier, say from sheep to cattle. Recent farming practices of using offal from sheep or cows in manufactured animal feeds are likely to have spread encephalopathies among cattle. Also, pituitary extract prepared by vets for injection into other cattle may have been a source.

What allows the prions of one species to infect some, but not all, other susceptible species? It may depend on how similar the primary structure of different prion proteins is. If the size and amino acid sequence of prions of two species are sufficiently similar, then they may infect both.

Extension: controversy about the prion theory

In 1997, Stanley Prusiner of the University of California at San Francisco was awarded a Nobel Prize for work on his **prion theory of encephalopathies**. Other workers have argued that an as-yet-undiscovered virus was involved. However, the infectious agent for encephalopathies has been shown to remain infectious even after bombardment with radiation sufficient to destroy all DNA or RNA present. As proteins are not destroyed by radiation, whereas viruses (which contain nucleic acid) are, prions are most likely to be responsible. Nevertheless, much remains to be discovered about prions, normal and flipped.

Answers

Chapter 1

1 1.4 mm = 1400 μm; 660 nm = 0.660 μm.

Chapter 2

1 The image of *E. coli* (excluding the variable flagella) is 60 mm. As the magnification is × 25 000, the actual size is:

$$\frac{60 \times 1000}{25\,000}\ \mu m = 2.4\ \mu m$$

2

Eukaryotic chromosomes	Bacterial chromosomes
Within nucleus, enclosed by nuclear membrane	In cytoplasm, attached to plasma membrane
Linear structure, occurring in pairs, one from each parent	Single, circular structure
DNA helix supported by histone protein 'scaffold'	DNA without histone protein
Contains lengths of non-coding DNA, called introns	Without 'introns'
Each chromosome has a centromere somewhere along its length	No centromere

3 Pasteur used S-shaped glass flasks in a demonstration similar to that in Figure A.1.

an issue at the time: could living things suddenly be formed (spontaneous generation) from non living things?

the flasks Pasteur used

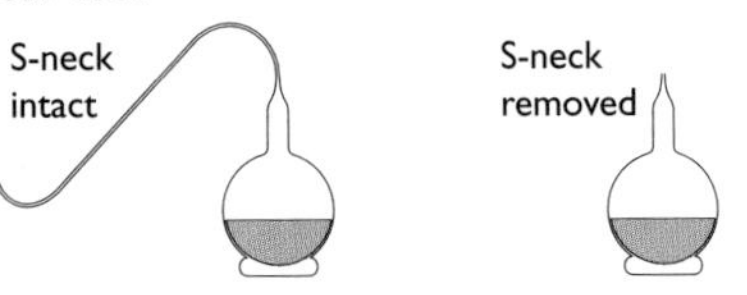

showing where microbes come from

tubes 2–5 were sterilised at outset of experiment

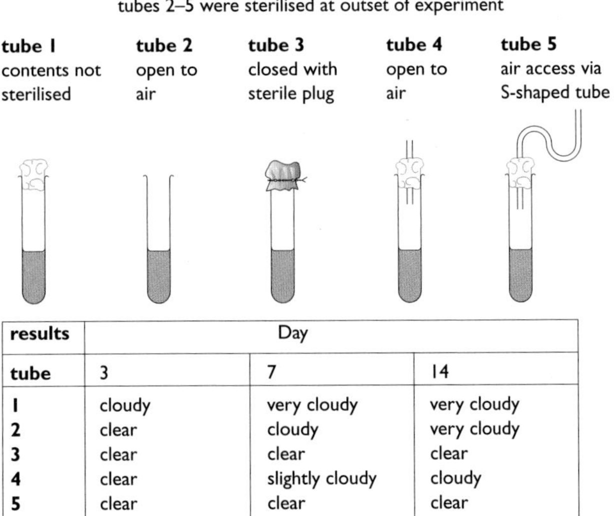

results	Day		
tube	3	7	14
1	cloudy	very cloudy	very cloudy
2	clear	cloudy	very cloudy
3	clear	clear	clear
4	clear	slightly cloudy	cloudy
5	clear	clear	clear

the number of microbes that developed (judged by cloudiness) was in proportion to the communication with the external air

Figure A.1

4 In C squares of the counting chamber, the average number of bacteria per square was:

$$9 + 4 + 7 + 6 + 8 + 6 + 5 + 4 + 7 + 4 = 60 / 10 = 6$$

The C square has a volume of 0.00025 mm^3, therefore the estimated number of bacteria in 1 mm^3 of the culture was:

$$6 \times 4000 = 24\,000 \text{ bacteria}$$

(because 0.00025 mm^3 is 1/4000 of 1.0 mm^3).

5 Pyruvic acid (as pyruvate) is an intermediate of respiration common to the aerobic and anaerobic pathways.

6 The significance of microbes' ability to fix atmospheric nitrogen is given on page 36.

Chapter 3

1 Magnification = the number of times that an image is larger than the specimen, i.e. size of image / size of specimen

Resolution is the ability to see structures that are close together as separate. If two tiny objects are so close they appear as one structure, then magnification will not resolve them.

The resolution of a light microscope is determined by the wavelength of light.

2

	HIV	TMV
Structure	Two single strands of RNA	One single strand of RNA
On entry	Converted to DNA helix by reverse transcriptase	Remain in the cytoplasm and 'read' at ribosomes
	Becomes 'provirus' in nucleus in chromosome of host T4 cell	Viral proteins formed Viral RNA and viral protein synthesised immediately
	Later, synthesis of viral protein and RNA, and new viruses assembled	New viruses assembled immediately
	Viruses escape and infect other cells	

3 Viruses may be cultured:

- *in vitro*, meaning the biological process is occurring in cell extracts (literally 'in glass'), or
- *in vivo*, meaning the biological process is occurring in a living organism (literally 'in life').

Chapter 4

1 The features of eukaryotic cells are listed in Table 2.1, page 3, and the features of hyphae are shown in Figure 4.1, page 24.

2 The connections between alcoholic fermentation and lactic acid fermentation, and how they differ, are shown in Figure A.2.

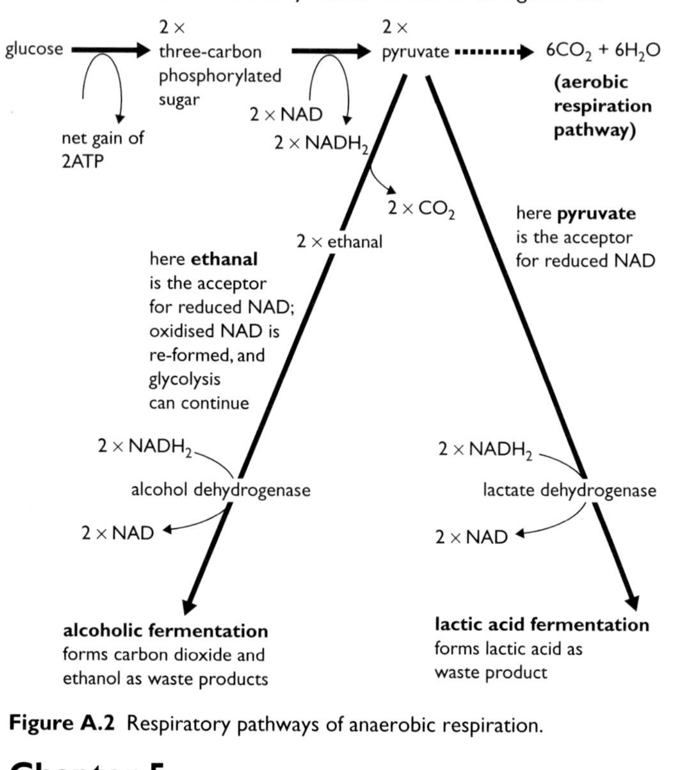

Figure A.2 Respiratory pathways of anaerobic respiration.

Chapter 5

1 The classification of feeding strategies, including heterotrophic nutrition, is shown in Figure A.3.

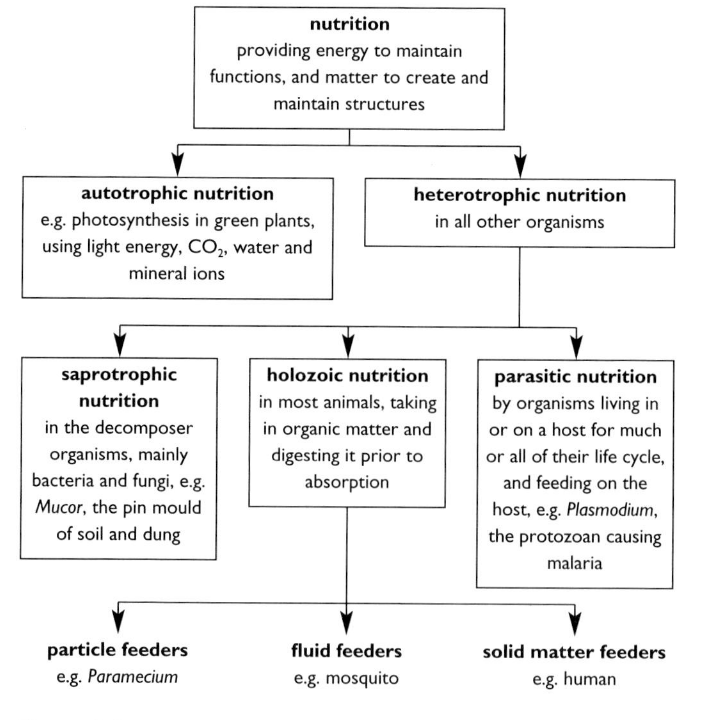

Figure A.3 Feeding strategies.

Chapter 6

1 Silage, like compost, is made from grass, but creation of silage is an entirely different process. It is preserved by fermentation with the exclusion of air. The results of your internet search under 'silage making' should cover issues such as: harvesting of 'leafy' grass; partial drying (wilting); additives that may be used; how anaerobic fermentation is achieved; pit silage and baled silage; later use of silage in stock feeding.

2 Hydrogen bonds are electronic attractions between an electropositive hydrogen atom and an electronegative atom in another group or molecule, e.g. hydrogen bond:

—O—H···O<   >N—H···O<

—O—H···N<   >N—H···N<

They are found in and between many molecules, such as water, proteins, DNA and cellulose.

3 In water rich in nitrates and phosphates, algae grow unchecked (bloom), but eventually they die and decay. Aerobic metabolism of the saprotrophic bacteria involved uses up dissolved oxygen, and anaerobic conditions develop. Organisms such as fish die. Organic matter now decays anaerobically, releasing H_2S (and other substances). Many remaining aquatic organisms are killed. An aquatic environment takes time to recover from this (eutrophication) process.

4 Combined nitrogen formed by *Anabaena* becomes available for use by plants and animals: **a)** due to some loss of ions into the medium from the cyanobacterium cells; **b)** when *Anabaena* is eaten by herbivorous animals; and **c)** on death and saprotrophic decay of the *Anabaena*.

5 Leguminous plants belong to the family Papilionaceae (Leguminosae).

Green vegetables of this family are peas, french beans, runner beans and broad beans.

Common wild flowers in this family include vetches and clovers, gorse and broom, trefoils, medics and clovers.

6 In electric storms, small quantities of nitrogen combine with oxygen, or with traces of hydrogen (present in the upper air), under the influence of the high-tension electric discharge. Traces of nitric oxide and ammonia are formed and washed down in the rain.

7 Greenhouse gases such as methane (plus water vapour and, to a lesser extent, CO_2), once in the upper atmosphere, trap infra-red radiation coming up from the earth as the soil is warmed by incoming solar radiation. Some of this heat is re-radiated from the upper atmosphere back to earth. We rely on the greenhouse effect to keep the earth warm, but a large enhancement of the greenhouse effect may alter living conditions unfavourably.

Chapter 7

1 An excess of ethanol is potentially toxic to humans because of its effects on:

- the central nervous system – as a narcotic drug it interferes in brain cell function and ultimately destroys brain cells
- the liver – by raising acetaldehyde levels in the blood (which inhibits heart and other muscle), and causes accumulation of fat (possibly leading to obstruction which may contribute to hypertension)
- the alimentary canal and diet – by enhancing the likelihood of cancer in the gut mucosa and associated organs (e.g. pancreas and liver), and by adding nutrient-free calories to our intake, leading to likelihood of obesity.

2 Milk from cows afflicted with the udder disease mastitis should be rejected until antibiotic treatments (directly into the udder via the teat) have been successfully completed. If antibiotic-contaminated milk reaches the dairy, the bacteria of 'starter cultures' used in yoghurt production may be inhibited or killed.

3 Fermenters are used:
- in batch culture in the production of alcohol in wine- and beer-making (pages 44 and 45); cheese and yoghurt production (page 46); industrial enzyme production (page 51); and antibiotic production (page 53)
- in continuous culture in the cow's rumen (page 32); the aerobic tanks and anaerobic digesters of sewage works (page 34); in biogas plants (page 42); and in Quorn production (page 48).

4 Water may be removed from spent medium by reverse osmosis, as illustrated in Figure A.4.

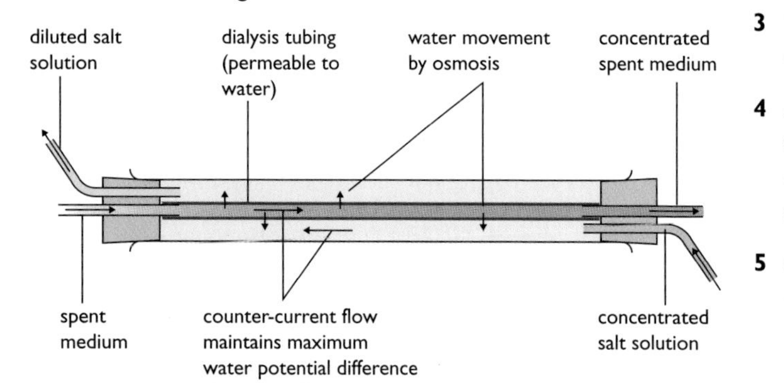

Figure A.4 A laboratory pilot plant for water removal by reverse osmosis.

5 Antibiotics are added to the feed of intensively reared livestock such as chicken and pigs. Here they reduce or prevent the incidence of many diseases, and the animals are shown to grow better with them. Possible dangers arise from the over-exposure of bacteria to antibiotics, encouraging the appearance of antibiotic-resistant strains of disease-causing organisms.

Chapter 8

1 The key steps by which a gene dictates the type of protein a cell manufactures are:
- transcription – part of the coded information of the chromosome is copied and delivered to a ribosome in the cytoplasm in the form of mRNA
- amino acid activation – amino acids in the cytoplasm are activated by reaction with ATP and with short lengths of transfer RNA
- translation – the information in the code is used to determine the sequence of amino acids in the protein, and this occurs as the ribosome passes along the mRNA, picking up activated amino acids in the right sequence, and allowing the formation of peptide bonds between amino acids.

2 Cultivated grasses (rice, wheat, maize) literally feed the world. If these plants could manufacture their own nitrogenous fertiliser (in effect), this would vastly improve productivity without demanding expensive nitrogen-rich fertilisers.

3 We can speculate on the possible genes and characteristics that might escape into disease-causing microbes, making them a greater threat to health.

Chapter 9

1 If plastics were manufactured from oils from current crops, this would greatly reduce the use of fossil fuels. These could be conserved for future use, and the atmospheric CO_2 would rise more slowly.

2 Sugar, salt, and drying of foods all prevent decay in foods exposed to the air, by causing any spores that land on the food to die by reverse osmosis.

Chapter 10

1 Preventive medicine seeks to combat heart disease and cancer by identifying and encouraging healthier life styles (involving diet and physical activity).

2 In cholera, chloride ions move from epithelium cells into the lumen of the gut by active transport caused by protein pumps in the cell membranes, driven by ATP. Water is moved by osmosis (a special case of diffusion), in which metabolic energy is not involved.

3 How resistance to the drug's action arises in microorganisms is described on page 52.

4 *Plasmodium* is present only briefly in blood plasma, and so is less at risk from the actions of the patient's immune system. In blood or liver cells the parasite is relatively protected, as well as having access to the nutrients it requires.

5 **a)** Parents' concerns:
- in very rare cases, MMR has given a child one of the three diseases it is designed to protect against
- occasionally there are (well publicised) cases of an adverse reaction to the injection
- occasionally the emergence of autism in a child has coincided with the time MMR was administered, so in the minds of some, the two events are connected (although the evidence suggests they are not).

b) Public health experts' concerns:
- memories of what it is like for children to get measles, mumps or rubella are fading fast, so vaccination may seem a pointless exercise
- when vaccination levels dropped in the USA there was an increase in cases, and 123 people died of measles between 1989 and 1991; vaccination levels are falling in areas of the UK, and the incidence of these diseases is now rising again
- unvaccinated children are at risk of all three disease, all of which may develop serious complications such as encephalitis (50% of children with this are left with permanent brain damage)
- later in life, adults can contract mumps or rubella, the latter with serious risks for pregnant women, for example
- separate vaccines (rather than MMR) ultimately give children the same level of protection (and the same risk of adverse effects?), but the process takes longer, and children are more likely to catch one of the diseases before they have had all three injections (injections start after 1 year; prior to this children are protected by their mother's antibodies).

6 How the segmented genome of the flu virus contributes to its genetic variability is shown in Figure 10.29, page 79.

7 The HIV virus and AIDS, the disease it causes, create particular problems for the people and economies of less-developed countries such as Zimbabwe or Zambia because it almost exclusively handicaps the young adult population at their time of greatest economic activity and family-building.

8 The insecticide DDT is harmless to humans at concentrations that are toxic to mosquitoes, but applied at these concentrations it rapidly collects in body fat and is transferred from prey to predator through the food chain, steadily increasing in concentration. It accumulates at the top of the food chain at concentrations that may cause harm. The stability of the DDT molecule means that it does not biodegrade readily, but persists (which is why it is so effective), and so remains a health threat.

Glossary

Entries are *aides-mémoire*, rather than formal definitions.

agar gelatinous polysaccharide, obtained from red algae
antibody a protein produced by blood plasma cells derived from B-lymphocytes when in the presence of a specific antigen, which then binds with the antigen, aiding its destruction
antibiotic organic compounds produced by microorganisms which selectively inhibit or kill other microorganisms
antigen a substance binding specifically to an antibody
aseptic technique a procedure delivering sterile media and equipment for culturing microorganisms
attenuated reduced pathogenicity in a microbe

bacillus a rod-shaped bacterium
bacteriophage a virus that parasitises bacteria
batch culture ingredients are added to fermenter at start, and after the reactions are completed the products are harvested
BCG bacille Calmette–Guerin, a TB vaccine named after the work of two French scientists
binary fission when a cell divides into two daughter cells
biodegradation recycling of dead organisms and waste matter by saprotrophs
biodeterioration when saprotrophs degrade economically useful materials
biotechnology the industrial and commercial applications of biology, particularly of microorganisms and enzymology
broth liquid medium for growing microorganisms

capsid protein coat of a virus
coenocytic fungal tissue not divided into individual cells
complementary protein protein in the body's defences against antigens that invade the body
chemosynthesis use of chemical energy from oxidation of inorganic compounds to synthesise organic compounds
chromosome, eukaryotic visible at nuclear division, each consists of a long thread of DNA packaged with protein
prokaryotic a circular strand of DNA in the cytoplasm, attached to the plasma membrane
clone a group of genetically identical individuals (or cells)
coccus spherical bacterial cell
continuous culture a culture of cells maintained in growth phase by supply of nutrients and removal of products/waste matter
commensalism a mutually beneficial association between two organisms of different species
cyanobacterium photosynthetic prokaryotes

decomposer organisms (typically microorganisms) that feed on dead plant and animal material, allowing matter to be recycled
detritivore an organism that feeds on dead organic matter
downstream processing alteration of a biotechnology product

endemic restricted to a certain region
endotoxin bacterial toxin that remains in/on the cell
eukaryote organism having cells with a 'good' nucleus
exotoxin bacterial toxin secreted by bacteria

facultative having the capacity to do something if necessary, or if the opportunity arises, e.g. be anaerobic, or be a parasite
fermentation anaerobic breakdown
fermenter vessel used for the growth of microorganisms
open a non-sterile fermenter system, open to atmosphere
closed an enclosed, sterilised fermenter system

gene probe prepared sequence of DNA made radioactive, coding for a particular amino acid residue sequence
genetic engineering change to the genetic constitution of individuals or populations

haemocytometer an apparatus for counting cells
helper cell type of T-lymphocyte which interacts with B-lymphocytes to stimulate plasma cell formation
hydrogen bond a weak bond caused by electrostatic attraction between a positively charged part of one molecule and a negatively charged part of another
hypha (pl. hyphae) the tubular filament 'plant' body of a fungus

immunisation e.g. inoculation/vaccination, the injection of a specific antigen, derived from a pathogen, to confer immunity
immunity resistance to the onset of a disease after infection by the causative agent
inflammation painful swelling and redness in response to infection or tissue damage
interferon proteins formed by cells of the immune system in response to virus infections

macrophage a large, phagocytic white cell
mucus a watery solution of glycoprotein with protective/lubrication functions
murein a cell-wall component of prokaryotes (a macromolecule built of amino sugar monomers)
mutation an abrupt change in the DNA of a chromosome
mutualism symbiosis in which both organisms benefit
mycelium a mass or network of hyphae
mycorrhiza a mutualistic association between plant roots and fungi

nitrifying bacteria chemosynthetic autotrophs using energy from exothermic chemical reactions
nitrogen fixation atmospheric nitrogen reduced to ammonia

opsonin type of antibody that attacks bacteria and viruses, facilitating their ingestion by phagocytic cells

pandemic worldwide distribution
pathogen a disease-causing microorganism
pili thin, short filaments protruding from some bacteria, involved in attachment
plasmid a small, circular length of DNA that is independent of the chromosome in bacteria
plastid an organelle containing pigments, e.g. chloroplast
prokaryote tiny unicellular organism without a true nucleus, having a ring of RNA or DNA as a chromosome, e.g. bacteria
Protoctista kingdom of the eukaryotes, consisting of single-celled organisms and multicellular organisms related to them

recombinant DNA DNA that has been artificially changed, involving joining together genes from different sources
restriction enzyme enzymes, also known as endonucleases, that cut lengths of nucleic acid at specific sequences of bases
retrovirus viruses that, on arrival in a host cell, have their own RNA copied into DNA which then attaches to the host DNA

saprotrophic organism that feeds on dead organic matter (i.e. saprotrophic nutrition) – also referred to as **saprobionts**
single-cell protein protein derived from unicellular organisms
slime layer sheath around some bacteria
sticky ends unpaired nucleotides at the end of a DNA helix, produced when the DNA is cut by certain restriction enzymes
symbiosis literally 'living together', e.g. parasitism

vaccination conferring immunity from a disease by injecting an antigen (of attenuated microbe or inactivated component) so the body generates antibodies and memory cells
vaccine a preparation of pathogenic microorganisms or their antigenic components which can induce protective immunity
vector an organism that transmits a disease-causing organism, or a device for transferring genes during genetic engineering

Index

Page numbers in **bold** represent tables

acid-fast bacilli 72
acidophiles 17
acquired immunity **76**
active immunity **76**
activated sludge 34
active transport 3, 76
acute disease 68
aerobic respiration 15, 82
agar plates 6, 7, 10, 23, 52, 57
Agrobacterium tumefaciens 60
AIDS *see* autoimmune deficiency syndrome
Alcaligenes eutrophus 63
alcohol
  fermentation 26
  from biomass 43
  see also brewing; wine making
alcoholic drinks 45
algae 28–9
alkalinophiles 17
Alnus glutinosa 38
Amanita muscaria 25
amino acid synthesis 36
ammonification 40
Amoeba 28
Anabaena spp. 16, 36, 37
anaerobic respiration 15
ancient bacteria 17
Anopheles mosquito 82
antibiotic resistance genes 57
antibiotics 52–3
  downstream processing 53
  prophylactic 52
  resistance 52
  scale-up process 53
  screening 52
  sensitivity 52
antibodies 67, 74
antigen presentation 75
Arabidopsis thaliana 63
Armillaria mellea 85
artificial immunity **76**
Ascobolus 31
aseptic techniques 6
asexual reproduction
  bacteria 4, 5
  yeasts 26
  plasmodium 83
Aspergillus 27
athlete's foot 84
autoimmune deficiency syndrome (AIDS) 20, 80–1
  prevention 81
  treatment 81
autotrophic bacteria 14
Azotobacter spp. 36

Bacillus anthracis 13, 14
bacteria 2–17
  ancient 17
  cell components 2–3
  cell division and reproduction 4
  chemosynthetic 14
  classification 14–15
  culture 6–8
  cytoplasm 2
  extremophiles 17
  genetic material 3, 4–5, 49, 56, 58, 60
  genetically modified 56, 57
  Gram-negative 15
  Gram-positive 15, 68
  growth curve 12–13
  heterotrophic 14
  lawn 52
  monitoring of growth 9–11
  nitrogen fixation 36
  nutrition 14
  parasitic 14
  photosynthetic 14, 16
  prokaryotes and eukaryotes **3**
  respiration 15
  saprotrophic 14
  sexual reproduction 5
  shape 14
  transgenic 56
  see also cyanobacteria
bacterial diseases 67–77
  causes of 67
  cholera 70–1
  food poisoning 68–9
  tuberculosis 72–3
bacterial growth 4
  calibration curve 9
  conditions for 7
  diauxic 13
  generation time 13
  growth curve **12**–13
  monitoring 9–11
  number of generations 13
bacterial toxins 67, 70
bacteriophages 21
  culture 23
batch culture 49
binary fission 4
biodegradation 62
biodeterioration 62
biogas 42, 43
biogeochemical cycles 31
biological washing powder 50
biomass
  alcohol from 43
  fuels from 42–3
biosensors 50
biotechnology 44–53
  antibiotics 52–3
  brewing 45
  fermented foods 46–7
  food 48–9
  industrial enzymes 50–1
  wine making 44
blood-clotting mechanism 74
blood-sucking insect 82
boot-lace fungus 85
bovine somatotrophin, genetically engineered 59
bovine spongiform encephalopathy 86
bread making 27
brewing 45
broth 9, 12–13
Brucella abortus 65

Candida albicans 84
canning 65
capsid 18
capsule
  bacteria 3, 67
  virus 20
carbon recycling 31
catalysts 41, 50
cell cycle 4
cell division 4
cell walls 2, 19, 49, 56, 72
cellulase 32
cellulose breakdown 32–3
  by non-ruminants 33
  by ruminants 32
  in soil 33
Ceratocystis ulmi 25
cheese making 27, 46
chemical preservation **64**
chemosynthesis 40
chemosynthetic bacteria 14
chitin 24
Chlamydomonas spp. 28
Chlorella 28
chlorination 35
cholera 70–1
  prevention 71
  treatment 71
cholera toxin 70
chromosomes 4, 55
closed fermenter systems 49
Clostridium spp. 36
Clostridum botulinum 65
commensalism 2
communicable diseases 66
complement proteins 67, 74
compost 31
conjugation 5
continuous culture 49
Coprinus 31
Creutzfeldt-Jacob disease 86
cryophiles 17, 31
culture of bacteria 6–8
  agar 6–7
  aseptic techniques 6
  broth 9, 12–13
  growth conditions **7**
  inoculation 8
  preparation of cultures 8
  turbidity change 9
culture media 6, 7, 9
culture of viruses 22–3
  animal and human viruses 22
  bacterial viruses 23
  fertilised hens' eggs 22
  tissue culture 22
cyanobacteria 16, 36
  in paddy fields 37
  in subsistence farming 37
cycling of nutrients 30–1
  carbon recycling 31
  decomposition of compost 31
  decomposition of dung 31
  see also nitrogen cycling
cytoplasm 2

DDT 83
decomposers 31
dehydration 65
denitrification 41
diarrhoea 68, 70
diauxic growth 13
dilution plate method 9, 10
diplococci 14
Diplococcus pneumoniae 14
disease prevention 66
diseases
  bacterial 67–77
  communicable 66
  encephalopathies 86
  fungal 84–5
  protoctistan 82–3
  viral 78–81
DNA 3
DNA viruses 18
downstream processing 53
droplet infection 72
drug resistance 73, 83
dung, decomposition of 31
duodenum 32
Dutch elm disease 25

electron microscope 18
encephalopathies 86
endoparasites 18
endospore formation 5
endotoxins 67
enteritis 68–9
enterotoxins 68
environment 30–43
  cellulose breakdown 32–3
  cycling of nutrients 30–1
  fuels from biomass 42–3
  nitrogen cycline 36–41
  sewage treatment 34
  water purification 35
enzymes, industrial 50–1
epidemiology 71, 79
Epidermatophyton spp. 84
epithelium 67, 68, 70
Escherichia coli
  genetic engineering 54
  structure 2
eukaryotes 1, **3**, 24
  genetic engineering in 59
exotoxins 67
extremophiles 17

fermented foods 46–7
  alcohol 26, 43
  cheese 46
  sauerkraut 47
  soya foods 47
  yoghurt 46
fermenters 32, 49, 51
filamentous algae 28
fimbriae 3
flagellae 3
fluid mosaic model 3
food infection 68
food poisoning 68–9
  incidence of 68
  prevention of 69
food preservation 64–5
  chemical **64**
  dehydration 65
  heating 65
  refrigeration and freezing 64
foot and mouth virus 23
fossil fuels 42
Frankia 38
freezing 64
fungal diseases 84–5
  humans 84
  plants 85
fungi 24–7
  formation on fresh dung 31
  heterotrophic nutrition 24
  hyphae 24, 27
  parasitic 24
  Penicillium 27
  saprotrophic 24
  spores 24, 27
  yeasts 26–7
Fusarium graminearum 48

gas vescicles 16
gel electrophoresis 55
gene copying 54–5
  genes from chromosomes 55
  messenger RNA 54
  working backwards from protein 55
genes 54
  antibiotic resistance 57
  switching on 58
genetic engineering 23, 54–61
  bovine somatotrophin 59
  engineered bacteria 57
  environmental and ethical issues 61
  ethical issues 61
  eukaryotes 59
  gene copying 54–5
  human growth hormone 58
  insulin 54–6
  plants 60
  prokaryotes 58–9
  switching on gene action 58
  vaccines 23, 61
  vectors 56

genetically modified organisms 54
genetically modified plants 60
genome 54, 79
global warming 33
glycolipids 3, 67
glycoproteins 3, 67
Gram, Hans Christian 15
Gram-negative bacteria 15
Gram-positive bacteria 15
greenhouse gases 43
growth rate constants 13

Haber process 50
Haef test 73
haemocytometer 9, 11, 12
haemoglobin 82
Halobacterium halobium 17
halophiles 17
heating 65
helper cells 75
heterocysts 36
heterotrophic bacteria 14
heterotrophic fungi 24
host defences 74–5
host organism 18, 83
human growth hormone, genetically engineered 58
human immunodeficiency virus 20, 80–1
hydrolytic enzymes 24
hyperthermophiles 17
hyphae 24, 27

immune system 74, 79
immunity 73, **76**–7
  see also vaccines
industrial enzymology 51
inflammation response 74
influenza 78–9
  inoculation 8
  epidemics and pandemics 79
  genetic variation in virus 79
  symptoms **78**
  treatment 79
  viral transmission and replication 78
insulin, genetically engineered 54–6
Irish potato famine 85
irradication 64

Jenner, Edward 76

Kluyveromyces lactis 50
Koch, Robert 1, 72
Koch's postulate 66
kuru 86

lactose operon 58
lag phase of growth curve 12
lambda virus 21
leaf-spot fungus 85
Leeuwenhoek, Anthony van 1
ligase 60
log phase of growth curve 12
lymphocytes 74, 75
lysogenic cycle 21
lytic cycle 21

macrophages 75
malaria 82–3
  control 82
  eradication of 83
  live cycle of *Plasmodium* 83
  transmission 82
memory cells 75
mesophiles 31
mesosomes 3, 16
messenger RNA 54
methane gas 42–3
Methanobacterium spp. 42
methanogens 42
Methylococcus spp. 14
micrometre (μm) 1
microorganisms 1
Microsporum spp. 84
milk, heat treatment 65
moquito 82
Mucor 24, 31
mucus 67, 74, 78
murein 2
mutualism 25, 38
mycelium 24
Mycobacterium tuberculosis 72
mycology 24
mycoprotein 48
mycorrhizal relationship 25
mycoses 84–5

natural immunity 73, **76**
nanometre (nm) 1
negative strand RNA virus 78
nitrates
  formation of 40
  in soil 40–1
nitrification 40
Nitrobacter spp. 40
Nitrococcus 14
nitrogen 36
nitrogen cycling 36–41
  combined nitrogen 36
  maintenance of soil nitrates 40–1
  nitrogen fixation 36–9
nitrogen fixation 36–9
  economic importance 37
  mechanism of 36–7, 39
  mutualistic micro-organisms 38–9
nitrogenase 36, 38
Nitrosomonas spp. 40
Nostoc spp. 36
nucleus **3**, 24, 26, 28
nucleic acids 18

open fermenter tanks 49
opportunist infection 81
opsonins 74
oral thrush 84
organic acids 27
Oryza sativa 37
osmosis 2

pandemic 79
Paramecium 28
parasitic bacteria 14
parasitic fungi 24
passive immunity **76**
Pasteur, Louis 1, 76
pasteurisation 46, 65
pathogens 66
peat 33
penicillin 27
Penicillium spp. 27
Penicillium chrysogenum 27
Penicillium roqueforti 27
Petri dish 7
phages *see* bacteriophages
phospholipids 3
photosynthesis 36
photosynthetic bacteria 14, 16
Phytophthora infestans 85
Pilaria 31
pili 3
Pilobolus 31
pituitary dwarfism 58
plants
  fungal diseases of 85
  genetically modified 60
plaques 23
plasma membrane 3, 81
plasmids
  antibiotic-resistant 57
  bacterial 3, 5
  recombinant 56
  Ti 60
  as vectors 56
  yeasts 26
Plasmodium spp. 28, 82–3
  life cycle 83
plastics, biodegradability 63
polio vaccine **77**
polyhydroxybutyrate 63
potato blight 85
preventive medicine 66, 81
prions 86
prokaryotes 1, **3**
  genetic engineering in 58–9
prophylactic antibiotics 52
protein-rich foods **38**, 48–9
  fermenters 49
  mycoprotein (Quorn) 48
  single-cell protein 49
proteins 3
Protoctista 28–9
protoctistan disease *see* malaria
protozoa 1, 27, 82
Prusiner, Stanley 86
Pseudomonas denitrificans 41

Quorn 48

Ramularia 85
refrigeration 64
replica plating 57
replication
  bacteria 4
  viruses 19
respiration 15
restriction enzymes 55, 60
retroviruses 20, 80
reverse transcriptase 20, 54, 80
Rhizobium 38, 39
ribosomes 2
rice growing 37
ringworm 84
RNA viruses 18, 78
root nodules 38, 39
Ross, Ronald 1
rumen 32
ruminants, cellulose breakdown by 32

Saccharomyces 26–7
Salmonella enteritidis 68, 69
Salmonella typhi 14
Salmonella typhimurium 69
saprotrophic bacteria 14
saprotrophic fungi 24
sauerkraut 47
scale-up process 53
scrapie 86
senescence phase of growth curve 12
sewage contamination 34
sewage treatment 34
sexual reproduction 5
sickle-cell trait 82
sigmoid growth curve 12
silage 31
single-cell protein 48, 49
size of microorganisms 1
slime layer 3, 16
smallpox 76–7
Snow, John 71
soil
  cellulose breakdown 33
  nitrates 40
somatotrophin, genetically engineered
  bovine 59
  human 58
Sordaria 31
soy sauce 47
soya foods 47
Spirillum minus 14
Spirogyra 28
Spirulina 17, 49
spores 24, 27, 31
staphylococci 14
Staphylococcus aureus 14
  binary fission 4
  enteritis 68
stationary phase of growth curve 12
streptobacilli 1
streptococci 14
Streptococcus pyogenes 14
subsistence farming 37
superoxide dismutase 59

T4 helper cells 80
temperate viruses 21
thermophiles 17, 31
Thiobacillus ferro-oxidans 17
thrush 84
Ti plasmids 60
tissue culture 22
tobacco mosaic virus 19
toxin 67, 70
transformation 5
transgenic crops 60
transmission electron microscope 18
Treponema pallidum 14
Trichophyton spp. 84
Trichophyton mentagrophytes 84
Triticum dicoccum 37
tuberculosis 72–3
  existing immunity 73
  multiple-drug-resistance 73
  prevention 73
  treatment 73
  vaccine 73

unicellular organisms 26
units **1**

vaccines and vaccination 76, **77**
  AIDS 81
  genetically engineered vaccines 23, 61
  malaria 83
  polio **77**
  tuberculosis 73
  whooping cough **77**
vaginal thrush 84
van Leeuwenhoek, Anthony 1
vectors
  insect 83
  plasmid 56
  virus 61
Vibrio cholerae 14, 70
viral disease 78–81
  AIDS 80–1
  influenza 78–9
virulence 21
viruses 1, 18–23
  bacteriophages 21
  classification 18
  culture 22–3
  foot and mouth 23
  genetically engineered vaccines 23
  human immunodeficiency virus 20
  lambda virus 21
  replication 19
  temperate 21
  tobacco mosaic virus 19
  as vectors 61
Volvox 28

water purification 35
water supply 35
whooping cough vaccine **77**
wine making 44
World Health Organisation 76, 83

yeasts 26–7
  alcohol fermentation 26, 43
  asexual reproduction 26
  bread making 27
  genetic engineering 59
  respiration 26
yoghurt 46

PLACES FOR CONTAINERS

CONTAINERS CAN be employed in a variety of ways to enhance the architecture of the house and the garden itself. In this chapter, all the possible situations for displaying containers are examined, and their particular attributes and constraints are discussed in detail. Patios and roof gardens, entrances and steps, paths and alleys, balconies and verandahs offer many design opportunities for large and small containers, and for container groupings. Choosing an appropriate container style, creating a balanced plant display that is in scale with the architecture of the house, and massing together color and shape in attractive and interesting ways are all key elements in achieving a successful scheme.

Before planning your display, it is important to assess the practical constraints of the situation as part of the overall design consideration – for example, you might want to train your plants to provide shelter on a windy roof terrace or employ tough plants as a screen for more tender specimens, or achieve a good vertical display of plants in a narrow space by using climbers or wall pots. This chapter provides useful guidelines on these more practical questions, as well as inspiration for attractive planting schemes.

LEFT: *The attractive gold-tinged leaves of the climber* Humulus lupulus *'Aureus'.*

ABOVE: *A decked patio is dominated by a container of foliage plants, including* Cornus alba *'Elegantissima' and variegated ivy.*

Entrances and Steps

First impressions are extremely important, and nowhere is this more true than for the entrance to a house or apartment. Very often the path to the front door or the steps leading up to it offer no actual space for the soil. Containers of plants make the perfect solution, providing a splash of color in what might otherwise be a monochromatic area of brickwork or paving. In addition, foliage and flowers can have the effect of softening the hard texture of the stone and concrete walls or paving, creating an altogether less forbidding approach to the building.

Entrances

The architecture of the house provides the backdrop to the container planting and you should take this into account in deciding what container to use. Brick, stucco, shingle, tiles and stone are all used for house walls, while the doorway itself can be in any number of styles from classical to rustic, gothic or colonial. Pick containers that blend with it – not only in shape, color and material, but in size as well so the scale is right.

The geometric simplicity of most townhouse doorways is enhanced if you position a container on either side, planted up with identical plants, whether in the form of, say, a pair of clipped box balls or large displays of ox-eye daisies (*Leucanthemum vulgare*). For a country home with a rustic door or porch, a looser style of planting combining climbing plants with some perennials would look charming.

Aromatic plants are particularly valuable for entrances. Sweetly scented climbers are always a good choice, as they provide such a wonderful welcome and take up relatively little of the available space on the ground. Among the most popular are honeysuckle (*Lonicera* spp.), jasmine and many forms of rose (although some are much more strongly perfumed than others – the deep pink 'Zéphirine Drouhin' is one, and many of the old-fashioned roses are also deliciously scented). You can plant climbers together in a large container and allow them to scramble through each other. Annuals such as sweet peas (*Lathyrus odoratus*) can also be induced to climb and scramble over existing

Informal grouping
RIGHT: *For a country-style entrance, opt for an informal grouping of flowering plants. Here climbing roses provide height and soften the doorway, while a mass of old-fashioned cottage perennials, including Canterbury bells* (Campanula medium), *pansies* (Viola *sp.*), *scabious* (Scabiosa 'Butterfly Blue') *and catmint* (Nepeta × faassenii), *create a restful atmosphere.*

Formal arrangements
ABOVE: *Pairs of clipped bay trees* (Laurus nobilis), *trained as standards with plaited stems, flank a front door, their rounded forms contrasting with the vertical lines of the architecture. Other good shapes include spirals of clipped* Buxus sempervirens *(left) and balls (right).*

shrubs in an informal cottage-style arrangement, although you will not get the same sized blooms as you would in a more open situation. In spring some of the scented bulbs are worth growing in pairs of containers – the small, highly scented narcissi, for example, or pots of hyacinths: the large cultivated forms combine well with grape hyacinths (*Muscari* spp.). For summer, large pots of heavily scented Regal lilies (*Lilium regale*) make a wonderful addition to any entrance. Plant them either side of a conservatory door, for instance, but make sure that you stake them as discreetly as possible (see pages 72–73). Nothing looks worse than to see handsome flowering plants whose appearance is marred by prominent and unattractive stakes.

The size of the container, and of the display, is as important as the style: a planted pot roughly one third of the height of the doorway seems to offer a comfortable balance between the architecture and the planting. Poky containers simply look dwarfed and excessively large ones dominate the doorway, although if in doubt go for bigger rather than smaller pots.

Color is important too. Try to avoid the obvious contrasts of, say, white stucco walls and red geraniums (*Pelargonium* cvs.) and opt instead for something more subtle – a scheme which is a mixture of soft apricot and pale blue flowers with silver-gray or variegated silver-and-green foliage. Alternatively use the color of the front door as a key and link the flower scheme with the building in this way. A deep purple front door could be flanked by terracotta pots containing blue rue (*Ruta graveolens* 'Jackman's Blue') mixed with heliotropes, for example, with a deep blue/purple *Viticella clematis* climbing over it. If you do not opt for a formal evergreen planting of, say, box (*Buxus sempervirens*) balls, which look good all year round, you will have to be prepared to change the planting with the seasons. You could maintain the same large central shrub or climber in the container, and then change the surrounding smaller plants seasonally (see Large Containers, pages 38–41). If you want to adopt a single-color theme, you could plant white crocuses or dwarf narcissus 'Bridal Veil' for the spring, followed by white pansies or *Sutera diffusa* (sold as *Bacopa* 'Snowflake') for early summer, and white busy lizzies (*Impatiens* cvs.) or tobacco plants (*Nicotiana* cvs.) for mid- to late summer.

Another important factor to bear in mind when planting up containers for entrance displays is the aspect, since there is often very little shade in front gardens. If you have a south-facing entrance, in full sun, you can feature Mediterranean-style plants, a number of which are distinguished by silvery or felted leaves, among them herbs like artemisia, santolina and

Pots beside steps
LEFT: *These big terracotta pots of marguerites* (Argyranthemum) *punctuate a flight of steps.*

Stepped display
ABOVE: *A flight of steps makes an ideal display area for smaller plants, such as these Ivy-leaved and Regal geraniums (*Pelargonium *cvs.).*

lavender, all of which can then be clipped into neat shapes for a formal look or left to sprawl lazily over the edges of the pots for a more country-style appearance. Mulching the tops of containers with a layer of pebbles in any sunny situation will help to conserve moisture (see page 89). For shady entrances, go for larger pots of big-leaved architectural plants, such as hostas or some of the euphorbias. The big evergreen *Helleborus argutifolius* (syn. *H. corsicus*), with its spiny, hand-shaped, glossy, green leaves, is a good all-year-round performer in sun or partial shade and bears large drooping clusters of lime-green flowers. When the clumps get too unwieldy, you can divide them up (see page 98), although they will take a year or two to re-establish as large clumps. If you do not like their behavior in winter, when the old stems droop as the new foliage emerges in the center, cut the old foliage down and move the container to a less conspicuous place. They need regular watering and plenty of food.

Steps

A flight of steps can look bleak and uninviting unless the brick or stone has weathered and softened. If the steps are wide enough, a container on each step can look attractive. This treatment works well provided the planting is repeated and the containers are similar in size and color.

You can opt for a formal approach with a neat geometric look using soldierly plants with an upright habit, or you can soften the whole effect using billowing shapes, like trailing geraniums (*Pelargonium* cvs.) or clouds of brachycome or lobelia.

If you do not want to put anything on the steps themselves, you can position a large container at the top and bottom of the flight, so that the plants either soften the angle made where the wall and steps meet, or accentuate the architecture, depending on the choice of plant. For the former, you could combine trailing silver-leaved helichrysums with petunias; and for the latter, foliage plants with an architectural growth habit like cordylines, as well as the Australian tree fern (*Dicksonia antarctica*), hostas and ferns.

For a less formal look for wide steps, plant up some fairly shallow containers using plants with a sprawling habit that will soften the edges of the steps, and help to give a lived-in feel to the hard surfaces. Scented plants, such as lavender or santolina, are ideal since, as you brush against them, they release their perfume. Another contender for this relaxed style of planting is the daisy-gone-crazy, *Erigeron karvinskianus*, with its starry, small, daisy-like flowers that fade gently from pink to white, or nasturtiums, which can be used to trail over the edge of the step walls. The darker red forms of the latter look particularly good with soft golden stone or with well-weathered bricks. If the steps are white, you can opt, in warmer climates, for the strong contrast of scarlet geraniums (*Pelargonium* cvs.), but under temperate skies softer blues or pinks look less startling. The pastel-colored, ivy-leaved geraniums will appear more in keeping with a cool climate, particularly when combined with one of the dark green, small-leaved ivies. Ivy-leaved geraniums have the additional advantage of aromatic leaves, particularly when bruised or crushed. If you have little space and are keen to grow herbs, provided the steps have a sunny aspect they offer a useful space for a collection of culinary herbs, such as basil, thyme, marjoram, rosemary and sage. You can grow them in individual pots, or combine them in a mixed display (see page 15).

Same color, different shape

ABOVE: *Pansies are used in all of these containers, but the effect of each is totally different. At the top of the page, large-faced pansies combine with petunias to give an informal display. To the left and right, small-flowered pansies grown in tiered pots produce topiary-like structures.*

Patios

WHETHER YOU HAVE A PATIO as part of a much larger garden or whether you have just a small paved area as your entire garden, containers of plants play an important part in decorating and furnishing these otherwise unprepossessing hard surfaces. If you are going to group displays of plants in containers on the patio, you need to consider how the container material blends or harmonizes with the hard surface where they will be placed. Big oriental ceramic pots, now available at low cost from the Far East, work well on decks or gravel; terracotta looks good on stone surfaces; and lead and wood combine well with brick. Do not mix the types of container too freely – keep to some kind of theme, either of shape or of material. There is nothing to stop you painting various less-than-lovely containers, such as plastic pots, to make them look more attractive, and to help unify the color theme (see pages 36–37).

For a patio the major points to bear in mind are not only the color, shape and texture of the container planting but also the size and style of the containers themselves. Try to avoid too many small containers dotted around the patio and opt instead for bolder groupings of large containers, ideally with a large-leaved tree or shrub as a focal point, and perhaps some small containers in a similar material surrounding it. Not only do large numbers of small containers make for hard work when it comes to watering, but they also bring a restless, busy look to an area which is used predominantly for relaxing.

Most patios are sited between the house and the rest of the garden, if any, and it is therefore important to think about the architecture and style of the house itself when planning the patio planting. If possible incorporate some fairly large plants to relate the scale of the planting to that of the house itself. You can do this in various ways: by positioning large containers close to the house and allowing climbing plants or wall shrubs to support themselves on the house walls, or by creating a group with a large plant as the central element – a datura (*Brugmansia*), perhaps, or tree fern.

Too much flower color and too little in the way of foliage interest will also create a restless look. Large-leaved climbers and shrubs make a good foil for some of the more brightly colored annuals, and you can opt for

Green and white
RIGHT: *This little patio in a formal town garden presents a well-orchestrated and themed container display. Handsome lilies (*Lilium longiflorum*), with their huge, waxy, white trumpets, are echoed by the lower-level planting of pure white petunias, and backed by neat, geometrically shaped clipped box (*Buxus sempervirens*) balls.*

SILVER-LEAVED PLANTS

The shrubs and perennials here are all suitable for a sunny patio. *Senecio*, *Helichrysum*, lavender and *Convolvulus cneorum* make medium-sized shrubs. The others are smaller perennials.

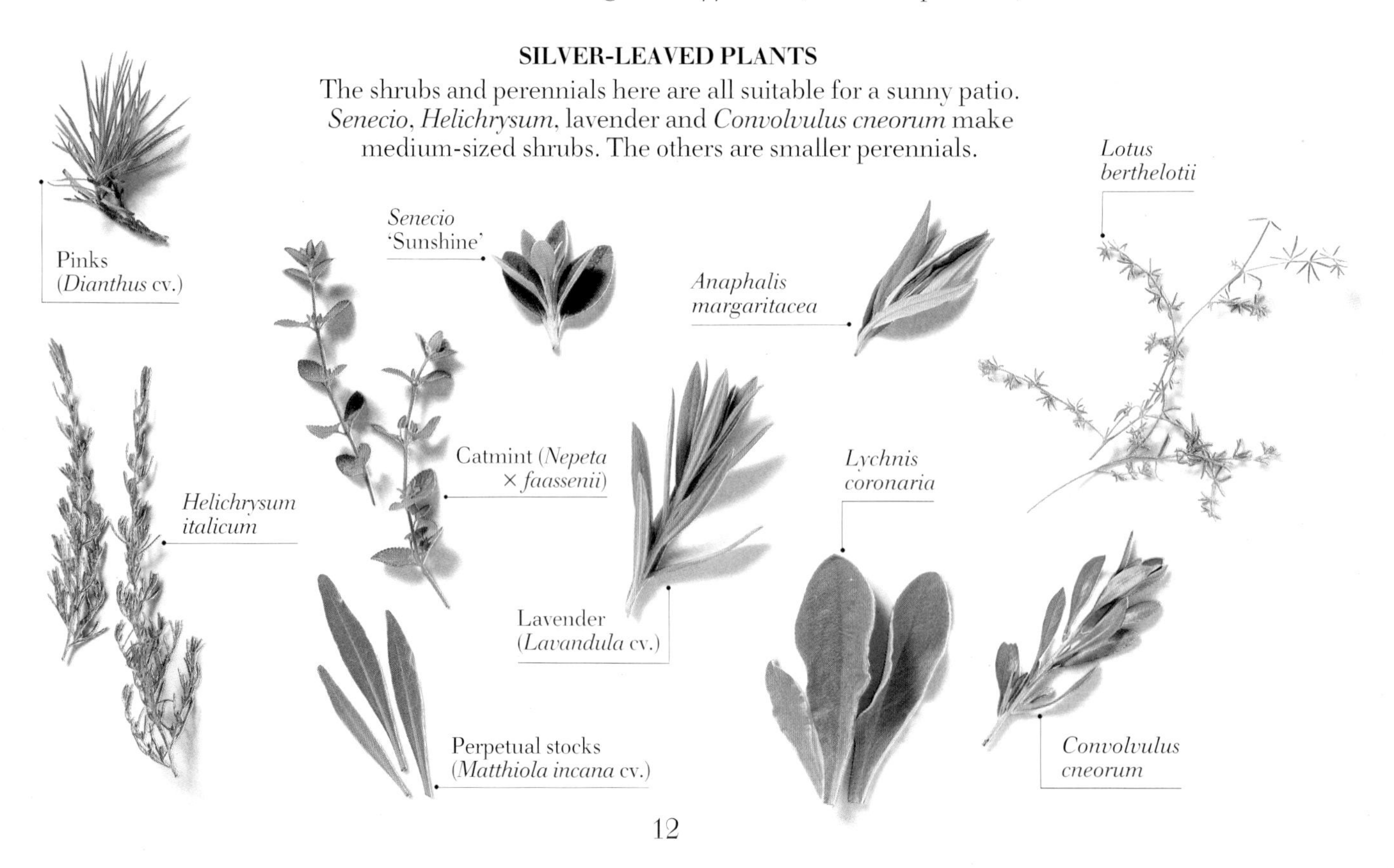

Stone terrace
ABOVE: *Containers are used to create interest on this sunny sunken stone terrace where form and color both play a part. Containers do not always have to be planted up – here, the Ali Baba pot is used as a focal point in its own right.*

impressive annuals, like the large-leaved giant tobacco plant (*Nicotiana sylvestris*), or architectural clump-forming perennials, like *Hosta sieboldiana* var. *elegans*, to create interest up to about 1.2m (4ft) in height.

A single-color theme, or some kind of continuity in the type or style of pots used, will give the area unity and harmony, but you do not need to be too purist to achieve a good effect. Keep the color theme cool (pastels in blues, pinks, whites or mauves) or hot (vibrant yellows, oranges, reds, mauves and purples) and avoid a largely pastel theme with odd outbursts of strong colors. The latter will kill the impact that the former makes.

Although most patios are built in sunny parts of the garden – some are actually built at the end of the garden rather than near the house in order to catch what sun there is – a shady patio can look extremely attractive. However in shade you will not be able to use the hot colors which tend to belong to sun-loving Mediterranean-type plants. A cooler green-and-white theme works well for shade, and looks particularly effective when allied to terracotta or stone paving. Ivies and ferns, in which the interest concentrates on leaf texture and shape as much as on color, with the contrast of pure white flowers – busy Lizzies (*Impatiens* sp.) are notoriously good in a shady situation – give a timeless classical appeal. If you can afford an attractive piece of statuary to set among the pots, so much the better, but go for quality, and aim for less rather than more so that your arrangement combines impact with elegance.

If the patio is large and sunny, you can opt for an informal look with sprawling mounds of silver-leaved plants, herbs and cottage-garden flowers in soft colors. Scented plants make a wonderful addition to the patio, particularly those with good evening scent, like night-scented stocks or tobacco plants, which drift across as you sit enjoying the last of the sunshine. Drifts of toning colors look particularly good: the range of pinks, from delicate pale tones to deep rich almost-reds, is an excellent choice. Hydrangeas will provide you with a long display of flower color and some of them, particularly certain mop-headed varieties, go on to give you wonderful russet colors in winter as the heads slowly fade and die. Combined with less permanent displays of summer-flowering annuals and perennials with a lax habit, like verbenas and helichrysums, in toning colors, these can make a wonderful grouped display.

Lead containers and stone troughs are another good solution for a patio area. Blue and yellow flowers both look good against these gray materials and you can change the planting seasonally to ensure a long display. Pansies and small hyacinths, such as *Muscari* spp., in shades of deep blue-mauve make a wonderful combination for a small stone trough, and deep blue flowers, such as those of *Stachys macrantha* 'Superba', scabious (*Scabiosa caucasica*), *Salvia patens* or giant larkspur (*Consolida ambigua*), look good in a trough combined with silver-leaved plants (see previous page).

Containers of plants can also be used to divide a patio or terrace from the remainder of the garden, often in an informally grouped display of shrubs and perennials, plus grasses or bamboos, or to give some shelter around a seating area. Evergreens, such as *Choisya ternata*, *Berberis*, *Elaeagnus* × *ebbingei* or *Prunus laurocerasus*, make a good medium-sized screen, to which flowering shrubs and perennials can be added to provide color.

Growing herbs on the patio

A sunny patio is the ideal place in which to grow a selection of culinary or medicinal herbs, although there is nothing to stop you growing them on a balcony or in a window box if you have limited space. Among the best herbs to choose are the ones with aromatic leaves or flowers, such as thyme, rosemary, lavender and sage, and you can group them in pots with specific themes.

Herbs look good grown in formal-style arrangements and in loose cottage-style plantings. Try to pick a good selection of leaf forms and types, as well as colors, since their virtue is primarily in their foliage rather than their flowers. Artemisia, with its wonderfully divided, silvery leaves, comes in many forms and is always a good standby, as is feverfew with its little daisy-like heads and feathery foliage, and lady's mantle (*Alchemilla mollis*) with its soft rounded leaves and cloudy heads of yellow-green flowers which have a frothy appearance.

Keep the kitchen herbs in a pot by the back door, handy for the cook. Basil and parsley both look effective grown individually in large terracotta pots, and parsley also makes a good foliage addition to flowering displays, particularly with blue or white pansies. Almost all herbs do best in light soil in a sunny situation, since the majority are from the Mediterranean region.

There is a growing interest in specialist collections of medicinal herbs, and certainly in those used for natural cosmetics. Although a container collection would be unlikely to yield enough for any major experiments in medicinal herb preparations, you can certainly grow enough for making infusions of leaves of certain herbs, such as mint for example. Mint is an excellent choice for containers since it tends to spread.

Fruit and vegetables

You can also grow some edible produce on the patio. Tomatoes in grow-bags (disguised in wooden troughs), strawberries in special planters, and perhaps runner beans in a large, deep container, supported on a wigwam of bamboo canes. Remember that any fruit and vegetables will need frequent watering, and plenty of feeding if they are to produce worthwhile crops.

MIXED HERB PLANTINGS

Herbs can be grown in any number of ways on the patio, either formally or informally. You can buy purpose-made herb pots, with specially divided sections for the different kinds, or you can simply plant them up yourself. If you wish, you can create your own special-interest herb pots using, for example, all culinary herbs (see below left), or all medicinal herbs (see below right), or several species of one genus, such as the different artemisias.

Culinary herb display
BELOW: *This small terracotta pot contains culinary herbs, including oregano* (Origanum vulgare), *lemon thyme* (Thymus × citriodorus) *and variegated lemon thyme* (T. × c. aureus), *sweet basil* (Ocimum basilicum) *and dark-leaved basil* (O. b. purpurascens), *hyssop* (Hyssopus officinalis) *and sage* (Salvia officinalis).

Medicinal herb display
BELOW: *The following medicinal herbs make an attractive grouping: wormwood* (Artemisia absinthium), *tansy* (Tanacetum vulgare), *golden feverfew* (Tanacetum parthenium *'Aureum'*) *and hyssop* (Hyssopus officinalis).

Herb pots
ABOVE: *Although herbs are usually combined in containers, why not grow the different forms each in their own pot, as here? The golden-leaved marjoram* (Origanum vulgare *'Aureum'*) *shown right and the variegated sage* (Salvia officinalis *'Icterina'*) *shown left are both good foliage plants in their own right. The different forms of basil* (Ocimum sp.) *also make good single plantings and will grow well on a sunny, sheltered patio. Bush basil* (O.b. *var.* minimum), *which has a compact, round habit making it appear naturally topiaried, is ideal.*

Balconies

Although balconies can vary in size from no more than a window ledge to areas large enough to hold a party on, there is usually relatively little space for plants, and you need to use a certain amount of ingenuity if you are going to create a decorative area of planting that you can also use for relaxation.

The solution is to include climbing plants, which take up remarkably little of the available floor space but add greatly to the finished planted effect. As Graham Rose pointed out in his book *The Low Maintenance Garden*: 'Six vigorous climbers, planted in a strip of land 30cm (12in) wide by 9m (30ft) long would soon provide a froth of foliage and colorful flowers over 56sq m (600sq ft) of wall. To provide a similar area of color and interest on the ground would require 20 times the area of cultivated land and would need between 100 and 200 times as many herbaceous plants.'

Climbers are therefore an important element in any balcony scheme, and, of course, containers are the only recourse for planting. The size and shape of the plant is a vital part of the overall effect, and you need to aim for depth without width, so that they do not encroach too much on the available floor space. One way around the problem is to construct your own wooden, window box-style containers, lined with plastic to make them more durable. Do not forget to incorporate drainage holes in the containers, and make sure that they are strong enough to bear the weight of the growing medium.

As with a roof terrace, you need to make sure that the balcony can bear the weight of the loaded containers. If you are in any doubt, call in a building contractor to advise you. You may also need to check that the surface is waterproof and that it drains adequately, as well as being durable enough to walk on.

Since balconies are likely to be exposed, particularly those on high-rise apartments, make sure that you have enough protection both for the plants and for the human occupants. Climbing plants that are fairly tough will create a good windbreak – griselinia is often recommended, and being evergreen it provides a

Mixed planting
LEFT: *This attractive corner shows how successfully plants of different heights and forms can be combined to provide a screen for an urban balcony. Bright pink geraniums (*Pelargonium *cvs.), hydrangeas, busy Lizzies (*Impatiens *cvs.) and white waxy lilies are backed by climbing and trailing plants – notably ivies and vines.*

Contrasting colors
ABOVE: *A brightly colored, mixed display of fuchsias, felicia and* Oenothera cheiranthifolia *in contrasting colors makes an excellent feature on a stone balustrade. If possible, organize the containers so that a matching pair flank an entrance or a flight of steps.*

permanent screen. If you want light to permeate in winter, when you do not use the balcony, opt for a deciduous hedging plant, like hornbeam (*Carpinus betulus*), kept clipped to the size required.

If there is no overhead shelter, you can create your own with vines trained over a supporting network of poles and wires, which provide excellent dappled shade in summer, and if you pick one of the fruiting kinds, you may even get a crop of grapes. For north- or east-facing balconies, climbers like *Clematis montana* or *Actinidia deliciosa* make good subjects for such situations.

Any climbing plants you grow will need to be trained over a support, unless they are of the self-supporting type, like ivy, which has sucker pads.

At the foot of the climbers you can grow trailing annuals and lower-growing perennials or small shrubs. Among some of the best for this purpose are catmint (*Nepeta* × *faassenii*), low-growing campanulas, nasturtiums and lobelias. If you are planning to grow any of these in the same container as a climber, you may have to choose plants that thrive in a poor, dry growing medium as a vigorous climber may well take most of the nutrients and moisture.

Displaying plants on balconies

When planning a display for a small area, think first about your own needs in terms of the space available and then decide how to organize what you have left for

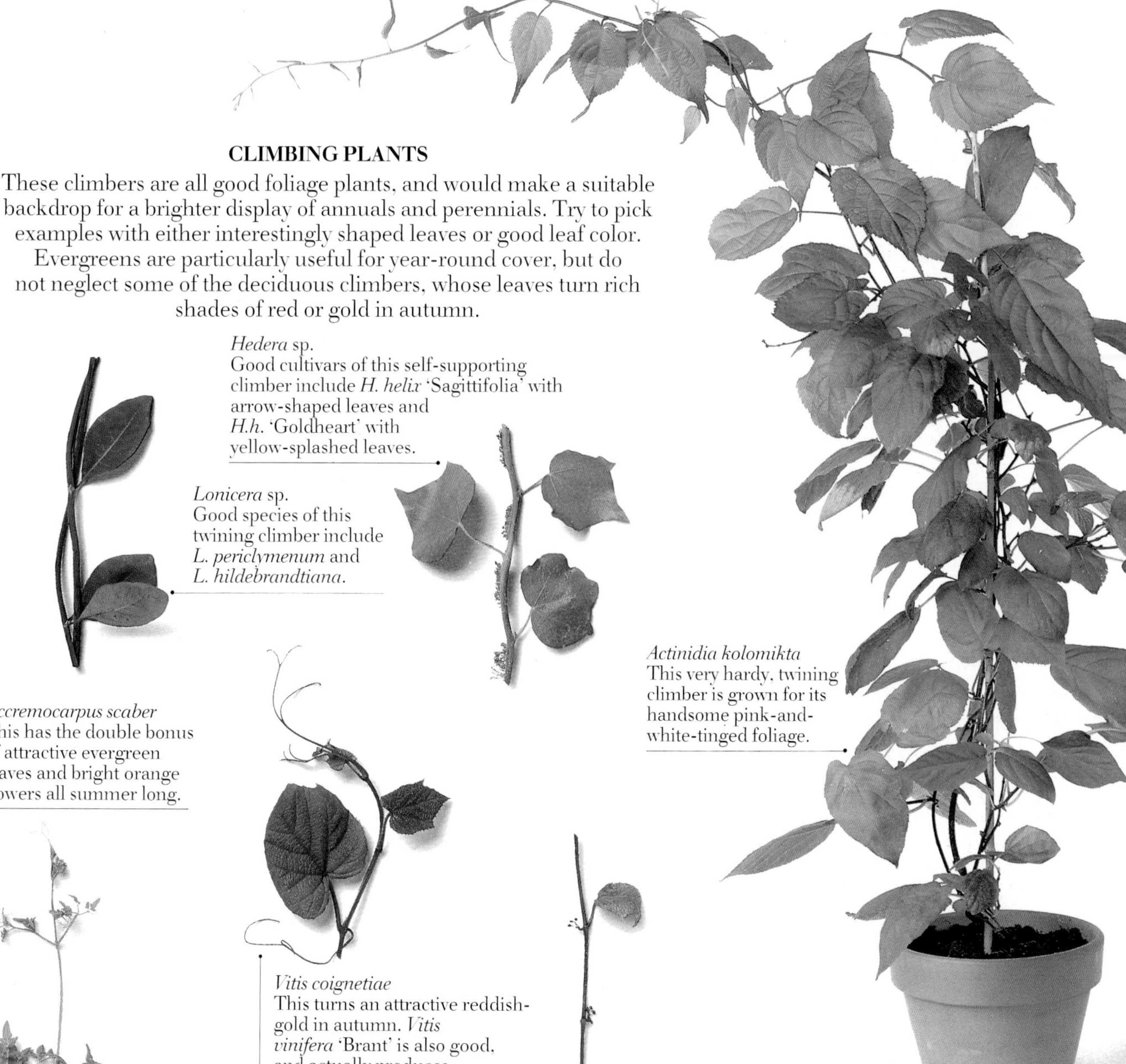

CLIMBING PLANTS

These climbers are all good foliage plants, and would make a suitable backdrop for a brighter display of annuals and perennials. Try to pick examples with either interestingly shaped leaves or good leaf color. Evergreens are particularly useful for year-round cover, but do not neglect some of the deciduous climbers, whose leaves turn rich shades of red or gold in autumn.

Hedera sp.
Good cultivars of this self-supporting climber include *H. helix* 'Sagittifolia' with arrow-shaped leaves and *H.h.* 'Goldheart' with yellow-splashed leaves.

Lonicera sp.
Good species of this twining climber include *L. periclymenum* and *L. hildebrandtiana*.

Actinidia kolomikta
This very hardy, twining climber is grown for its handsome pink-and-white-tinged foliage.

Eccremocarpus scaber
This has the double bonus of attractive evergreen leaves and bright orange flowers all summer long.

Vitis coignetiae
This turns an attractive reddish-gold in autumn. *Vitis vinifera* 'Brant' is also good, and actually produces edible grapes. Both varieties support themselves with tendrils.

Parthenocissus tricuspidata
This provides good color in autumn and is self-clinging.

the planting. Growing climbing plants is one solution; another is to pick plants that have either a narrow columnar or a trailing habit. Among the good tall and narrow plants are topiary spirals or columns made from clipped box, privet or holly (see page 8), topiary pyramids constructed from pansies (see page 11) or from grape hyacinths (see page 72), and dwarf conifers, although the latter tend to look too stiff and formal in a mixed arrangement, their pillar-like shape being at odds with the more exuberant flower planting. If you do use dwarf conifers, combine them with ivies and variegated-leaved plants, and perhaps a few pure white flowers – winter cyclamen have a similarly architectural shape that goes well with the formal outline of the evergreens. Anything that forms a rounded bushy shape may well encroach too much on the available space for seating.

Look carefully at the surrounding architecture, not only of the wall that forms the backdrop to the balcony but also the style of the retaining structure. This can take the form of plain or fancy iron railings, or posts and rails. Whatever the type, it is imperative that it is strongly constructed; if you have a post-and-rail structure, the rails must be nailed to the inside of the post, not the outside, for extra strength. You can, if you wish, hang window boxes on the balcony supports, but make sure that the material and color tones well with them.

Cottage-garden planting
ABOVE: *Small containers are massed together on a wooden decked balcony to create a cottage-style display. Climbing plants are used to soften the trellis, and also act as a foil.*

A simple repeating planting scheme that acts as a ribbon at the base of the balcony, with some plants trailing through the railings, is often the most successful. Red works well against most brick houses, as does brilliant blue or pure white, but not all three together! For colored stucco walls in pastel shades choose softer colors – pinks, mauves and lemons, for example. For town houses the more sculptural flowers tend to look best: stiff and soldierly scarlet tulips, perhaps, or neatly clipped evergreens.

Larger balconies and terraces

With more space at your disposal you can opt for more elaborate displays, and in this instance it probably pays to focus the planting in a single, large, massed display, perhaps around or backing a pillar or support. Make sure that there is at least one large specimen to act as the central core – a brugmansia (formerly datura) with its exotic, tubular, bell-shaped flowers and strikingly handsome leaves would be a good choice, although it is frost-tender. Attractive flowering climbers, such as jasmine or solanum, could be trained to grow up any supports, and a range of interesting perennials could

Clematis
'Madame Julia Correvon'

Clematis
'Victoria'

FLOWERING CLIMBERS

Of all the flowering climbers, clematis is among the most popular, with a species or cultivar that flowers in almost every month of the year. The small-flowered *Clematis viticella* and some of the small-flowered hybrids make really good candidates for a container, as they enable you to concentrate on their particularly delicate beauty.

Clematis texensis
'Princess of Wales'

Clematis viticella
'Margot Koster'

Clematis viticella
'Purpurea Plena Elegans'

Clematis
'Caroline'

Clematis
'Jackmanii'

Clematis
'Marie Boisselot'

Clematis montana

play the subsidiary role, being moved around as they come into flower. There is a wonderful range of geraniums (*Pelargonium* cvs.), including those with deliciously scented leaves and smaller, but no less pretty flowers, which have a long flowering season. Also useful are the bigger marguerites (*Argyranthemum*), which form generous clumps or which can be trained as standards, as indeed can fuchsias, honeysuckle and even wisteria if you wish to have the flower color high off the ground (see page 74). Try to ensure that the group is backed with a couple of good foliage shrubs – like *Choisya ternata* with its whorls of glossy green leaves or a palm like *Trachycarpus fortunei* – to give some depth and substance to the planting. In limited spaces use a more slender climbing plant for this purpose, such as the golden-leaved hop (*Humulus lupulus* 'Aureus', see pages 6 and 64).

If you are lucky enough to have a terrace with a broad stone balustrade, you can use it to display large handsome containers with striking plant combinations. Make sure that the containers, and the plants they contain, are bold enough for the situation: anything too small will simply look niggardly. Again, some degree of symmetry will help unite the scheme, whether by repeating the flowers, the shapes or the container types, although too great a concentration on symmetry starts to look like a late-Victorian bedding-out display. For stone terraces, concentrate on large containers with sprawling plants like senecio and euphobias, which make good year-round performers, with summer interest in the form of big lilies, tobacco plants or fuchsias. Spiky-leaved plants make a focal point: try yuccas, with their sword-shaped leaves and cordylines, with their elegant spray of slender evergreen leaves.

Roof Gardens

A ROOF GARDEN, even a tiny one, provided it is well screened from the elements and from neighboring apartments and houses, can make a lush oasis of flowers and foliage in a hostile city environment. By virtue of the situation you have no choice but to garden in containers, and the size and type will be determined to some extent by the shape and load-bearing capacity of the roof. It is a good idea to get an expert to check these structural elements before you plan out the roof garden, because the weight of a container laden with plants and growing medium is often considerable. You also need to make sure that the roof drains adequately and you should take out third-party insurance just in case a portion of your garden should take off in a high wind and cause damage.

Since roof gardens are almost certainly very exposed, the first consideration is to set up some kind of screen, either manmade or in the form of plant material. Stout trellis is ideal, since it provides a support over which to grow an exciting range of climbing plants, which then create a far better microclimate for less hardy plants under their protection. Make sure that any trellis you put up is solid enough to support the weight of the plants you have in mind and, even more importantly, that it is solidly anchored.

If you are lucky enough to have a roof terrace that is surrounded by trees, use these as the backdrop for your planting scheme. These will enable you to dispense with some of the screening devices needed for more exposed terraces.

If you are laying a floor for the roof terrace, consider putting in an irrigation system at the same time, since it will greatly reduce your watering needs at a later point, and it may be inconvenient to cart containers of water on to the roof. If you are organizing piped water to the roof, think about installing a pool or fountain; you can

Built-in containers
LEFT: *Large trees and shrubs flourish in built-in containers, while pots of summer-flowering annuals, set against a backdrop of foliage plants, including a yucca and a chamaecyparis, provide color at floor level.*

Creating impact
ABOVE: *Handsome flowering plants – the African lily (*Agapanthus africanus*) and the giant tobacco plant (*Nicotiana sylvestris*) – are grouped together for impact.*

create an effective small water garden using a large half-barrel filled with a variety of plants chosen for providing different habits and foliage (see page 54).

Exploit the trellis to grow some good architectural-leaved climbers, such as variegated ivies, hardy vines or creepers, plus flowering, and ideally scented, climbers such as honeysuckles, jasmines and clematis. *Clematis armandii*, for example, has attractive evergreen leaves and its flowers are scented like almond blossom.

In smaller tubs or boxes around the perimeters of the roof you can grow whatever takes your fancy, including some fruiting plants if the roof garden is large enough. Try to position the containers so that you avoid the obvious central square with plants dotted around the perimeters and, if the roof terrace is large enough, you could grow some screening plants so that they partially mask the areas. Bamboos in large containers would be excellent for this purpose.

Remember that too many different colors will make a small space appear smaller, so opt instead for a single-color theme, or at least a harmonized one for the different seasons. A few large-leaved plants, curiously enough, will make the space seem larger, while too many small plants will reduce its apparent size. Go for large handsome pots with equally large, handsome plants, but remember that there will be virtually no overhead shade, unless you construct some kind of pergola, so take care to pick sun-loving varieties.

Because of the exposure, any plants of uncertain hardiness will need winter protection. If you have a penchant for Mediterranean-style plants, you may need to cover the roots with sacking (see page 103). Although you do not use the roof terrace in winter, try to think about its appearance at this time of year, and group any containers of plants that need lagging in this fashion behind an evergreen screen, if possible. Pick sufficient evergreen foliage plants to give the terrace some degree of winter interest, and choose at least one good autumn performer such as a Japanese maple (*Acer palmatum*).

The Japanese theme works very well on roof terraces, as it usually looks good as an extension to the living room and you can use purpose-bought reed screens to provide protection and shelter. The key to Japanese-style design is simplicity, so opt for less rather than more, and go for strong shape and form rather than color. Among the popular plants for a Japanese look are small-leaved evergreens, such as the species rhododendrons and azaleas, which make attractively rounded small shrubs, box (*Buxus sempervirens*) and privet clipped into ball shapes, bamboos, ferns and hostas. A couple of handsome oriental stone statues or earthenware pots will add to the atmosphere. For the surface of

TRAINING A CLIMBER UP A TRELLIS

Climbers grown in pots need some form of support. You can start them off with a simple trellis which is freestanding, and then allow them to climb over other plants once they are established. Alternatively, train them over an existing wall to which wires or trellis have been attached.

1 *Cover the bottom of the pot with pebbles. Insert the base of the trellis and add multi-purpose growing medium to sit the plant below the rim.*

2 *Insert the plant, first teasing out the rootball, and fill the container with growing medium to within 2.5cm (1in) of the rim.*

3 *Carefully remove any existing supporting canes and start to spread the shoots of the plant against the trellis.*

4 *Tie in the shoots with plastic ties, making sure that they are not so tight as to damage the stems.*

5 *As the plant matures, tie in any new shoots to the support when they are still young and pliable.*

The established plant, a Parthenocissus, *tied in to the support. Notice the way that the leaves turn outwards and upwards towards the sun.*

Cottage-style display
LEFT: *A lush mixture of flowering and foliage plants creates a cottage-like effect in the corner of a small roof garden. Tables are useful for raising the smaller flowering plants to create color at a different level, as here, where small containers of geraniums (*Pelargonium *cvs.) are backed by taller cannas.*

the roof terrace, combine decking – constructed from durable hardwood, such as teak – with gravel, which can be raked into a variety of fascinating, formal Zen-type designs. Solid bench seats or plinths on which to stand the containers can be constructed from recycled railway sleepers, if the roof terrace will hold the weight of the bases, pots and plants after a good watering. Make sure that the garden is screened appropriately from prevailing winds, using bamboo or reed panels. This will not only ensure that your roof garden is pleasant to sit in but also will provide year-round protection for your plants so that they do not become battered and damaged.

Using trellis

When you grow climbing plants in containers, you will need to provide support for them. To attach trellis to a wall, fix it to a wooden frame nailed to plugs inserted in the wall. If you mount the trellis on battens, away from the wall, you allow air to circulate around the plants.

SMALL SCREENING PLANTS

Low-growing bushy plants can be grouped closely together in order to form low-level hedging, which is useful for dividing space on a roof terrace into compartments. Suitable plants are suggested below.

If you want the plants to form a screen, nail the trellis to the back of a wooden container, making sure it is fixed low down on the container to keep the trellis stable. Use screws rather than nails, and fix at the base and again closer to the top for maximum support.

Bought trellis normally comes in preservative-treated softwood, but there is nothing to stop you making your own trellis, if you wish, by screwing together lengths of timber. Opt for slightly thicker timber battens for a roof garden, to allow for greater strength. The spaces between the trellis battens allow some air to circulate around the plants, and lessen the risk of the entire screen taking off in a high wind.

Mirrors can be useful as a means of increasing the apparent space and can be incorporated into *trompe l'oeil* features on one of the trellis walls.

Hedging plants for roof terraces

Among the screening plants you can use for roof terraces, or balconies for that matter, are both tall and short, deciduous and evergreen forms. For low hedges to divide up the space, consider using box which forms a slow-growing, dense, evergreen barrier. There is a golden-variegated cultivar, *Buxus sempervirens* 'Aureovariegata', which looks attractive grown with the plain green form. Box can be struck easily from cuttings. For taller hedges consider using griselinia, yew, thuja or ivies, the latter grown over trellis. (See below for small- and medium-leaved screening plants.)

ENCOURAGING BUSHY GROWTH

Screening plants, such as the griselinia shown here, which resists pollution, need to be encouraged to bush out at the base to form a dense screen.

In the first season after planting, trim off any leading shoots down to an outward-facing bud, using a sharp pair of secateurs/ pruning shears. Removing the tips of the leading shoots encourages lateral shoots to form, and will make the shrub bushier.

Choosing a suitable flooring

Bear in mind that all the materials needed to create your roof garden will need to be carried up to the roof and that access can be very tricky. For this reason, it can pay to choose flooring materials that are as lightweight as is possible. Conventional paving, such as stone slabs may be too heavy for the load bearing capacity of your roof so choose lightweight tiles or perhaps chippings instead. The latter can be spread thinly over the surface of the roof to reduce weight further.

MEDIUM TO LARGE SCREENING PLANTS

For a more exposed situation you may need to create a windbreak for the terrace, in which case you will need some bigger plants to make a hedge. You can plant the hedge in containers, ideally in long, reasonably deep troughs rather like window boxes. Suitable plants are suggested below.

Walkways and Paths

CONTAINERS OF PLANTS can be used to great effect in areas of the garden, such as walkways or paths, where there is either little soil or where there is limited sunlight. On the one hand, they allow you to create interest at specific points in the landscape by drawing the eye across a narrow area towards a focal point, such as a piece of statuary; while on the other hand, they enable you to bring plants that prefer higher light levels into shady areas for short periods of time in order to give the area a lift of color and interest.

Gardeners dealing with borders in shady areas need to establish whether the soil is dry or damp before selecting appropriate plants, but as you control the environment in the container, this does not apply.

On the whole it pays not to mix plants that prefer different kinds of situation, in part because there is no way they would grow together in nature. You can relax the rules, but look at foliage color and leaf shape to see if they blend with the surrounding planting.

The hard surfaces of both paths and walkways need softening with good foliage shapes and there are plenty of climbing plants and perennials to choose from, but in shady areas you will have to sacrifice brilliance of flower color and concentrate the interest on foliage.

In general, shade plants have large leaves in order to present as much available surface to the light as possible for photosynthesis to take place (see page 80). Variegated leaves are inclined to revert to their natural, entirely green state if the light levels are too low.

Walkways

The narrow area that runs alongside a house or the narrow passage between houses is often fairly heavily shaded, and yet, for town dwellers, it represents a

Clashing colors
RIGHT: *The brilliantly colored* Begonia × tuberhybrida *and* Lobelia *'Sapphire' make a striking combination. Limit the palette to two contrasting colors for impact.*

PLANTING A WALL POT

Planting up a wall pot is similar to planting up a hanging basket but you need to organize the plants so that they spill over the front rim. Keep taller plants for the back, but make sure that they will make a balanced crown shape rather than a pyramid.

Toning colors
RIGHT: *This little wall pot has been planted up with a mixture of blue marguerites (variegated* Felicia capensis*), scaevola and variegated ground ivy* (Glechoma hederacea *'Variegata'). Delicate plants with small flowers suit the smaller size of most wall pots, rather than the more blousy-looking petunias so frequently seen in hanging baskets.*

PLANTS FOR WALL POTS

Among the suitable plants for a small wall pot are the following:

Blue Display
Small-flowered pansies, e.g. *Viola* 'Azure Blue'
Grape hyacinths (*Muscari* spp.)
Small-leaved ivies, such as *Hedera helix* 'Sagittifolia'

Pink display
Diascia (such as *D.* 'Ruby Field')
Verbena (such as *V.* 'Rose du Barry')
Silene (such as *S. pendula*)

1 *Cover the base of the pot with a layer of gravel to promote drainage. Fill the wall pot by approximately one-third with multi-purpose growing medium.*

2 *Start to plant up the pot – here the variegated* Felicia capensis *is being inserted into one side of the pot – spreading the roots out well as you do so.*

3 *Firm in well, then add the second plant to the other side of the pot, draping the trailing stems attractively over the edge.*

4 *Plant up the central part of the container with scaevola. Add growing medium to within 2.5cm (1in) of the rim and press down well. Water well.*

significant part of the space available for gardening. Walkways are probably among the most difficult places to plant successfully because of the limitations of space and light, and yet these very limitations can be made to work for instead of against you provided that you plan the planting scheme carefully.

The key is to keep the planting simple, and to use repeating groups of plants, which will help to magnify the available space. It is important that you choose the plants from among those that naturally do well in semi-shade, with perhaps the occasional sun-loving plant brought in to add flower color and cheer the scheme up a bit, provided it is not expected to flourish for long in low light levels. Geraniums (*Pelargonium* cvs.), for example, although preferring sun, will actually perform reasonably well in semi-shade for a while, but will eventually grow spindly and straggly in their search for light. You can avoid this problem by moving the plants periodically into better positions.

The expanse of wall surrounding the walkway will need some kind of adornment. You can use climbing plants or you can fix wall pots to the vertical surfaces, if you prefer. If you opt for the latter, keep the pots plain and make sure that their material and color tones with the wall surfaces. Higher up the walls light levels normally improve – the darkest shade will be at the foot – and you can therefore grow similar plants to those you would select for a hanging basket: begonias and busy Lizzies will do well in semi-shade and they come in a good range of colors. Try to keep the color theme co-ordinated to some degree, and perhaps lighten the impact by using silver-splashed or colored foliage – *Helichrysum* 'Sulphur Light' has almost fluorescent foliage, and there are plenty of ivies that have gold- or silver-splashed leaves. To grow up the walls themselves you can use an ornamental hop (*Humulus lupulus* 'Aureus') with its attractive gold leaves, any of the ivies, ideally those with large leaves like *Hedera colchica* 'Sulphur Heart', or the Dutchman's pipe (*Aristolochia durior*), again with large handsome leaves. Some honeysuckles will also do well in semi-shade, in particular the late Dutch honeysuckle, *Lonicera serotina* 'Honeybush', which is also very fragrant and therefore particularly suitable for walkways.

At the foot of the wall grow ribbons of ferns or clumps of hostas, for example, both of which enjoy semi-shade. Smaller woodland perennials, such as tellima and tolmiea, will also do well in poor light. Of the medium-sized shrubs, cotoneasters will cope successfully with these conditions, as will pyracanthas, which carry an added bonus of attractive berries in autumn that range in color from scarlet to yellow.

SHADE-LOVING PLANTS

There are a number of shade-loving plants, whose charms, though less obvious than those of their sun-loving cousins, nevertheless have an appeal for the discerning eye, mainly because of their attractive foliage.

Damp shade
LEFT: *This small container, planted up with a fern (*Athyrium filix-femina *'Victoriae') and underplanted with* Alchemilla mollis, *is ideal for providing foliage interest in a damp shady area.*

Dry shade
RIGHT: *In an area of dry shade this small pot of Japanese anemones (*Anemone × hybrida *'Honorine Jobert') and* Tellima grandiflora *makes a strong impact.*

Try to group the containers so that they break up the length of a long narrow passageway, while still allowing sufficient access. Trailing plants will help to soften the look of any hard surfaces – ivies, which tolerate poor light levels, are always useful.

Paths

Paths tend to get ignored in gardens, their utilitarian purpose in leading you to a specific point of interest tending to discourage any further planting ideas. You can, however, give them a complete change of emphasis and additional interest by placing containers of plants judiciously at particular points.

One of the more successful ideas is to foreshorten a long path by positioning pairs of containers, planted with the same plant, on either side at intervals along it. This is particularly successful in an area of the garden which perhaps lacks any real structure or form. Be careful, though, to pick varieties that blend sympathetically with the surrounding planting in your garden. If you go for those that prefer the same site and soil conditions – such as dry shade or damp soil – you are less likely to make an obvious mistake. Do not plant sun-loving plants along a woodland path: they will not thrive because of the low light levels. A container of Japanese anemones and ivies, however, would be ideal.

SHADY PLANTS

Among the best plants for shady places are those below. The fatsia is a shrub, while the hosta, fern and euphorbia are all perennials. The tobacco plant is an annual and there are various forms.

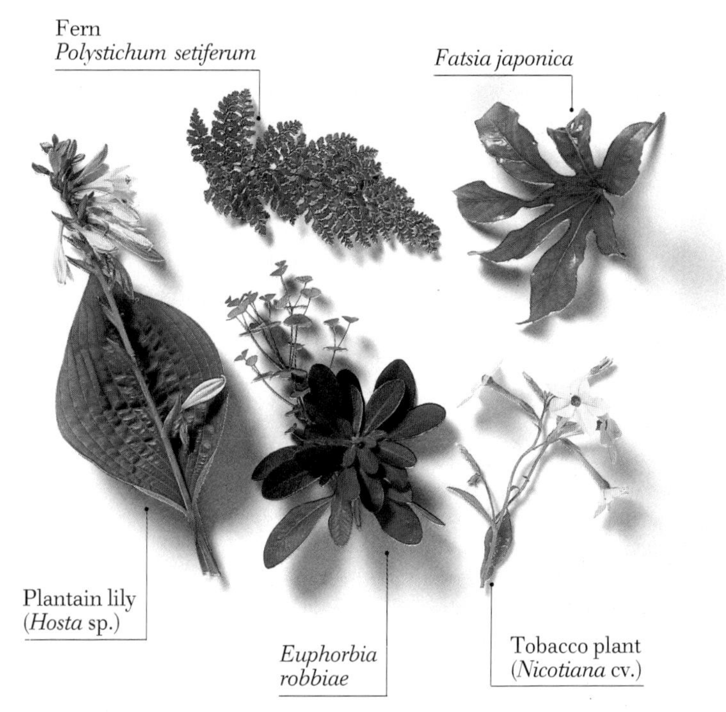

Pots on paths
LEFT: *These neatly marshalled terracotta pots of alliums, marking the path down an almost jungle-like corner of the garden, not only bring a touch of formality to an otherwise informal area, but also provide a bridge between the stone path and the loose planting on either side. At other seasons of the year you could replace the alliums with other flowering plants, keeping the color theme and plant type the same to give continuity to the scheme.*

DESIGNING WITH CONTAINERS

CONTAINERS COME in many different shapes, materials and styles, and choosing the most appropriate form is one of the first elements to consider. Your next task is to decide which are the most attractive and appropriate plants for them. This chapter is packed with ideas.

The first section concentrates on the various types of container available. A fully illustrated directory of container forms is featured, with advice on choosing which type is best suited to the purpose you have in mind. This is followed by advice on how to plant up the different styles to best effect – with information on hanging baskets, window boxes, small containers, large containers and unusual objects. At the end of each section, seasonal planting schemes show how to plant up the same container for spring, summer and winter to achieve all-year-round color. (When planting up containers in winter, check that the listed plants are hardy in your area – if in doubt, check in the Plant Directory on pages 106–155, or in an appropriate reference work.)

LEFT: *The toothed leaves of the creeper* Parthenocissus henryana *are marked with silver-white veins.*

ABOVE: *Two handsome stone urns on a York stone-flagged patio are filled with small mounds of purple pansies* (Viola *cv.*).

Choosing Containers

Choosing appropriate containers, both for the setting and for the plants you buy or grow, is a key element in successful container gardening. It also represents quite a large cash outlay, so it is important to be as well informed as possible on the range of containers available to you, and their differing virtues and, occasionally, drawbacks.

Materials for containers

Containers are available in a wide range of materials, and there is something suitable for every taste. Terracotta, wood and metal are the most popular and commonly used natural materials, while concrete, plastic and fiberglass lead the synthetics. All these materials are used to make containers in a wide variety of sizes and shapes to suit all types of garden.

Terracotta containers

The most popular material for containers is terracotta – literally 'cooked earth'. Made from clay and fired in a kiln, these containers range from hand-thrown pots that are works of art in themselves to mass-produced, mass-fired pots created from standard molds. The most beautiful examples are undoubtedly the hand-thrown

POTS AND URNS

By far the most popular containers, pots and urns come in a wide variety of shapes, sizes and materials to suit your needs.

ones, and some of these are antiques – weathered and worn to wonderful, soft, washed-out colors that blend superbly with almost any kind of planting scheme. Imported hand-thrown versions can now be bought at low cost, but some of the molded pots currently being made to good traditional designs are excellent, and much less expensive.

If you do not like the natural reddish-brown of terracotta, or it does not suit your particular planting scheme, you can always paint the pots to achieve just the right color. This is particularly useful for mass-produced pots which are often too strongly colored when they are new and have not yet been exposed to the elements. You can paint and distress them to look instantly aged and attractively weathered in a good range of softly tinted colors – the classic French blue-gray always looks good and works well with most planting schemes, as does a weathered deep green.

Shapes of terracotta container are wide-ranging, from classic flower-pot shapes to rimless, slightly conical pots, urns, baskets and troughs, plain or molded in a variety of different ways to complement a variety of styles and themes. There are some excellent replicas of traditional Georgian and Victorian styles and shapes, if you want a period look to match the architecture of your house, and most designs come in several sizes, so you can have uniformity of design with disparity of size, ranging from tiny flower-pots to huge tubs for large plant displays. It is normally best to use just one style of container. An assortment of varying containers in one display is likely to distract attention away from the planting scheme.

Large terracotta seed pan

Terracotta 'egg' pot
35cm (14in) in diameter, 35cm (14in) high

Terracotta swag pot (plain)
40cm (16in) in diameter, 33cm (13in) high

Fiberglass 'lead-effect' urn with verdigris finish

Terracotta pot with glazed rim
32cm (12½in) in diameter, 28cm (11in) high

Terracotta seed pan with lattice effect

Stone basket with lattice effect

Metal containers

There is an excellent choice of metal containers suitable for planting up, from inexpensive galvanized buckets and tubs to antique lead troughs and cisterns, and including a new range of ornate twisted-wire or filigree containers from the Far East which make excellent hanging baskets. Inexpensive metals will quickly rust unless they are treated with some form of preservative. Painting is the best solution, and you can create some very attractive effects using some of the new decorative paints on the market, such as the verdigris-style kits currently being manufactured by specialist paint companies (see page 37 for further information).

If you are lucky enough to own a lead container, such as an antique cistern, these look best left in their natural state, planted up in sympathetic colors. Blue, violet or orange flowers go well with the natural blue-gray of the lead. Nasturtiums, in mixed shades of reds and oranges, look wonderful too.

Again the range of sizes and shapes is unlimited, although many of the traditional metal containers are geometric in shape and fairly formal in style.

Wooden containers

Wooden containers range from Versailles tubs, the rectangular box-shaped containers used for topiary

TROUGHS, TUBS AND WINDOW BOXES

Troughs, tubs and window boxes are available in a wide variety of materials – from wood and terracotta to lead and stone. An important consideration when choosing window boxes is that the supporting sill or ledge will hold their weight. Fiberglass containers – which come in a range of styles – are extremely lightweight.

Slatted wooden window box

Terracotta plain window box 40cm (16in) wide

Terracotta swagged trough 30cm (12in) wide

Wooden half-barrel 40cm (16in) in diameter, 30cm (12in) high

Versailles tub (untreated), 35cm (14in) wide, 43cm (17in) high

Metal filigree window box, with terracotta pots

Wooden window box 80 × 28 × 28cm (32 × 11 × 11in)

shapes, through half-barrels to window boxes. You can leave the wood in its natural state, simply coated with a horticulturally safe wood preservative, or you can prime it and then paint it in whatever color you prefer. Although not as long-lasting as terracotta, for example, most wooden containers, even if made from softwoods, have a reasonable life expectancy if looked after properly, and those made from more expensive hardwoods are extremely durable.

If you are worried that the wood has been preserved using a less-than-horticulturally friendly chemical, line the container with plastic to be on the safe side. As in all containers, there must be drainage holes in the base.

Stone containers

There is a wonderful range of stone containers, from rough-hewn sandstones to polished marbles, the colors of which can really enhance the variety of hues in the planting. If your patio has a York stone surface, for example, be careful to choose containers that blend successfully with it. Highly polished marble would jar in these circumstances, but would work well on a smooth tiled surface. Of all container materials, natural stone is the heaviest but also the most durable. In addition, natural stone ages attractively, being enhanced by weathering and the onset of greenery. You can, if you wish, create an imitation stone effect by

Peachstone trough
45 × 28 × 23cm (18 × 11 × 9in)

Cedarwood window box
60×20×20cm(24×8×8in)

Lead trough
74 × 25cm (29 × 10in)

Terracotta chimney pot
30cm (12in) wide, 50cm (20in) high

Lead tub
30cm (12in) in diameter, 25cm (10in) high

Adam-style fiberglass window box
60 × 21.5 × 28cm (24 × 8½ × 11in)

Square terracotta pot
23cm (9in) wide, 17.5cm (7in) high

WALL POTS AND HANGING BASKETS

When grouping hanging baskets and wall pots, take into account that the plants will probably need frequent watering. If you position them up high, make sure that you can reach them easily – either with a watering can or hose. Attach hanging baskets using a strong rope or chain, suspended either from a hook or metal bracket, solidly fixed to the supporting structure.

Terracotta straight-edged wall pot
25cm (10in) wide,
20cm (8in) high

Metal filigree hanging basket
34cm (13½in)
in diameter

Terracotta 'shell' wall pot
33cm (13in) wide,
25cm (10in) high

'Honey-pot'-style wall pot

Terracotta wall pot

Terracotta lattice-effect wall pot

covering an existing container with a mixture of sand and cement and painting yogurt on to the surface to attract algaes and lichens, in the manner of natural stone (see page 37). This is frequently done with old ceramic sinks to turn them into attractive stone troughs, and is surprisingly effective.

Man-made materials

Plastic, fiberglass and concrete are commonly used for containers, in addition to natural materials, but by and large they are not a patch on the latter. Plastic containers are practical, but the material is totally unsympathetic to any planting scheme – the contrast between the dead and lifeless manufactured material and the living ingredients being too great for comfort or aesthetics. If you have no alternatives, for reasons of economy, paint the plastic pots in a subdued color that harmonizes with the foliage of your chosen plants – dark green or slate blue would be a good choice. At all costs avoid plain, white plastic pots. They stand out too prominently and draw attention away from the plant towards the container. The lightness of plastic containers, which makes them so practical, contributes to their insubstantial air.

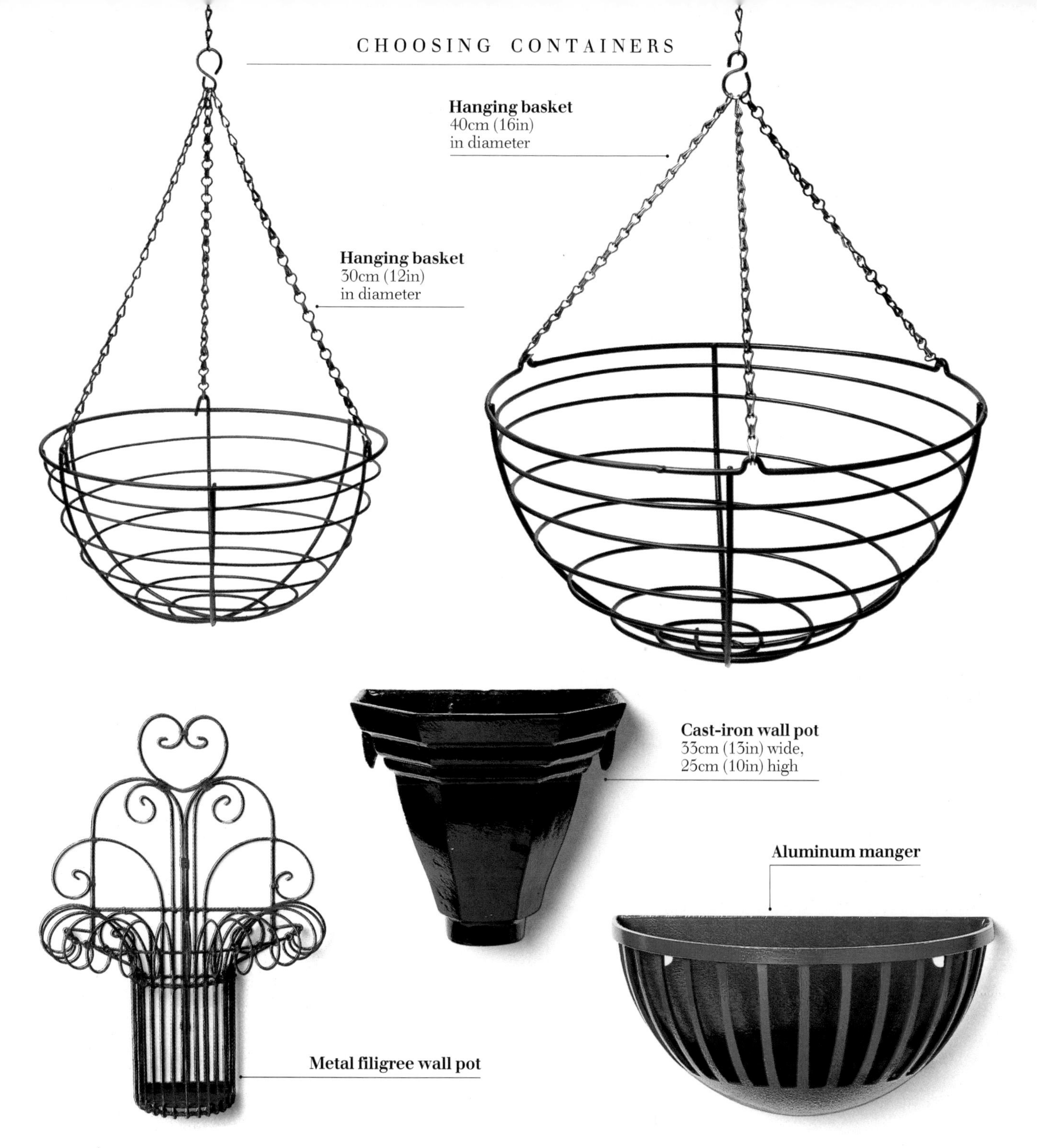

Fiberglass is a more recent innovation for containers, and it is light, strong and long-lasting. Like plastic, however, it is an inert material and is not particularly attractive. Concrete is used as a substitute for stone, and is molded to form pots and planters. It is extremely heavy, but is as durable as stone when reinforced and is therefore often favored for large containers.

Container size

Containers come in a whole range of sizes from massive, specially constructed, raised beds, sometimes made from railway sleepers, or from brick or concrete, to the smallest terracotta flower-pots. When choosing an appropriate size and shape of container, the first important consideration is the ultimate position it is going to occupy. If it is for a roof terrace or balcony, for example, weight is bound to be an important consideration, since a container full of growing medium is extremely heavy and the timbers supporting the area may not necessarily be able to hold a considerable weight. If you are in doubt, it is a good idea to get the load-bearing capacity of any such area properly checked by a building contractor before buying and installing any containers.

For certain situations you may be better advised to have the containers specially made up or to make them yourself. Wooden window box-style containers are quite easy to construct from reasonably heavy-duty timbers, but remember that the screws must be long enough as the wet growing medium will exert a considerable amount of lateral pressure on the joints.

Container shape

Containers come in a variety of shapes from small squat tubs to elegant Ali Baba pots, from square Versailles tubs to pedestal urns. The style of the container plays an important part in the overall scheme, and although container shapes and sizes can be mixed very successfully, it is usually better to keep one element the same. In other words, if you are going to combine a number of different shapes, keep the container material uniform. A group of terracotta pots in different sizes and shapes provides the basis for an attractive display, and the variation in size allows you to plant a range of different-sized plants, from small trees to alpines.

Positioning containers

One of the main aims with container displays is to be able to get a good vertical array of color and texture. You can do this by incorporating some large pots, for taller climbers, trees and shrubs, and also by using a wall, if you have a suitable one, to hang wall pots and hanging baskets. Smaller containers can be positioned at the front for annuals and little perennials. It always pays to group containers in one large display, rather than scattering them piecemeal along a balcony or around a patio. A single large container can form a focal point on a patio or in a small garden or back yard, particularly if planted with a large perennial or handsome shrub or small tree. This will help to lead the eye into your planting area and give it a sense of structure.

Any wall-mounted containers will need a good-quality bracket from which to hang, which must be securely screwed into the supporting masonry or timber structure, as they are a considerable weight when fully planted and particularly after they have been watered. Make sure that the supporting chains or ropes are

PAINTING CONTAINERS

You can achieve attractive and original results by hand-painting bought containers. Either paint the entire pot in one uniform shade or apply two or more colors in a geometric design. Stripes are particularly easy to create and can look very effective. For an even finish use masking tape as a guide; for a random appearance, paint the stripes roughly by eye – the rather wavy lines you create have a lively, decorative appeal.

Painted pots
LEFT: *This little terracotta container, with a juvenile box* (Buxus sempervirens) *topiary, has a distressed paint effect in duck-egg blue.*

Choosing paints
BELOW: *There is a good range of colors suitable for painting containers. The best colors tend to be those with white or black added.*

strong enough to take the weight of the basket or other container, and that any hooks are long enough and are well screwed into the supporting wall for security.

Decorating containers

There are many ways to brighten up containers, or to change their appearance to suit a particular scheme. You can opt for simple painting, using either flat emulsion and matt varnish for a non-glossy finish, or oil-based paint for a shiny appearance. Painting will help to transform disparate containers into a unified scheme, by using a range of toning colors or different shades of the same color.

If you wish, you can mix your own paint shades by adding artist's oil colors to oil-based paints, and acrylics to water-based paints. You usually need only a very little of the artist's oils or acrylics to color the paint, so proceed very cautiously and mix the colors first with a little white paint before adding to the whole tin. You can paint the entire pot, or you can choose to decorate it with surface patterning. Varnish protects the paint.

WEATHERING A POT

Most terracotta pots look more attractive when they have been weathered. You can speed up the natural weathering process by painting the surface of the pot with a layer of yogurt, which will attract the growth of algae and lichen.

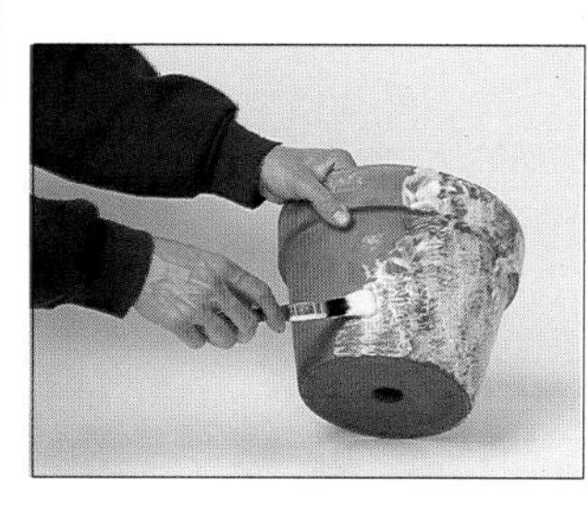

Coat the pot with a thickish layer of yogurt and leave outside to dry for a week or so before planting it up. It will gradually acquire an attractive greenish patina.

CREATING A VERDIGRIS EFFECT

The attractive bluish-green sheen that copper acquires when left in the open can be imitated using special paint effects. Kits are available containing two colors of paint, but you can make up your own colors using white acrylic paint, tinted with artist's acrylic colors.

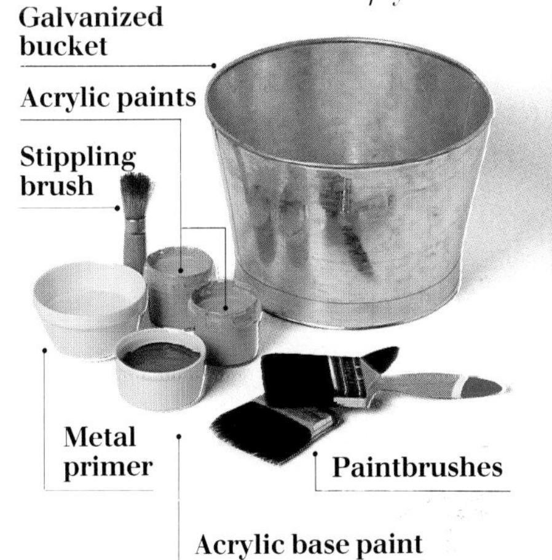

1 *Clean the bucket thoroughly. Apply a coat of primer to the bucket with a soft-bristled paintbrush and allow to dry.*

2 *Apply the base coat of acrylic paint to cover the surface of the bucket, using a soft-bristled brush.*

3 *With a stippling brush, dab on the first shade of green in patches. Allow to dry, then repeat with the second shade.*

4 *Allow to dry and then coat the bucket with acrylic varnish to prevent the surface from peeling or flaking.*

The finished container *(planting details on page 26).*

Large Containers

WHAT CONSTITUTES a large container is a matter of proportion rather than actual size, but for the purpose here it is roughly 45cm (18in) or more in height and 45cm (18in) or more in width. Some large containers are so handsome that they make a feature in their own right, and look better unplanted. Ali Baba pots are a typical example: they are such an appealing shape that they can be used unadorned (see page 14).

By virtue of their size really large containers are imposing, and this attribute needs to be considered when planting them up. It does not mean that the planting has to be equally large, but it does have to complement the size and shape of the pot in some way. Nothing looks worse than a large container full of small straggly-looking plants that do not echo its structure or form, and nothing looks better than a large-leaved hosta in an equally large and handsome terracotta pot. If in doubt, err on the side of simplicity.

Apart from any other consideration, big containers give you the chance to grow large and imposing plants that would quickly protest at the straitjacket of a less generous pot. Small trees, such as the Japanese cut-leaved maple (*Acer palmatum* var. *dissectum*) or the delicate weeping birch (*Betula* 'Trost's Dwarf'), will do very well in large pots. So will big palms, such as *Trachycarpus fortunei*, whose leaves resemble the sails of a windmill – hence the common name, windmill palm – or the Australian tree fern (*Dicksonia antarctica*). Keep the busier mixed plantings for smaller containers, such as window boxes or hanging baskets, and opt for bold statements with your large containers.

Cool colors
RIGHT: *A large terracotta pot makes a focal point in a courtyard, planted with cool mauves, grays and whites, in the form of melianthus, artemisia and* Malva sylvestris *'Primley Blue'. Silvery-gray leaves, like those of melianthus or artemisia, are the perfect foil for both cool- and hot-colored flowers.*

HOT COLORS

An old washtub has been painted up and varnished before being planted with a hot-colored summer display of annuals and herbaceous perennials. These have a long season – the nicotianas, the cuphea and the pansies, in particular, last well. If you cut back the main spike on the lupin when it starts to fade, secondary spikes will form in its place.

Large pots offer you the chance to show off some of the more exciting tender plants, with huge leaves or remarkable flowers. The banana plant (*Ensete* sp.), with its long, thickly ribbed leaves, would be a good choice, as would angel's trumpet (*Brugmansia* sp.), with its large leaves and huge scented flowers.

Seasonal displays

If you want to change the planting in big containers to give variety throughout the season, it is a good idea to place one permanent, tall, shrub-like plant in the middle of the container and then change the lower-level planting each season. You can also leave any trailing plants, such as ivy, around the rim of the pot to soften the planting.

Depending on the situation, you can keep the color theme the same throughout the year by simply altering the plants – white hyacinths and narcissi in spring, white busy Lizzies, lupins, geraniums (*Pelargonium* cvs.) and petunias in summer, and white chrysanthemums and asters in autumn. Alternatively you can create a different look for each season: yellows and blues for spring; pinks and reds for summer; and mauves and gray-greens for autumn and winter.

SPRING PLANTING

This large terracotta container is planted for spring with narcissi underplanted with pansies, which are surrounded by ivies. If you opt for the taller types of narcissi, you may need to provide some kind of support. A rectangle made of split bamboo canes, arranged horizontally and tied together at the corners, is an effective method of staking them.

ALTERNATIVE PLANTING SCHEME

You could keep the ivy to trail around the pot but replace the daffodils with tulips. The lily-flowered or parrot forms look exotic and are taller than narcissi. You could use the same colors for the tulips as for the narcissi to keep the yellow/white theme, which always looks fresh and springlike. The Lily-flowered tulip 'White Triumphator', with its pointed, pure white petals, is a good choice. As an alternative to pansies, plant a collar of primulas around the edge of the container.

WINTER PLANTING

A large container in winter is best planted with a central architectural specimen plant, and underplanted with smaller winter-flowering plants. Here ornamental cabbages and ivies are used around the rim of the container.

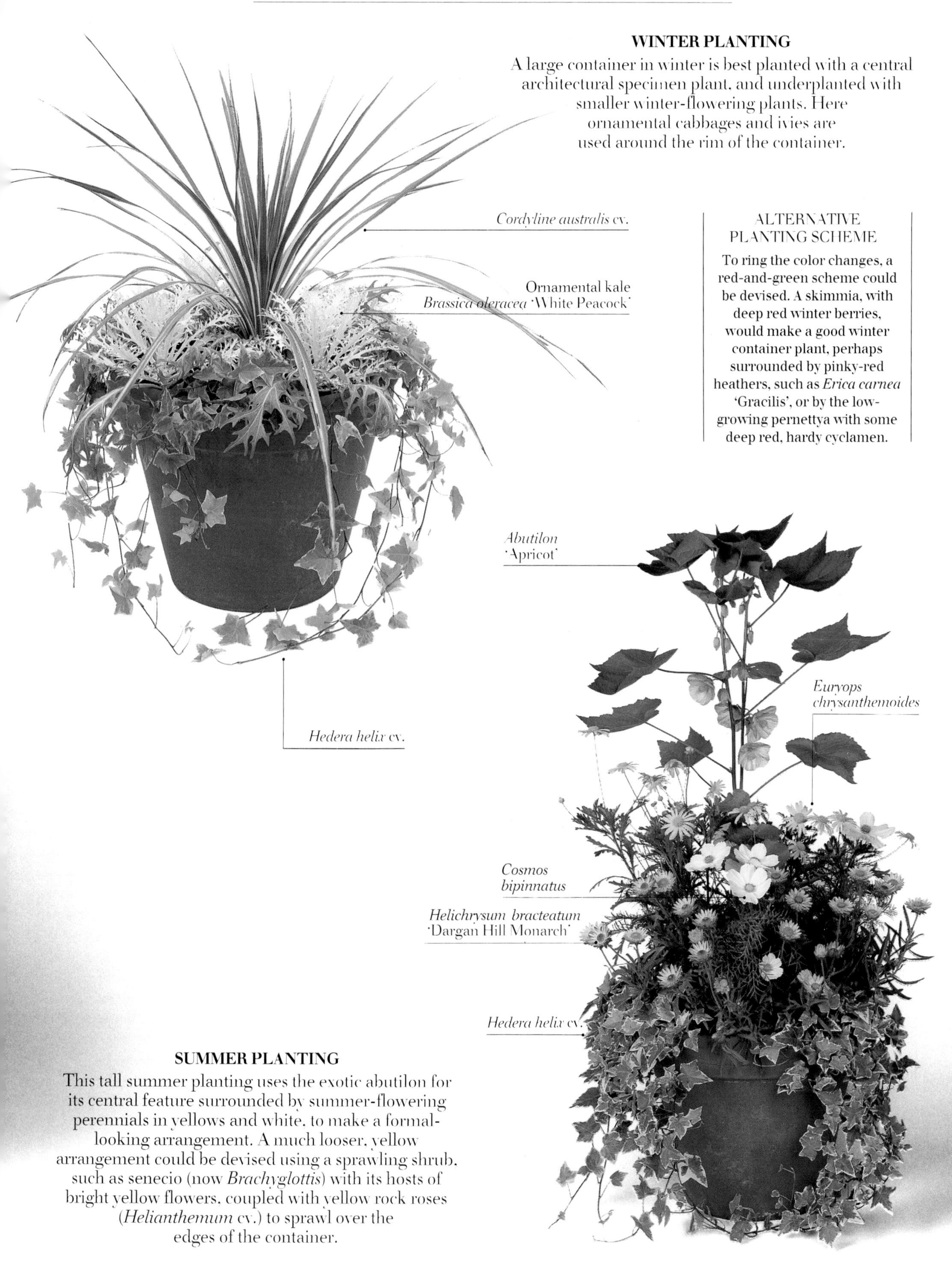

ALTERNATIVE PLANTING SCHEME

To ring the color changes, a red-and-green scheme could be devised. A skimmia, with deep red winter berries, would make a good winter container plant, perhaps surrounded by pinky-red heathers, such as *Erica carnea* 'Gracilis', or by the low-growing pernettya with some deep red, hardy cyclamen.

SUMMER PLANTING

This tall summer planting uses the exotic abutilon for its central feature surrounded by summer-flowering perennials in yellows and white, to make a formal-looking arrangement. A much looser, yellow arrangement could be devised using a sprawling shrub, such as senecio (now *Brachyglottis*) with its hosts of bright yellow flowers, coupled with yellow rock roses (*Helianthemum* cv.) to sprawl over the edges of the container.

Small Containers

CONTAINERS, at their smallest, can hold a single plant. Grouped together, these can make as effective a contribution as a single larger and showier pot. The small size tends to show off the attributes of the plant, particularly when it is a single specimen in a container, so that it makes a special impact. A group of four containers, say, each holding a single, particularly attractive pansy, or a neat little clipped ball of box (*Buxus sempervirens*), can be used on each window sill in place of a window box.

Small containers need positioning so that their charms are not overlooked. They are ideal for decorating a flight of steps, with a single repeating pot on each step. In Europe scarlet geraniums (*Pelargonium* cvs.) are used in this way, and the simplicity of the scheme is its chief virtue. You can vary this, for example, by creating a black-and-white theme, placing alternate containers of black pansies and white bacopa along a walkway.

Some of the most attractive plants are the little alpines that can be found in their natural state in the mountainous regions of the world. They have a delicacy and charm, and often bear singularly beautiful flowers in glowing colors. Small groups of them can be planted in shallow containers, provided you pick those which marry well together. Do not forget that the foliage is an important characteristic, and try to mix those that have a clump-forming habit with those with a looser habit, to vary the shapes they make. These little troughs and pots have all the charm of small jewel cases, provided you put them where they can be seen easily, such as at the front of a container arrangement or along a flight of steps. Among the best plants are the saxifrages, sedums and dwarf cultivars of dianthus.

In general, alpines need a growing medium containing plenty of grit or coarse sand, so that it drains freely. In ordinary growing medium their crowns will get wet

PLANTS WITH STRONG FLOWER IMPACT

The plants shown here are just some of those which look good planted up alone in small containers in order to focus attention on their enchanting, jewel-like flowers. Position them where they can be seen easily.

Single color displays

LEFT: *A pot planted up with a single form of flower, in this case the almost-black-flowered pansy,* Viola *'Molly Sanderson', makes a far stronger statement than a mixed-color display, and the attractively rounded form also helps to draw the eye towards it.*

Snap dragon (*Antirrhinum* – dark red cv.)

Pansy (*Viola* 'Molly Sanderson')

Busy Lizzie (*Impatiens walleriana*)

Potentilla fruticosa 'Alba'

Sweet William (*Dianthus barbatus* Nigrescens Group)

Marguerite (*Argyranthemum* sp.)

and they will rot off. You need to ape their natural growing conditions of poor free-draining soil if they are to succeed. They usually prefer sun to shade, and do best on an open sunny terrace.

Many of the spring bulbs are an ideal choice for small containers, in particular the dwarf cultivars of daffodils (*Narcissus* cvs.) and tulips, such as the small species tulip, *Tulipa clusiana*, with its star-shaped open flowers. They tend to look best planted up in single-species displays rather than mixed together, and it often pays to plant up a pair of containers to increase their impact. Other suitable small spring bulbs are the dwarf forms of iris, which could also be planted up alone to make an effective display in a small container. They need a free-draining growing medium and most of them require a sunny site, although the tiny *Iris verna*, which has lilac-blue flowers, will tolerate light shade.

Pay particular attention to achieving an attractive marriage between the pot and its contents. With small containers it is important to ensure that the container and the plants it holds have some particular merit – of either shape, color or form – or its smallness will mean that it will be overlooked. Single-color themes, or handsome architectural shapes, will help create impact. You can use ageing and distressing techniques (see pages 36–37) to alter the color of the pots, if you wish, so that they harmonize better with the planting.

Do not be afraid to use plants that are often used as underplanting for mixed schemes on their own in containers. The humble lobelia, particularly the dark blue form, looks extremely attractive planted up alone in a small container, as does the little *Sutera diffusa* or the simple daisy, *Bellis perennis*. By planting it on its own in a small container you persuade the onlooker to notice a plant's individual, and often overlooked, virtues. Even the rather undistinguished alyssum can make an attractive mound of flowers, which trail spontaneously over the rim of the pot.

For an autumn alternative to the foregoing spring and summer plants, you can plant small containers with dwarf chrysanthemums, or with autumn-flowering crocus. Good candidates for small pots for winter color are cyclamen, with their small reflexed flower heads in white, pink or red, and their attractively

Small-flowered plants

LEFT: *Plants with a mound-forming habit, like the Sutera diffusa* (sold as Bacopa *'Snowflake') shown here, make good subjects for small containers. You can move them out of the spotlight once they finish flowering, or replant the container with something different to flower later in the year (see pages 44–45).*

SMALL CONTAINERS

Alternative plants for small containers could include some of the following

Pink and red
The small form of sweet pea (*Lathyrus odoratus*) 'Knee Hi')
Busy Lizzie (*Impatiens* New Guinea hybrids)
Iberis umbellata
Schizanthus Hit Parade
Stocks (*Matthiola* Brompton Series)

White
Dimorphotheca pluvialis
Gypsophila elegans
Lavatera trimestris 'Mont Blanc'

Yellow
Limnanthes douglasii
Tagetes 'Gold Coin'
Coreopsis tinctoria

Blue and mauve
Brachycome iberidifolia
Felicia bergeriana
Anchusa azurea 'Blue Angel'
Lobelia erinus 'Crystal Palace'
Convolvulus tricolor 'Blue Flash'

mottled, dark green leaves, and the bright pink heather, *Erica carnea* 'Gracilis', which forms an attractive bushy shape. Small balls or pyramids of box (*Buxus sempervirens*) can also be used as a backdrop for a winter display. As the box is slow-growing it will be some time before it needs potting on. It is relatively easy to propagate from semi-ripe cuttings.

Herbs are also good candidates for small pots, especially the low-growing forms, such as thyme and parsley. These also make good companion plants for flowering ones: blue pansies look attractive combined with the curly leaves of parsley, for example. Little pots of santolina can be used to line a path or steps.

Container shape and size

Small containers can be narrow and deep or shallow and wide. Pick the plants carefully to suit the size of the container. Small narrow pots are ideal for more formal-looking plants like little clipped evergreens, or tiny standards, such as *Coprosma* × *kirkii*, or a small canna lily. The shallow wider ones are best for looser displays of sprawling flowering plants, like diascia, verbena or brachycome, with their delicate flowers and floppier stems. Equally good are small shallow containers of architectural sempervivums.

Small spring flowers
LEFT: *These little crocuses, in mixed mauves and whites, make an attractive display when grouped together in small matching pots. Covering the growing medium with a layer of moss not only helps the display, but also prevents the birds from attacking the crocus bulbs.*

Seasonal plantings

Small containers are ideal for seasonal plantings, using bulbs in spring and annuals in summer, with small-flowering perennials for winter. Ideally the planting scheme should be kept very simple, with just one or two colors at most. For impact, plant up small containers with the same scheme, and line them up along the edge of a patio, alongside a path or on a window sill.

WINTER PLANTING

A bowl-shaped terracotta container measuring approximately 15cm (6in) tall by 23cm (9in) in diameter has been planted with winter-flowering primulas in two colors and heather, which serves as a subtle backdrop and a perfect foil to the brightly colored primula flowers.

ALTERNATIVE PLANTING SCHEME

In winter there is rather less to choose from when it comes to finding subjects for a small container, but cyclamen and heathers are a good choice. You can opt for an all-pink or all-white display, mixing the two plants together, with perhaps the heather positioned centrally in the pot and the small cyclamen surrounding it, with ivy trailing over the sides. Pernettya, with its bright pink, waxy berries, is another good winter choice, as is the ornamental kale (*Brassica olearacea*) which comes with either white- or pink-tinged leaves.

SPRING PLANTING

A single planting of one species or cultivar – in this case *Cardamine trifolia* – is particularly effective in a small container, especially since you can then replace it with another plant for later in the year. The creeping perennial used here is ideally suited to container-growing as it can be invasive if allowed to spread freely in a border.

Cardamine trifolia

ALTERNATIVE PLANTING SCHEME

You can plant the following on their own or you can mix some of them together: for example, grape hyacinths with narcissi. Crocuses and snowdrops tend to look better in single plantings, because the flowers are so delicate.

Grape hyacinth (*Muscari* sp.)
Snowdrop (*Galanthus nivalis*)
Scilla spp.
Crocus cvs.
Dwarf tulips (such as *Tulipa clusiana*)
Dwarf iris (such as *Iris gracilipes*)
Dwarf narcissus (such as 'Tête-à-Tête' or 'Bridal Veil')

SUMMER PLANTING

Among the most attractive flowering plants are the genus *Mimulus*, with their snapdragon-like flowers that are borne profusely in summer. They are best planted on their own, as they like a moist growing medium, whereas many of the summer-flowering annuals prefer drier soil. *Mimulus* Malibu series, grown as an annual, comes in shades of yellow, red and orange.

Mimulus sp.

ALTERNATIVE PLANTING SCHEME

The following plants look attractive in a single planting scheme. *Brachycome* has a bushy habit with daisy-like flowers, while *Lewisia* is a clump-forming perennial.

Lewisia spp.
Saxifraga cvs.
Sedum spp.
Gentiana spp.
Brachycome spp.
Erigeron karvinskianus

Hanging Baskets

A HANGING BASKET attracts attention in a way that no other form of planting does. This has both benefits and disadvantages. The main benefit is that you can plan the hanging basket more or less in isolation; the disadvantage is that any error or shortcoming is only too obvious.

Today there is a wide range of containers suitable for hanging baskets, made of galvanized wire, rattan or plastic (see pages 34–35). The most attractive are generally the least obvious; the first aim in any hanging basket should be to disguise the container.

Beware the very bright and jazzy planting. It dazzles the eye to such an extent that you can quickly sicken of its brilliance. The most successful plantings tend to be those with a limited color palette and some attractively textured contrasts, using leaf color and form as well as flower power for effect.

For a change of pace, consider planting an all-green hanging basket. Ferns are a good choice because they make a naturally symmetrical, rounded shape, looking attractive from all angles. They are ideal for cool shady areas since the plants themselves need relatively little light to survive. As a result they are the obvious choice for shady patios or walkways.

Unlike other containers, which can be used to provide all-year-round interest, hanging baskets are predominantly a summer affair. Their shape and position normally requires a great deal of foliage to disguise the actual structure of the basket, and this is not generally available in the winter months, although you can make relatively successful winter plantings with, say, several ivies to cover the base, and winter-flowering pansies for color. The summer, in particular, gives you the best choice of trailing plants for hanging baskets (see page 49). At this time of year it is possible to create a really luxuriant display of flowers complemented by foliage plants that give a variety of subtle colors and textures. Alternatively, if you use a container

Architectural displays
ABOVE: *Ferns, such as the lush green* Polypodium vulgare *'Cornubiense', are an excellent choice for hanging baskets. You will need an attractive container since the fern will not disguise the base.*

Pastel mix
RIGHT: *The colors of this gently toning hanging basket have been carefully selected to offset the golden stone of the house behind. Begonias, lobelia and helichrysum are the principal ingredients.*

PLANTING UP A HANGING BASKET

To make the rounded basket stable before you plant it, rest it on a bucket or similar container. You will need enough sphagnum moss to line the basket, a black plastic bag as a liner and enough multi-purpose growing medium to fill the container.

1 *Push the sphagnum moss down into the basket and around the sides to cover the whole of the frame effectively.*

2 *Insert a black plastic bag, with drainage holes pierced in the base. Cut three holes in the sides and push the plants through from the outside of the basket.*

3 *Fill the interior with a layer of growing medium and add plants to the middle and upper sides, making sure that their roots are covered by the growing medium.*

4 *Build the central part of the scheme so that it rises to a slight peak. Make sure that the planting is well balanced on all sides.*

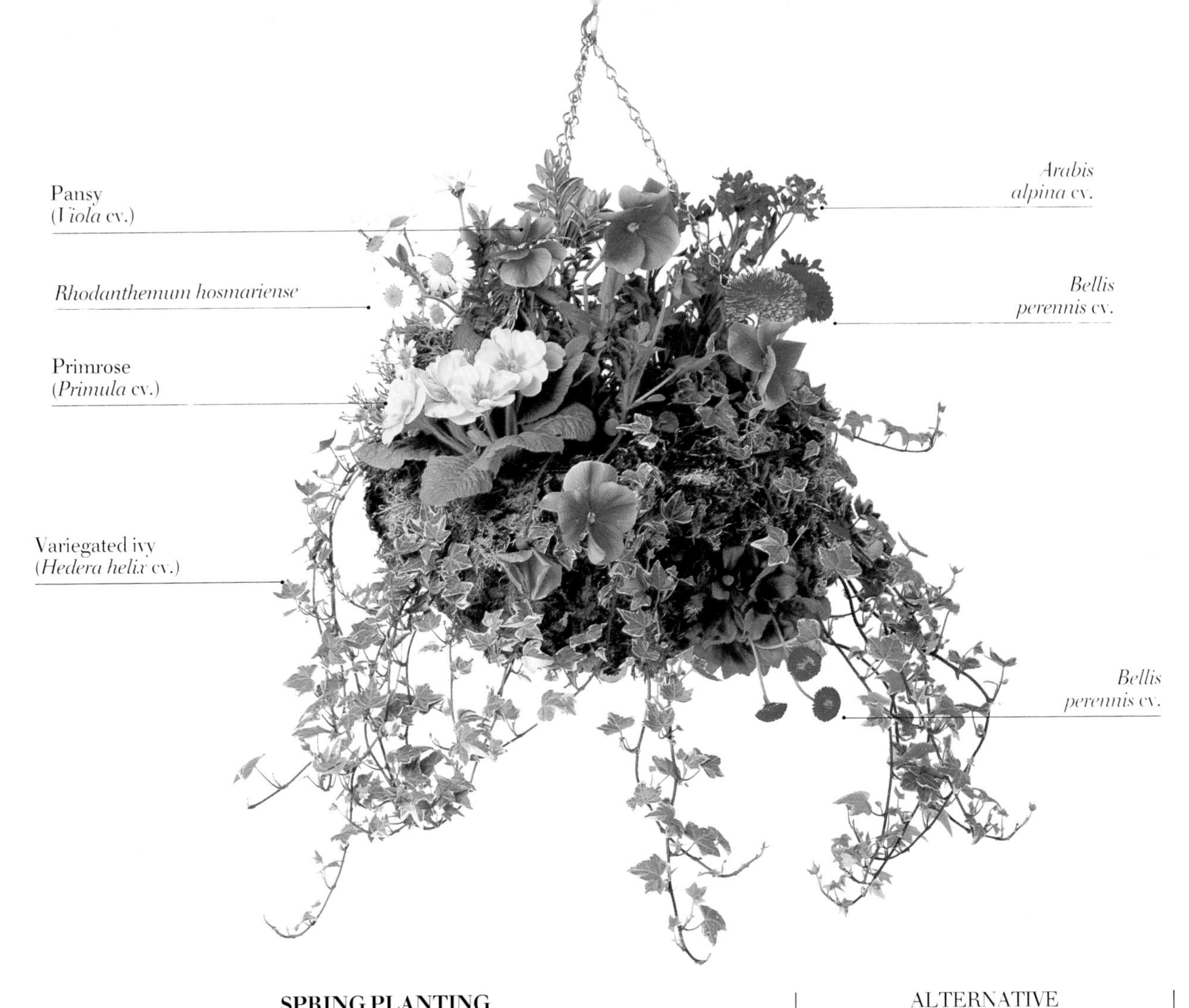

SPRING PLANTING

This spring planting centers on pansies, spring bulbs and ivy with a principally mauve and yellow theme. Primroses, pansies, small narcissi and grape hyacinths can be used for an early spring planting, with daisy-like flowers replacing the bulbs for early summer. It is always possible to modify the planting as the season progresses, replacing some, but not all, of the plants to keep it constantly looking good, and in full bloom. Ivies are an excellent choice as they trail over the rim and edges of the basket (see page 65).

ALTERNATIVE PLANTING SCHEME

You could change the color scheme to blues and pinks, with scillas, primulas and pink or blue pansies, if you wish, but keep the foliage interest the same.

such as a metal colander (strainer) or a rope basket, you are less constrained by the need to cover the base, and you can turn this type of hanging container into an all-year-round feature more easily.

Planting up

It is important to remember that the hanging basket is viewed, at least in part, from below and that any planting must be encouraged to disguise the sides and base of the container as much as possible. Nothing looks worse than a hanging basket with a brilliantly colored top and brownish dead-looking moss forming the base and sides.

To cover the basket completely, it helps to plant into the sides as well as the top. Plants bought or raised in Jiffy pots are ideal for this purpose, as the whole pot can be pushed into any available space in the container. Juvenile plants will grow quickly to fill the container and make it look lush, but it pays to leave the container, resting on a bucket or similar support, for a couple of weeks to mature before hanging it in its final position. For a standard hanging basket, about 12in (30cm) in diameter, you will need six to seven container-grown plants, which will rapidly fill the container in about three or four weeks.

For most displays some trailing plants are a must (see page 51 for suggestions). They will drape over the sides of the container, effectively masking its construction from view. Among the many suitable plants are verbenas, nasturtiums, helichrysums, diascias, pansies and begonias, fuchsias and lobelia, and geraniums (*Pelargonium* cvs.). The different kinds of ivy (*Hedera* cvs.), *Helichrysum petiolare* and *Tolmiea menziesii* make good foliage trailers for hanging baskets.

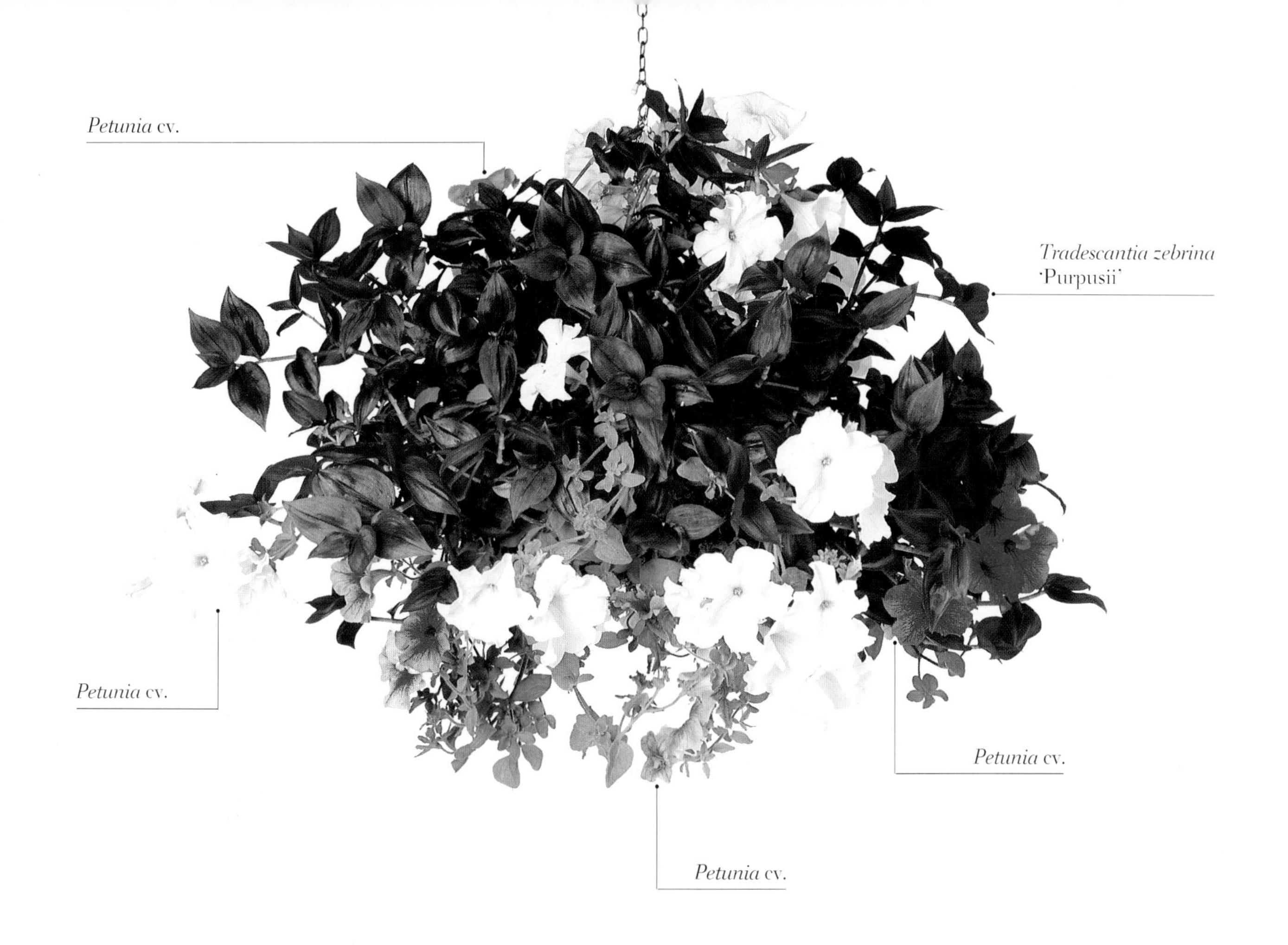

SUMMER PLANTING

Here some strongly contrasting colors have been used for impact, the rich dark leaves of *Tradescantia* making a perfect foil for the large-flowered pink and white petunias. The plants have been inserted through the sides of the basket as well as at the top to make a good rounded shape. An alternative plant would be lobelia, although it tends to dry out quite rapidly. You can also use tender house plants, such as kalanchoe, for summer plantings. Leaf interest is largely created by ivies, helichrysum and tolmiea.

ALTERNATIVE PLANTING SCHEME

There is a much wider choice of plant material in summer, principally because many of the trailing plants suitable for hanging baskets are summer-flowering. Among the best are trailing geraniums (*Pelargonium* cvs.), nasturtiums, felicia, diascia, verbenas and fuchsias.

Caring for hanging baskets

By virtue of its exposure to the elements a hanging basket dries out extremely quickly, to the point where in hot sun it will need watering at least twice a day. Since it is almost certainly going to be positioned above head height, for the purposes of display, this presents a logistical problem with watering. The answer is to use a commercially available extender, which you fix to your can or hose, or to tie a cane to the hose (see page 89).

The weight of a fully saturated hanging basket is considerable, so make sure the container has proper support in the form of a securely mounted bracket. To encourage growth, feed once a fortnight with a multi-purpose fertilizer. It will take two to three weeks for the container to fill out, but if there are any unexpected gaps or holes, you can insert Jiffy-pot-grown plants at a later date (see right).

FILLING GAPS

It is notoriously difficult to get a hanging basket to form an even shape, as some plants grow and spread faster than others. You can correct the outline by adding Jiffy-pot-grown plants at a later date, pushing them through the sides of the container.

Cut a small cross in the black plastic liner and push the Jiffy pot, base first, into the growing medium. The roots will force their way through the container into the growing medium.

Window Boxes

THE LONG RECTANGULAR SHAPE of most window boxes has a restricting effect on the planting, not least because you also need to consider the way in which the planting frames or complements the window behind it. There is an excellent choice of window box styles available today, including some particularly handsome, hand-thrown terracotta ones, which look good in almost any situation (see pages 32–33). You can also paint or distress existing window boxes to blend better with the surrounding architecture. Even the most inexpensive, white plastic containers look infinitely more elegant if you give them a coat of dark green or mid-gray paint.

It pays, when planting up a window box, to think of the color scheme in the room of which the window forms a part. You do not want a pastel drawing room dominated by window boxes full of purple petunias or scarlet salvias. Equally, look at the outside architecture when determining the style and size of the window boxes. If they are too short in length for the window sills, they can look rather mean. If the planting is too high, it may dominate the window and also prevent light from coming into the room behind.

Whether you opt for formal plantings and single-color themes, or much more informal, massed displays of, say, herbs or small-flowered cottage-style plants or annuals, is largely a matter of taste. However, since most window box plantings are seasonal, you can change their style, if you wish, as well as their color from season to season.

If preferred, you can keep a couple of structural elements in the window box as permanent plantings – evergreens such as trailing ivies, for example, to soften the edges of the box or small hebes to form height in the background. You can then simply replace the surrounding planting – cyclamen in winter, pansies or bulbs in spring, geraniums (*Pelargonium* cvs.) and petunias in summer, to name some familiar and popular plants. Most of the plants listed in the chapter on small containers are suitable for growing in window boxes, but it is a good idea to incorporate some trailing plants as well, to soften the effect. Those shown right are just some of the ones available to you. As with all planting, think about foliage color and form in window boxes, and try to make sure that the foliage colors blend well with the flowering color scheme you have in mind: bright blue pansies and/or scillas look good with silver-leaved plants, for example, while yellows and oranges can be combined successfully with some dark green ivy for contrast.

White and green is a popular combination, but tends to look dull when backed by white paintwork. In this instance you might be better selecting a color scheme of greens and yellows, perhaps, or picking subtle mixes of

Clashing colors
LEFT: *An informal planting of summer-flowering plants, including* Osteospermum *cvs. backed by the attractively variegated leaves of coleus, is framed by pale lilac wisteria, soft peach roses, and lilac-colored* Agapanthus *flowers.*

A two color scheme
RIGHT: *The dark color of the window box itself echoes that of the painted shutters, allowing attention to focus on the planting – of geraniums (*Pelargonium *cvs.), petunias and ivy – in tones of pink and white. The generous width of the window box and the exuberant nature of the planting make a good contrast to the formality of the color scheme.*

TRAILING PLANTS

To soften the edges of a window, include trailing plants. There are quite a few to choose from, some of which are shown above. Other good ones are the small creeping campanulas, columneas, *Tradescantia zebrina*, nasturtiums (*Tropaeolum* cvs.) and *Lysimachia nummularia*.

pastel shades. Leave the red geraniums for brick or yellow stone. Their contrast with white paintwork can look crude, unless you live in a climate with brilliant sunshine and deep blue skies, where the natural sunlight makes them look less dominating.

Caring for window boxes

It is, unfortunately, all too easy to forget to water and feed the plants in window boxes, and nothing looks worse than a browned and dying display in such a prominent position. Remember to water well – it is surprising just how much water is required, especially on hot days in summer, since the plants transpire rapidly in these conditions, and the depth of planting medium in any window box is not great. Feed the plants regularly (once a fortnight on average) throughout the growing season with a multi-purpose liquid feed; and, if you live in a city, clean the leaves as well in periods of very dry weather to remove dust and grime. Your window boxes will only look at their best if they are full and healthy – skimpy-looking plants with dying or yellowing leaves are hardly welcoming or inspiring.

Fastening window boxes

As with all pots and containers used at high level, you must ensure that they are securely fixed otherwise you may find yourself liable if they blow down in a gale and damage someone else's property or hurt someone.

POT FEET

These are placed underneath the window box – usually one at each corner – in order to raise the container off the window sill. This process prevents water from becoming trapped underneath the container, which can cause the window sill to rot, and encourages air to circulate freely around the window box, thus reducing disease.

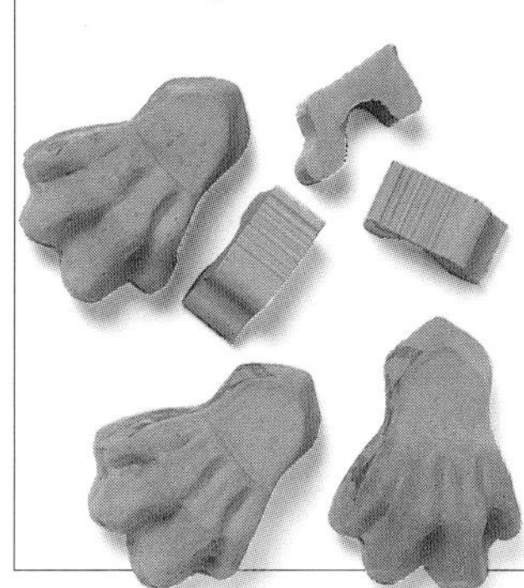

Unusual Containers

AS WELL AS the traditional terracotta, lead or wooden containers, you can make use of a range of other items, from wheelbarrows to purpose-made herb pots, from colanders (strainers) for hanging baskets to half-barrels as water gardens, and from chimney pots to old sinks. Recycling containers in this way is satisfying because it is economical and ecologically friendly, but make sure that you do not end up with a container display that is ill-assorted.

Good sources of unusual containers are junk shops and architectural salvage companies, where all sorts – from old chimney pots to drainpipes – can be obtained, often at low cost. You can also find interesting containers in potteries, but beware of the overly cute. You may be amused initially by a clay replica of an old boot, planted with flowers, but the novelty palls.

Interesting decorative features are a rather different matter, and it is certainly worth looking out for attractively molded containers, some of which are now of considerable value. For example, old lead water cisterns are extremely valuable garden ornaments, so if you come across a container of this sort in the process of renovating a house or barn, get it valued and insured.

Practical elements

The main point, when using unusual containers, is that they are suitable for the purpose – in other words they need to be frostproof and self-draining – and that you find a place for them in which their unusual appearance is an asset to the overall scheme. The subsidiary point is that you need to find ways of using them that either enhance their intrinsic oddity or disguise it completely – in other words play up their best features and play down their worst ones. Three or four old oil drums in disparate sizes, painted and distressed in, say, verdigris green, will harmonize well, while an old chimney pot, that is attractive in its own right, can be positioned to stand alone as a focal point.

The water garden
RIGHT: *A half-barrel makes an ideal water garden, with a selection of plants with differing foliage and form. Here pickerel weed (*Pontederia cordata*) and* Acorus gramineus *'Variegatus' are planted with water avens (*Geum rivale*).*

ALTERNATIVE PLANTING SCHEME

If you want to create a water garden in a large container, try to combine different forms of plant for maximum interest. Here are a selection of plants suitable for water-gardening:

Sword-shaped leaves
Acorus calamus 'Variegatus'
Butomus umbellatus
Irises:
I. ensata
I. pseudacorus
I. pallida,
Typha latifolia

Rounded leaves
Calla palustris
Caltha palustris
Mimulus guttatus
Sagittaria japonica

Floating aquatic plants
Aponogeton distachyos
Hydrocharis morsus-ranae
Nymphaea pygmaea

Oxygenating plants
Elodea canadensis
Lemna minor
Potamogeton crispus

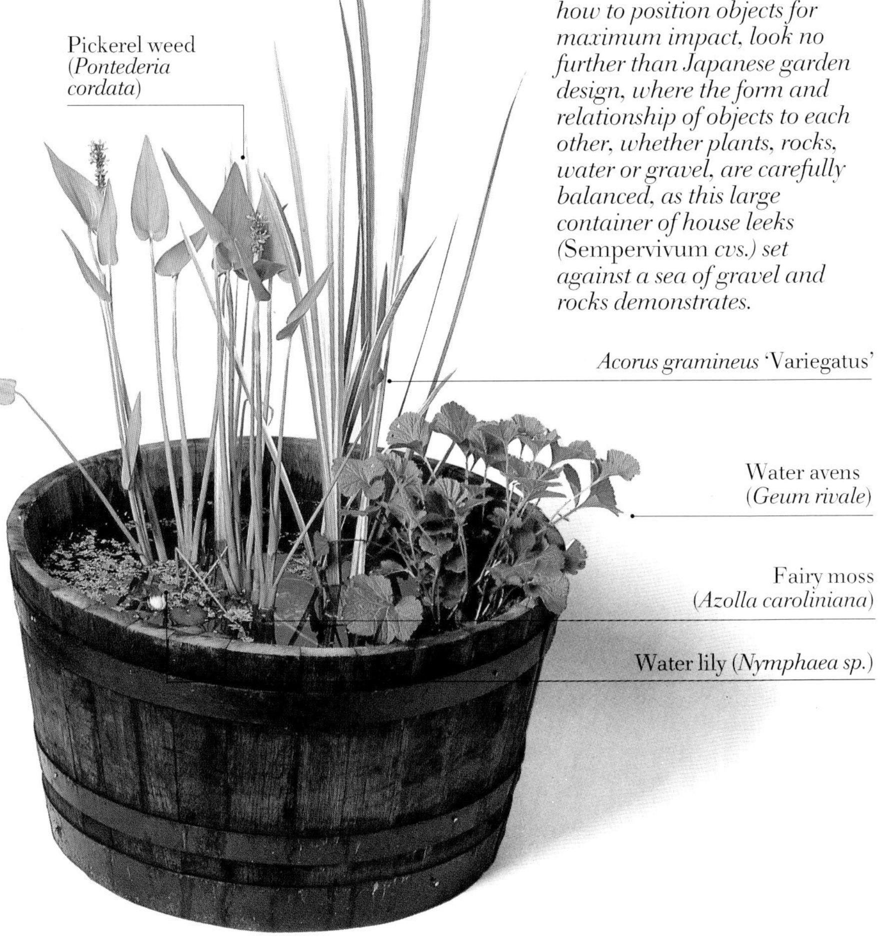

Oriental style
RIGHT: *If you want to find out how to position objects for maximum impact, look no further than Japanese garden design, where the form and relationship of objects to each other, whether plants, rocks, water or gravel, are carefully balanced, as this large container of house leeks (*Sempervivum *cvs.) set against a sea of gravel and rocks demonstrates.*

Whatever the containers, keep the theme simple and do not make a collection of curious shapes or different colors. They will simply look odd rather than interesting, unless you are lucky enough to have an artist's eye for combining disparate objects.

Planting up

The form the planting takes will depend on the style of the container. For a wheelbarrow or some former farm implement, for example, you need to devise a planting that is suitably informal in style – either herbs or small-flowered species plants with a trailing habit that will cascade over the sides. You can fill the center of the container with humble plastic pots of different sizes, since the sides will mask the pots from view. You can then change elements of the display as they come into or go out of flower.

Water gardens and bog gardens

Creating a small water garden greatly increases the scope of your planting, since most containers are better suited to plants that are drought-resistant rather than moisture-loving. The porous nature of most containers causes the growing medium in them to dry out more quickly than in a conventional border, so unless you are prepared to spend a great deal of time watering, your opportunities for growing moisture-loving plants in containers is limited.

If you want to go for large unusual containers, why not create a small water garden in an old half-barrel, which is large enough to provide a home for a good selection of water-loving plants, including the dwarf forms of water lily? If you decide to do this, you will need to soak the barrel first (see page 103) to make sure that the wood expands enough to make it watertight. You can then fill it with water and plant it up with some suitable water plants, such as *Pontederia cordata*, with its attractive blue flowers and arrow-shaped leaves, water-loving irises, such as *Iris laevigata* or *I. ensata*, water lilies and reeds. The main point to bear in mind when planting up a barrel or other container as a water garden is to include oxygenating plants, such as those listed in the box on page 54, otherwise the water will become stagnant.

You will need to use special baskets for planting in water – these have an open mesh construction which allows plant roots to penetrate the sides of the basket. If necessary, stand baskets on bricks to raise the height of the plants, especially those that prefer just to have their feet in the water. You can disguise ugly baskets with

Traditional troughs
BELOW: *A marvellous toning display of soft deep pinks and dark reds – dominated by petunias and* Verbena *'Sissinghurst' – sets off an old stone trough to perfection.*

fairy moss, scattering it around the rims. Always use heavy clay growing medium for water gardens, placing a layer of stones top and bottom to weight it down. Lining the planting basket with turf will also help to retain the growing medium.

Bog gardens look particularly good in the confines of patios or backyards which incorporate a great deal of hard surface, and help to counteract the busy impression created by the overuse of small-leaved flowering plants. Large green leaves help to give this kind of garden a lush feel, and since most of the large-leaved plants are shade-lovers, they are ideal for small backyards which receive very little sunshine as a result of high boundary walls. Among the plants you can grow in them are moisture-lovers, such as ligularias and rodgersias, whose virtue is their large attractive leaves. You can make a bog garden from a barrel. Simply line the container with plastic, remembering to puncture a few drainage holes in the base to promote aeration, and fill with rich organic growing medium which should be kept permanently moist. If necessary, rig up some kind of drip watering system.

Alpine troughs

The old square ceramic sinks, known as Belfast sinks, make ideal alpine troughs if you coat them with strong adhesive, to which you then apply a fairly stiff mixture of peat, builder's sand and cement using a trowel. Cover with damp sacking and leave it to harden off, out of the rain, for a few days.

PLANTING UP A SMALL SHALLOW CONTAINER

If you intend to plant a small container with alpines, you must use a free-draining growing medium and the container must have adequate drainage holes. The plants used here – dianthus, sedums and sempervivums – like the same conditions.

1 *Fill the middle of the container with a mixture of growing medium and coarse sand or fine gravel in a 50:50 ratio.*

2 *Start to add the plants, in this case a sedum, taking care to spread out the roots gently without tearing.*

3 *Continue to add the plants, sprinkling a little extra coarse sand or fine gravel around the crown of each one to encourage good drainage.*

4 *Complete the planting, cover the surface with a fine layer of coarse sand or fine gravel and water well.*

The large, low, square shape of these rough sinks is ideal for small jewel-like alpines. If you can get hold of an original animal drinking trough, so much the better. You will need to make sure that the growing medium has plenty of coarse sand to make it free-draining since alpines rot quickly in waterlogged soil. For an alpine display on a much smaller scale, an old clay roof tile makes an excellent container (see page 57). As most alpines are slow-growing you should plant them close together, and cover any gaps in the planting medium with gravel to prevent moisture loss.

Objets trouvés

An *objet trouvé*, such as an interestingly shaped log (see below and right), can also provide a home for small alpine-style plants, if you gouge out a little of the wood to make space for the growing medium. You can replant the container each season to ensure that it looks good all year round, with cyclamen in winter, perhaps, scillas or dwarf daffodils in spring, anemones in early summer and little daisies in high summer, followed by dwarf chrysanthemums in autumn.

Paint pots and oil drums can also be used as containers, but they will need to be painted after being cleaned thoroughly of their original contents. Large plastic pots can be employed in the same way, but remember to insert drainage holes if the container has none. You can either drill holes in the base or, if the container is plastic and you have no drill, sear them with a red-hot skewer. Cover the handle to avoid burning your fingers.

Household receptacles

A colander (strainer) can make a successful small hanging basket, since the holes will act as drainage holes. To draw attention to the unusual nature of the container, opt for a single-theme planting – perhaps using variegated ground ivy (*Glechoma hederacea* 'Variegata) or dark red nasturtiums. If you wish, you can paint the colander in an appropriate shade – dark green, perhaps, or dark blue.

Discarded household equipment, such as old iron cooking pots or an old washtub, make other possible containers, again painted or distressed in a sympathetic color. They can look good with a toning color scheme.

SPRING PLANTING

This wooden log is filled with an assortment of fairly shallow-rooting plants as there is very little depth of growing medium available. Natural woodland plants are more in sympathy with the style of the container than very brightly colored alpines, although the latter have suitably shallow roots.

SUMMER PLANTING

Delicate small flowers and variegated leaf color, in golds and greens, make a good combination for summer. Equally successful would be the small daisy *Erigeron karvinskianus*, with its pink and white flowers, or, for a blue display, one of the small campanulas, such as *C. portenschlagiana* 'Aceana', which has silvery small-leaved foliage.

WINTER PLANTING

For winter you can use either a winter-flowering pansy – ideally one with very small flowers, such as the dark 'Molly Sanderson' shown on page 42 – or cyclamen, as here. Small-leaved cultivars of ivy and ferns make the ideal accompaniment for the jewel-like flowers.

DESIGNING WITH PLANTS

THIS SECTION takes the two principal attributes of plants – their flowers and their foliage – and looks at the different ways these can be employed in grouping and arranging container displays.

Foliage tends all too often to be ignored as a feature in its own right, but the large number of really attractive foliage plants give you wonderful opportunities. This chapter shows you how to exploit their natural attributes, either by displaying them as single specimens or by using them as a backdrop to more brightly colored flowering specimens. In addition, information is given on topiary and staking techniques.

The beauty of flowers needs little introduction, but learning how to combine colors successfully in both harmonizing and contrasting schemes is an important element in container gardening. This section is packed with ideas for displaying flowering plants to best advantage – either by limiting the display to a single color, or by combining toning or contrasting shades in a mixed planting scheme.

LEFT: *The deep russet-purple leaves of* Pelargonium *'Vancouver Centennial' are edged with brilliant green margins.*

ABOVE: *The sword-shaped, silver-green leaves of* Astelia chathamika *are set off to perfection in an elegant stone urn.*

Displaying Foliage Plants

FOR MANY PEOPLE, foliage has no significance at all. It is flower color that seduces them. After a little while, however, the eye starts to sicken and tire of too much flower power and longs for a quieter, more relaxed palette. One of the reasons why nature attracts us is that our attention is caught by its rich tapestry of foliage – the myriad colors and shapes of its plant leaves – and yet as gardeners we are sometimes very slow to take advantage of these qualities.

Look closely at any photographs that particularly attract you in gardening books and you will find that the shape and color of the leaves, as well as the flowers, plays a significant part in the overall effect. In fact you do not really need flower color at all, since nature, and plant breeders, have provided us with a wonderful variety of plants that have terrific foliage – the huge, thickly ribbed leaves of some hostas, the glossy divided foliage of Mexican orange blossom (*Choisya ternata*), or the softly silvered, filigree-like leaves of wormwood (*Artemisia* spp.).

Far from being a background accompaniment to flower color, leaf form, texture and color can, and should, play an important role in any planting scheme. Look for plants that have naturally good form, such as *Hosta sieboldiana* var. *elegans* with its wide-spreading dome shape of huge, blue-gray, ribbed leaves, or create the form yourself by clipping a neat-leaved evergreen like box (*Buxus sempervirens*) into clean geometric shapes, or train a climber like ivy over a frame.

Some of the best foliage plants for container gardening are climbers. These are especially valuable for any gardener who is limited to growing plants in containers since, grown over a wall-mounted trellis, for example, or perhaps an arch or pergola they provide interest at a higher level that you would normally get in a garden from shrubs or trees. *Actinidia kolomikta*, with its heart-shaped green or variegated leaves, is always a good choice if you want a climbing plant with luxuriant and interesting foliage, as is the Dutchman's pipe (*Aristolochia durior*), again with heart-shaped leaves. Other good foliage climbers are vines, such as *Vitis coignetiae* which turns a rich ruby-red and adds color and interest to your planting scheme in autumn, and of course the huge range of ever-popular ivies, which have a marvellous range of leaf forms and colors, from the big white-splashed leaves of *Hedera colchica* 'Sulphur's Heart' to the tiny, neat, arrow-shaped, dark green leaves of *Hedera helix* 'Sagittifolia'. They are ideal for providing a subtle, attractive backdrop for flowering plants all year round.

Foliage for hedging and screening

A number of foliage plants can be used for container-grown hedges, which are useful for screening an exposed balcony or roof terrace and protecting other, more delicate, plants. Any of the small-leaved evergreens are a good choice, such as box, yew or privet. Privet has a particular advantage in that it will grow a lot

DARK-LEAVED PLANTS

There is an exciting range of dark-colored foliage at your disposal for container gardening, including some deep purples. Other good purple-leaved plants, apart from those shown here, include certain cultivars of the shrubs *Cotinus* and *Berberis* and of the perennial *Heuchera*, whose cultivar 'Palace Purple' is particularly effective.

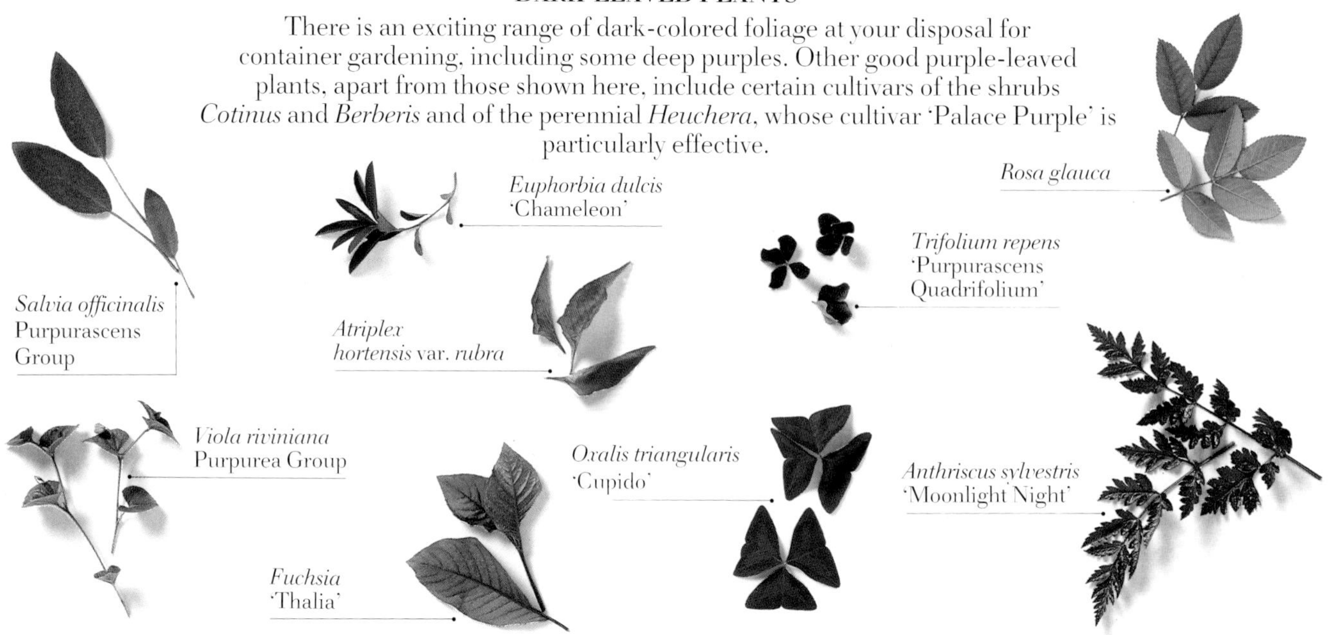

LEAF COLOR AND SHAPE

Leaves come in a wide range of shapes from tiny rounds perhaps 1cm (½in) in length to massive hand-shaped ones, like those of the giant *Gunnera*, which can measure up to 1.4m (5ft) across. Alternatively they can be composed of many tiny leaflets, like those of the false acacia (*Robinia pseudoacacia*), or they can be long and slender, like those of irises. Color can vary from simple green to all its shades, from yellow tones to silvery-blue, through variegated (splashed with other colors, as shown below) to deep purples (below left). Creating an attractive mixture of shape and color is an important element in successful container gardening, and it is perfectly possible to compose a superbly attractive planting scheme solely of foliage.

more quickly than yew or box, and there is an attractive golden-leaved variety. Less formal hedges can also be made from leafy climbers – which can be trained to scramble over trellis or wooden supports – such as vines, the golden-leaved hop, ivies, parthenocissus or Dutchman's pipe (*Aristolochia*).

For screens you can use large ceramic Chinese pots of bamboos. In fact bamboos are best grown in a container of some kind, since many of them spread uncontrollably if grown in the garden. You will have to remove them from the pot every few years in order to divide them, otherwise they become root-bound.

Foliage for backgrounds

Foliage plants in pots make a good background for a more colorful, flowering display and help to add structure and form to the overall appearance. Try to vary the leaf shapes and colors as well as the heights of

Contrasting colors
ABOVE: *This small terracotta pot is planted with a simple combination of attractive foliage plants: the large blue-green leaves of* Hosta *'Halcyon' and the yellowish-green palm-shaped leaves of* Tellima grandiflora.

VARIEGATED FOLIAGE

There are plenty of variegated plants to choose from, most of them preferring a bright, sunny position. When growing variegated shrubs, take care to cut back any shoots that revert to their plain, non-variegated form.

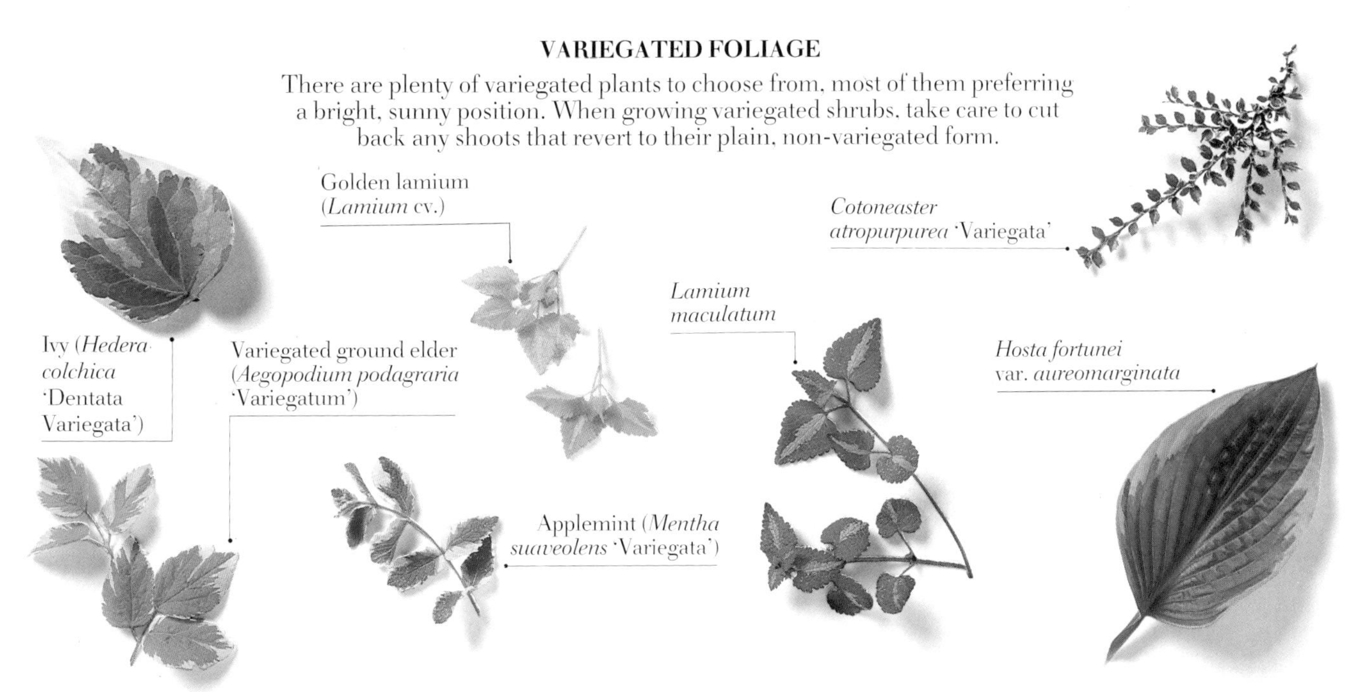

Combined displays
ABOVE: *The golden hop* (Humulus lupulus *'Aureus'*) *is trained over a birch-twig frame. A companion planting of curly parsley* (Petroselinum crispum) *and a variegated hosta* (Hosta fortunei *var.* albopicta) *completes the display.*

the plants to give an attractive display. Plants with strong forms – in other words, those with neat well-shaped outlines – are always good for this purpose. Hostas, ferns, dicentra, zantedeschia and clivia make excellent lower-level foliage background, as do neatly clipped balls of box or privet. Taller plants for this situation include *Choisya ternata*, small well-formed trees like the cut-leaved maples, *Rheum palmatum*, *Macleaya cordata*, palms like *Trachycarpus fortunei*, and the tree fern (*Dicksonia antarctica*). Similar to the tree fern in form are the sword-shaped plants like yuccas and cordylines from America and Australasia, which need a reasonably sheltered position. Cordylines will do well in partial shade; yuccas need full sun.

Foliage for color

When you display foliage plants you quickly realize that the leaves are far more colorful than you previously thought. The reason for this is that the brilliant color of many flowers tends to overwhelm the cooler, more subtle leaf coloring, but without the distraction of the flowers you begin to notice the quality and coloring of the foliage. In shady areas you are going to be largely dependent on leaves for interest, and it is often as a result of gardening in shade that you begin to appreciate foliage properly.

Colors range from silvery-gray to rich deep purples, and many plants have leaves that turn brilliant colors in autumn, in particular some of the Japanese maples, vines and creepers. Although it is hard to find a clash in nature, you can balance foliage colors better if you

TRAINING AN IVY PYRAMID

To train a small ivy pyramid you will need a suitable-sized container and a frame – the pot shown here is 23cm (9in) in diameter while the conical frame is 60cm (2ft) tall – plus two to three small-leaved ivy plants.

You don't need many materials in order to create your own stunning ivy pyramid.

1 *Plant up the ivy and place the frame inside the rim of the pot. Then criss-cross garden wire between the struts.*

2 *Unwind the shoots and train them up the frame, fastening them loosely with wire or plastic ties.*

The fully grown pyramid

consider the various shades of green. Be careful not to overwhelm more subtle, gray-green plants such as *Hosta sieboldiana* var. *elegans* or *Senecio* 'Sunshine' by combining them with too much grassy-green foliage. Aim for a mixture, but position the plants carefully, like a painter would arrange a palette, so that the colors harmonize. Green is a mixture of blue and yellow, but some greens have far more yellow in their make-up, while others have much more blue.

Texture and shape

Leaf texture and shape will also affect the way you perceive foliage colors and it is worth taking this into account when planning a display. Some leaves are rich and glossy – the waxy evergreen leaves of camellias, magnolias and choisya, for example – while others are softly felted, such as those of lambs' ears (*Stachys byzantina*) or *Lychnis coronaria*, both silvery-gray in color. You can try to create a Mediterranean feel using the silvery-gray felted or divided leaves that are customary in plants from sunnier parts of the world, or you can conjure up the atmosphere of an urban jungle with the large, thick, glossy leaves found in more humid regions. Be careful not to mix plants that come from areas of wildly differing climate – not only will they look odd together, but you will also find it more difficult to maintain them, since some will need copious watering while others will survive with minimal watering only. Over- or underwatering can spell rapid death to container-grown plants (see pages 88–89).

Variegated foliage – splashed with silver, white, gold or yellow – also helps to break up the monotony of an all-foliage arrangement, and will certainly lighten and lift the heavy effect produced by too much dark evergreen foliage. Set a couple of pots of a silver-leaved variegated plant like coprosma as standards against a

TRAINING AN IVY SPHERE

Rather than training a plant over an empty frame, you can use a wire frame filled with sphagnum moss. The plant can cling to the moss as it grows.

1 *Plant in the usual way (see opposite), allowing the ivy to grow over the mossy frame. Make sure that you keep the moss damp at all times.*

2 *Once the ivy has started to cover the support, trim off any untidy shoots as they appear using a pair of sharp secateurs/pruning shears.*

The fully grown and trimmed ivy sphere.
To maintain an even shape, trim any straggly pieces on a regular basis and turn the pot from time to time. Feed occasionally with proprietary foliar feed (see page 86) to promote healthy plant growth.

CHOOSING IVIES

Among the most useful foliage plants are ivies (*Hedera* cvs.), which come in many forms and which are invaluable for trailing over window boxes.

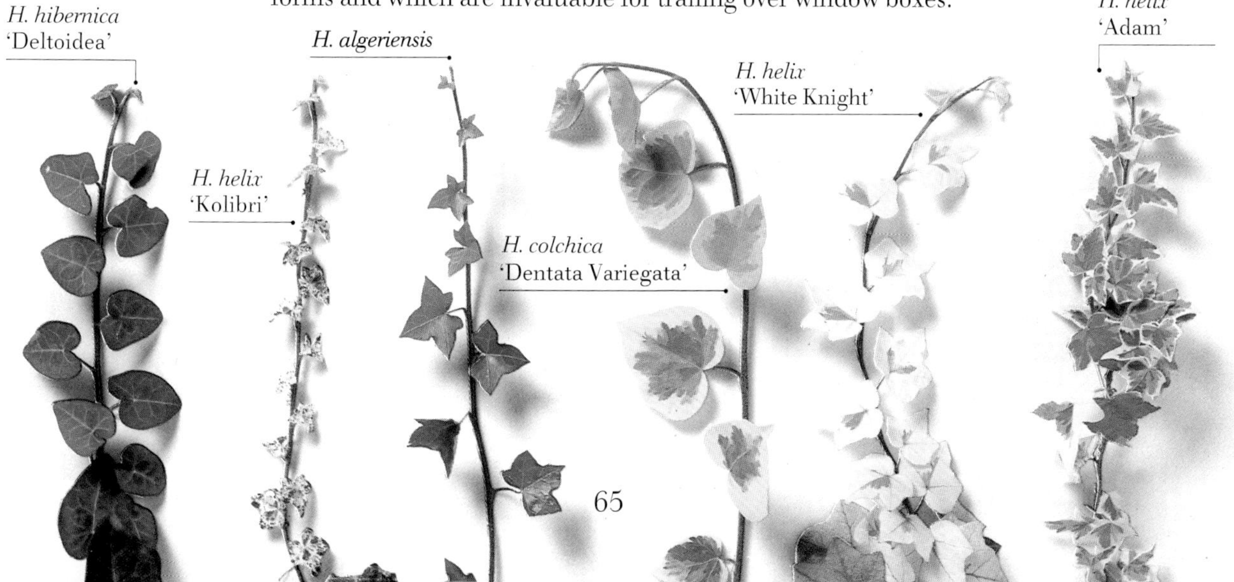

background of clipped evergreen privet or yew and you will see how the dense dark texture of the one provides a wonderful foil for the lightness and light-reflecting qualities of the other. Flower color is unnecessary, and if you keep the color palette simple in this way, your eye is immediately drawn to the form of the plants – the ideal way to display a standard, for example, where you want to focus attention on the sculptural shape.

Formal arrangements
ABOVE: *Neatly clipped box balls and spirals in terracotta pots create a formal look for this small patio garden, enclosed by clipped yew hedges. Clipped box borders around the two geometric beds emphasize the formal simplicity.*

Seasonal foliage

One of the nicest characteristics of foliage is the way it changes color throughout the seasons, as the plant matures. The new growth can sometimes be pink-tinged or a paler shade of green, and as the foliage dies in autumn you get some marvellous tints and hues, from rich golds to russet reds. Among the best plants for autumn color are the Japanese maples *Acer palmatum* var. *dissectum* with their delicately cut filigree-like leaves. There are some cultivars (see page 108) which have wonderful autumn color. As they are not particularly large or vigorous, and most of them have an attractively branched habit they make excellent container plants, and are useful for providing a focal point by a seating area on a patio, for example. The very light shade cast by their leaves gives protection, without restricting what can be grown below their canopy.

It is always worth including a good mixture of foliage shapes, and among the best plants for this kind of contrast are any of the sword-shaped leaves, particularly those of the big architectural cordylines, yuccas, agaves and dracaenas, as well as the smaller perennials with strappy leaves such as irises, *Montbretia*, *Crinum* × *powellii* (which has attractive starry pink flowers) and the rather tender but very spectacular *Clivia miniata*. Long leaves look particularly good when contrasted with the rounder, softer leaves of perennials like *Alchemilla mollis*, the crinkled leaves of *Tellima grandiflora* and the spathe-shaped leaves of many of the hostas, which are wonderfully varied in color, size and texture. Of all the foliage plants for containers, hostas are arguably the best. Since they are particularly prone to slug damage, growing them in containers certainly helps reduce the risk of pests. *Hosta sieboldiana* var. *elegans* is a prince among foliage plants, making a splendid large dome-shape about 1.2m (4ft) high of immense, thickly ribbed blue-green leaves.

Topiary

For the purpose of creating sculptural shapes and structures, foliage comes into its own. The art of clipping trees and shrubs into geometric shapes has been common garden practice since the 17th century, and was at its height in the formal garden designs of the French designer André Le Nôtre. If you look at paintings of this period, you will see that the landscape is punctuated by small formal clipped trees in what are known as Versailles tubs (see page 32) – rectangular pots whose geometric shape complements the neatly clipped plants contained within them.

The plants that are most successfully used for topiary are the slow-growing, small-leaved evergreens, such as box (*Buxus sempervirens*) or yew (*Taxus baccata*). Other evergreens can also be used for topiary, including privet (*Ligustrum* sp.), myrtle (*Myrtus communis*) and holly (*Ilex* sp.), but the latter lack the density that the first two offer. Because the two former plants are slow-growing they will take some time to create a topiary shape, so if you want quick results, opt instead for privet. However, if you are happy to create small topiary balls or pyramids, box can be grown successfully into these shapes in about two or three years, provided you buy reasonably large container-grown plants to start off with. Simple shapes, for example, basic geometric ones, can be trained and clipped freehand, but more complex shapes require a wire frame to act as a template for the design (see right).

Training topiary shapes

If you lack the patience to train and clip box into topiary shapes – it will take about three years to get a bush about 60cm (2ft) high from a box cutting and box takes quite quickly and well from softwood cuttings simply pushed into the growing medium – opt instead for creating similar-shaped forms with a climbing plant like ivy. You can buy a wire frame over which to train the ivy, and if you set several plants around the pot, the shape will thicken up and grow fairly quickly. You should have a well-filled frame within 18 months, and it actually looks quite pretty while growing and before it has completely filled out. Ivy grows easily from cuttings: just snip off short lengths with some roots attached and plant them around the edges of a container, feeding and watering them regularly to encourage them to establish quickly.

USING TOPIARY SHAPES

You can buy a variety of wire and metal structures for training topiary, including cones, pyramids, spheres and spirals. They can be used either as a clipping guide for formal topiary, placed over the young plants which then grow up through them, or to train climbing plants, such as ivy, to mimic topiary (see pages 64–65). For a less formal look create a frame yourself, like the one shown on page 64 (top), by bending and tying young birch twigs together.

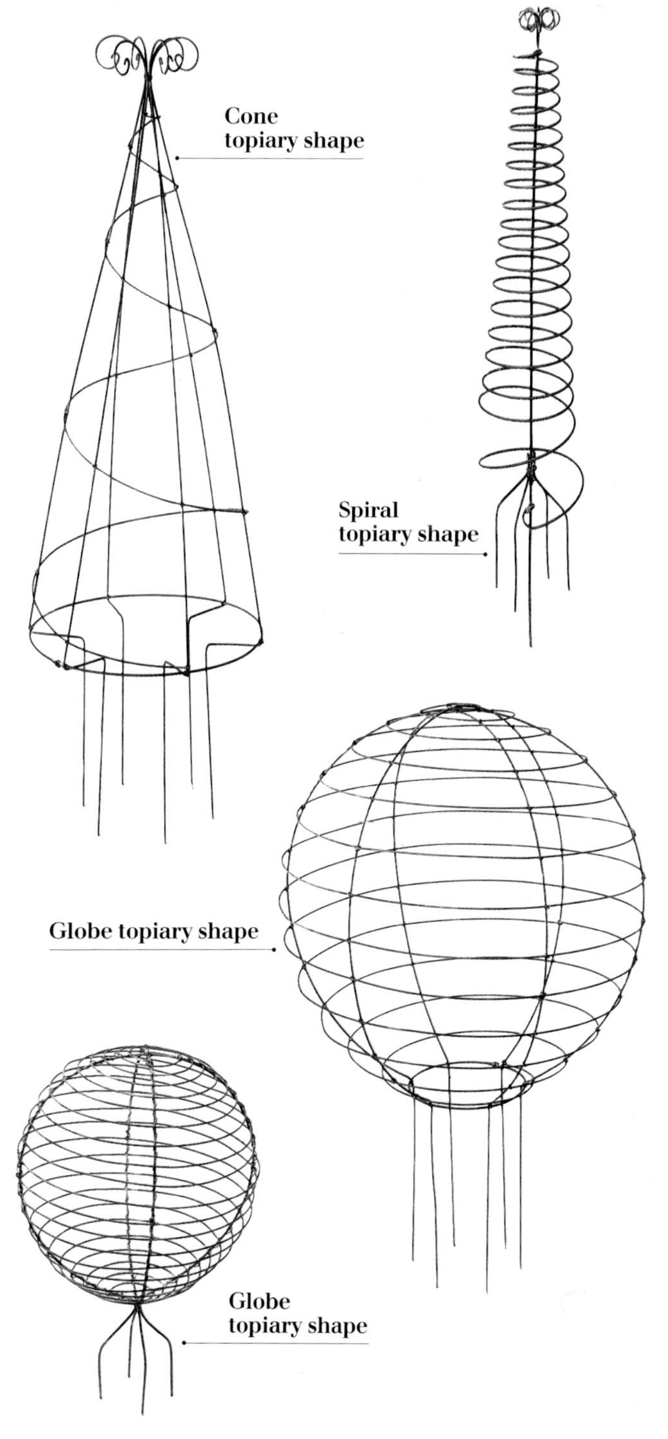

Displaying Flowering Plants

FOR MANY PEOPLE, flower color is the most exciting aspect of container gardening and they use containers, quite rightly, to create splashes of brilliant color on rather bland, hard surfaces, such as patios, gravelled terraces and steps. The secret, however, is to be able to orchestrate different flower colors so that they harmonize and fit in successfully with any other planting. Nothing looks worse than brilliant flower color that dominates or overpowers all the remaining planting, standing out from it like a rash on an otherwise smooth complexion.

Strong clashing colors can work extremely well, but they need to be carefully integrated into the whole plan of the garden and are often best confined to limited areas: on a flight of steps, perhaps, or in window boxes. The standard display of a myriad different containers, from hanging baskets to variously shaped plastic pots, filled with a jostling mass of petunias, geraniums and nasturtiums, is certainly eye-catching and has an enthusiastic following. Yet once you have learned to appreciate the more subtle charms of carefully blended, harmonizing schemes, and the sense of the delight that they bring, you may well start to think rather more selectively about flowering displays.

One of the first lessons is to look at flower form as well as flower color, and to notice how one very large, strong-colored bloom will overwhelm and detract from the charms of the quieter smaller flowers. Cottage gardens work well in this respect because they are a mass of small-flowered species plants which create a tapestry of color without any one type of flowering plant drawing attention away from the rest. Once their big brassy hybrid cousins muscle in, the charm is lost. Take heed from this in planning your container plantings, and try to choose plants that complement each other in form or color, rather than those which compete for attention and thereby spoil the potential harmony of the entire scheme.

There is a wealth of wonderful flowering plants that will grow successfully in containers, so try to experiment with some of the less obvious choices. Nonetheless, geraniums (*Pelargonium* sp.) are probably the most popular flowers of all for containers, closely followed by fuchsias and petunias. Take time to research what is available, particularly in the case of geraniums and fuchsias, as there is a mind-stretching array of different species and hybrids available, some of them with subtle but insidious charm. The appeal of scented geraniums,

TONING COLOR RANGES – COOL COLORS

Colors which have an element of blue in their make-up appear much cooler than those which have yellow or red. These pink flowers have a bluish tone, and will mix well with white, blue and mauve flowers to make a pastel display.

Violets are blue
ABOVE: *The cooler tones of mauve and deep purple harmonize in this container of petunias and begonias, backed by a display of pansies. Limiting the color range to tones of one color unifies a mixed planting.*

for example, with their ivy-leaved foliage perfumed with anything from lemons to peppermint, should not be overlooked (see page 142 for more information).

Using flower color in containers

One of the advantages of flowering plants in containers is that you can move the containers around to get the best from a particular flowering display, bringing a plant that is coming into flower to the forefront while concealing one that is past its best in the background. By using different-sized containers and plants of varying height, you can create large and eye-catching displays in a relatively limited space.

To exploit this to the full, you need a clear scheme in mind when you plant up the containers. As you would when planning a border, think about how the elements work together and try to provide a backdrop for the display that both frames it and provides a foil for the color. A dark yew or box hedge makes an excellent backdrop for containers, but so do a couple of large-leaved foliage plants in pots acting as the chorus to the more prima-donna-like flowering forms. A fence or

Pretty in pink
RIGHT: *Subtle variations in shades of pink are used in this attractive combination of hellebores, hyacinths and primroses in a terracotta container, whose deep earthy color sets off the warm pinks beautifully, giving the planting substance and depth.*

TONING COLOR RANGES – HOT COLORS

These plants have a great deal of yellow and red in their color make-up and will combine well with other yellow, red or orange plants to create a toning display.

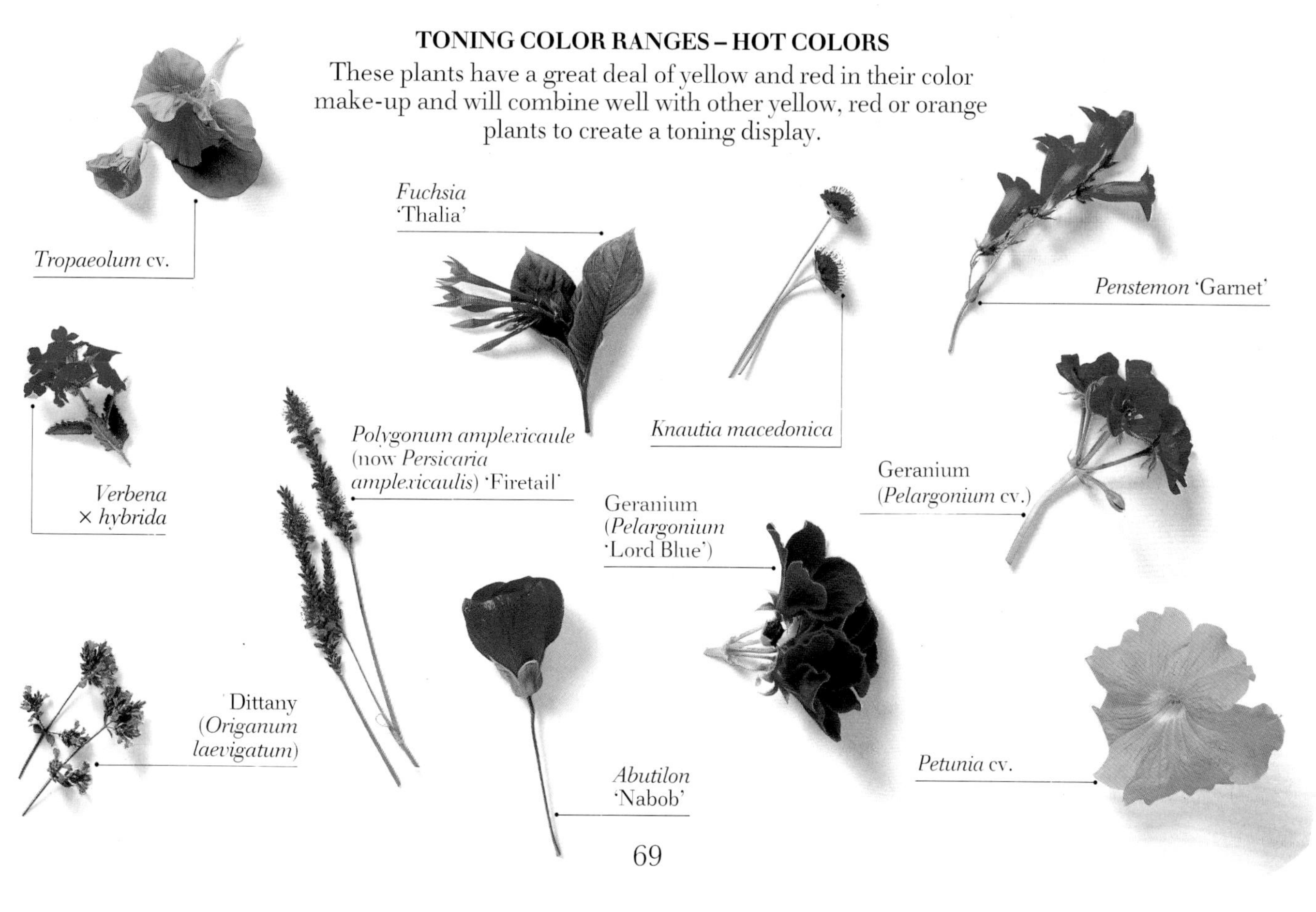

Toning colors
LEFT: *Here the pinks* (Dianthus *sp.) and a geranium (*Pelargonium *cv.) in the smaller container echo the color of the old-fashioned rose 'Roseraie de l'Haÿ' in the pot behind, both blending with the pink* Verbascum phoeniceum *hybrids in the background.*

Shades of pink
ABOVE: *In greater contrast, but still in toning colors, these Regal, Angel and Scented-leaved geranium cultivars illustrate how successfully you can vary the tonal range within a colorway to provide a scheme which harmonizes but does not lack interest.*

Color harmonies
RIGHT: *This deep pinkish-red mop-head hydrangea, whose rounded form contrasts well with the elongated spires of the white buddleja, has been planted in an ageing verdigris copper container which successfully complements the color of the flowers.*

wall provides an equally acceptable foil, but think about its color when planning the overall scheme. Enliven a dark brick wall with a gold-and-white planting scheme or echo the color of the front door in a porch display.

To achieve color at various heights, you can also use flowering climbing plants and shrubs. Remember, though, that these flower only for a limited period of time, so plan the display carefully for the maximum impact and, if possible, choose plants that offer the bonus of not only good foliage form but also an attractive flower display. Most flowering evergreens are good value from this point of view, particularly camellias with their glossy green leaves and spectacular flowers in a great range of colors, sizes and forms. Because they flower early in the year, they can get damaged by sun which scorches the frosted buds, so don't site them in an east-facing position. Like rhododendrons, camellias need a lime-free growing medium. Skimmias, hydrangeas and viburnums are all good flowering shrubs, with *Skimmia japonica* 'Rubella', for example, also offering the bonus of scent, as do some of the viburnums such as *V.* × *bodnantense.*

Among the more spectacular flowering climbers are the obvious choices such as roses and clematis, but also the less usual ones such as the brilliant orange *Campsis radicans* and *Solanum crispum* 'Glasnevin', with attractive blue flowers, *Akebia quinata* with its unusual chocolate-colored blooms, and *Rhodochiton atrosanguineus*, a tender annual with large, tubular reddish-purple bells and toothed leaves.

Some of the flowering climbers are ideally suited to scrambling over and through other plants, thereby maximizing the potential flowering display. One such plant is the flame creeper (*Tropaeolum speciosum*), aptly named for its brilliant scarlet flowers produced freely between summer and autumn. It looks particularly attractive grown over a dark evergreen, such as yew. Clematis can also be used successfully as scrambling climbers, sometimes mixed with roses. For example, the combination of a pink rose, such as the scented *Rosa* 'Zéphirine Drouhin', mixed with a pink clematis like 'Hagley Hybrid' are particularly successful. The 'Albertine' rose also makes a good weeping standard, but it is inclined to mildew if grown against a wall, so train it instead over a free-standing support.

Toning color schemes

One of the most successful ways of creating a harmonious color scheme is to opt for toning flower colors in a limited range – say, pinks and reds, or blues and mauves depending upon whether you favor a hotly colored display or a cool, serene one. Not only will toning colors unite the planting scheme, they will also focus attention on the different forms of flower. For this reason, pick candidates for this kind of scheme with quite different flower styles: big and blousy for example, combined with small and jewel-like. For a spring display a range of small-flowered narcissi could combine with yellow and cream wallflowers, backed by the yellow-splashed leaves of euonymus and *Aucuba japonica*. For summer the scheme could change to blues and mauves, with brachycome, lobelia and scabious in the smaller foreground pots, backed by powder blue lacecap hydrangeas, green-flowered euphorbias and the large-leaved hosta, *H. sieboldiana* var. *elegans*.

Focal points

As an alternative to a grouped scheme you can use some of the larger and more exotic flowering plants alone to create a focal point on a terrace, or at the front of the house. Angel's trumpets (formerly known as *Datura*, but now as *Brugmansia*), with their handsome leaves and huge, heavily scented, trumpet-shaped flowers, are a good choice, as are the big showy lilies like *Lilium regale*, again magnificently scented.

Help flowering plants that bear smaller blooms in great profusion to gain impact by training them into standard forms (see page 74) – fuchsias and honeysuckles are good examples, and even wisteria can be trained as a standard. Not only does this style create a neat architectural shape which looks good in town gardens and patios, but it also concentrates the color in one place, adding depth to the display and height where it is often most needed.

Any plant used as a focal point needs to be kept in prime condition. This involves regular dead-heading (see page 94), which will also increase the quantity of blooms and prolong the flowering season. A buddleja, for example, that is dead-headed will produce flowers from the secondary buds, so that the season is prolonged by several weeks.

Large-headed flowers

Flowers with large heads, like lilies and angel's trumpets, will need staking to ensure that they look their best, otherwise the weight of the flower heads will cause the supporting stems to bend and droop. It is important,

Tall and small
RIGHT: *The exotic flowers of* Lilium *'Pink Perfection' are partnered with the small clouds of pinkish-white flowers of the tiny daisy* Rhodanthemum gayanum. *The composition gains life and vigor from the contrast between the formal perfection of the lilies and the sprawling form of the daisies, unified by the harmonizing color scheme.*

Formal pyramid
ABOVE: *A tiny pyramid of blue grape hyacinths (*Muscari *sp.) makes a formal shape without the severity of clipped topiary. It is constructed by building a tower of pots planted with bulbs and then covering this with netting and moss.*

when staking a container plant that is used for display, to conceal the stakes as best you can, and you should choose the staking method not only to support the plant but also to ensure that it is not too obvious.

It is also vital with such key performers to remove any dead or dying flower heads and to keep the foliage in good condition. It is quite pointless to concentrate your attention on growing spectacular flowering plants if the foliage is yellowing through neglect.

Bushy flowering plants, like chrysanthemums or geraniums, will also benefit from discreet staking and supporting, and you can use canes, link stakes or brushwood for the purpose. Even quite small plants benefit from a few small twiggy stakes pushed down between the stems, to persuade the flowers to drape themselves more attractively (see page 95).

Small-headed flowers

Big is not necessarily beautiful in the world of flowering plants, and you can gain a great deal of pleasure from the sight of clouds of exquisitely formed, small flowers. Among the most enchanting are the small-flowered species clematis which are becoming increasingly popular. They make a wonderful background to a display of other delicate flowers, such as penstemons, diascias and tobacco plants (*Nicotiana* cvs.).

Among the attractive small-headed flowers are most of the old-fashioned species plants so beloved of cottage gardeners, including many of the daisies, as well as those with spires of small flowers that make, in their own way, as big an impact as some of the plants with showier blooms. Larkspur, lupins and stocks, as well as some of the bellflowers (*Campanula* spp.), fall into this category. On the whole their flower color is softer and less intense, but a display containing a mixture of these plants gains from an orchestrated palette in mauves, pinks, blues and whites, or in creams, golds, oranges, reds and purples. You can either grow the plants together in a large container (see pages 38–41) or make a border-style planting in miniature if you group containers each with a single-theme planting. The latter gives you more scope to move the containers around as the plants come into full flower.

In spring there is a wealth of small-flowered plants to choose from, including primulas and pansies, and the ubiquitous spring bulbs.

Growing roses in containers

Roses are a must for many gardeners, and there is nothing to stop you growing them successfully in containers – that is, provided you look after them properly and position the containers where they will be in the sun for at least half the day.

STAKING LARGE-HEADED FLOWERS

Flowers with large blooms and relatively weak stems require some form of staking if they are to look their best and remain upright. Try to insert the stakes as early in the growing season as possible, when you are less likely to damage the rootball of the plant. The following steps show how to plant up and stake bulbous, tall-stemmed plants – in this instance, Casa Blanca lilies.

1 *Select a large pot – one about 35cm (14in) in diameter should be sufficient to hold three bulbs – and start by covering the base thoroughly with a layer of pebbles or broken pots.*

2 *Partially fill the pot with growing medium, add a layer of coarse sand or fine gravel and plant the bulbs two and a half times their depth from the top of the pot, spacing them evenly.*

3 *Insert four or five canes, 1.2m (4ft) long (or slightly shorter than the eventual height of the plant) around the pot, deep enough to ensure that they are stable.*

4 *Fill the pot with growing medium to within 2.5cm (1in) of the top. When the stems reach 15cm (6in), tie them in to the wires.*

'Casa Blanca' lilies
These are fully grown. For tall varieties, tie in the stems as the plant grows.

Unfortunately the foliage and stems of many of the older roses (which usually have the best scent) are not particularly exciting to look at and the fact that they are in a container tends to lift the least attractive elements of the plant – the bases of the stems – into too prominent a position where they can detract from the display. Since they have a rather short flowering season, except for those varieties that are repeat-flowering, you will be left for quite a long time with a less-than-beautiful plant. The solution is to position some other, slightly lower-growing plants in front of such roses, or to grow the much smaller Patio (or Dwarf Cluster-flowered) roses which have been developed in recent years, and move them out of the focal position once they have bloomed. The latter bear attractive and abundant flowers in a range of colors, but unfortunately many of them have very little scent, which is one of the chief glories of the rose.

If you plant roses in containers, you are going to have to ensure that they are in prime condition, free of disfiguring black spot and mildew, and certainly flowering as profusely as possible, and pruned regularly to this end. They are not the ideal choice for the *laissez-faire*, low-maintenance style of container gardening, so if this is your style you would actually be much better off opting for drought-resistant, easy-to-care-for flowering plants like geraniums (*Pelargonium* cvs.) which will cope with most forms of neglect apart from over-enthusiastic watering or deep shade.

One of the best ways to deal with roses in containers is to grow them as standards. This limits the unattractive stem element to a single supporting stem and concentrates the flower color where you really need it – at the top of the plant. in one large mass. There are various roses that do well as standards, including the aforementioned 'Zéphirine Drouhin'. Another is 'White Pet,' with its abundance of small white flowers. You can also have a weeping standard, which is created by grafting a rambling cultivar on to a non-climbing rootstock, in which the flowering stems droop attractively towards the ground from a height of 1.5m (5ft) or so, or you can train a climbing rose with the addition of a stake to support the base of the stem. These then create a more bushy standard form.

Growing clematis in containers

The other two great loves of flower gardeners are clematis, dealt with here, and fuchsias, discussed opposite. Most of the clematis insist on having their roots in cool shade in order to thrive, which means that you need to be careful where you site the container and how you deal with the surface of the pot. An array of large pebbles over the soil surface will help to prevent the sun scorching the roots. The choice of clematis for containers varies from the large-flowered hybrids to the small-flowered species, and there are devotees of both kinds. Since the different forms flower at different seasons, you can pick a clematis to go with a flowering

TRAINING A STANDARD

Both foliage plants, such as the variegated Coprosma *shown here, and flowering plants can be trained as standards. A single stem is supported to a specific height and then encouraged to form its bushy shape.*

1 *Allow the leading stem to grow, and trim off side shoots to a couple of leaves. Wait until the leading stem has reached the required height.*

2 *Cut off the growing tip of the leading stem in order to encourage the secondary buds to develop and begin to bush out.*

3 *Support the stem of the plant with a cane pushed into the growing medium and tied loosely in to the main stem with green plastic ties.*

The young Coprosma × *'Kirkii Variegata', with the top growth beginning to fan out into a neat standard shape.*

display from spring through to autumn. Among the early-flowering clematis are those of the *C. montana* group, with many of the larger hybrids flowering in summer, followed by some of the viticella group in late summer (see also page 19).

Growing fuchsias in containers

The different cultivars of fuchsia are also a rich source of inspiration for gardeners. Their attractive habit and long flowering season, which extends right through summer and well into autumn, makes them ideal for growing in containers as they can be relied upon to create a delightful and lengthy display. Like roses they can be trained with a single supporting stem to create extremely effective, weeping standards. A pair of standard fuchsias in containers is an ideal choice for positioning on each side of a porch or entrance. You create the standard in the usual way from one of the upright bush-type fuchsias. Support the leading stem and pinch out any emerging side shoots to one pair of leaves until the leading stem has reached the required height – normally about 1m (3ft) or so. When two or three pairs of leaves are produced above the required height on the leading stem, pinch them out and leave the plant to develop naturally, so that it creates a bushy shrub on a slender supported stem (see also below left).

The different colors of fuchsia range from white, such as *F.* 'Annabel', through pale pinks, like *F.* 'Pink Galore', to pinky-mauves, such as *F.* 'Leonora', to deep orange-reds, like the attractively elongated flowers of *F.* 'Thalia'. Mixtures of these colors can be found also, such as the mauve-and-white flowers of *F.* 'Estelle Marie' or the large double flowers in red and purple of *F.* 'Dollar Princess'.

Flowering standards

LEFT: *Among the range of flowering plants that can be encouraged to form standards are daisy-like marguerites* (Argyranthemum frutescens), *which are trained so that their natural bushiness is allowed to form only after the main stem has already grown to about 1.2m (4ft) in height. Here the plant's neat formality is complemented with an elegantly striped container in green and white.*

CARE AND MAINTENANCE

THIS CHAPTER covers all the basic practical information you require in order to maintain both plants and containers in good condition – from choosing healthy plants to recognizing and treating common ailments.

Sound, sensible practice in looking after container plants is an important element in keeping a display in peak condition, and this section covers every aspect of care, including how to pot plants, which growing medium to use, how to feed and water plants, how to keep them tidy through staking and dead-heading, how to prune them in order to promote healthy growth and how to increase your stock of plants through propagation – either from seed or from cuttings. In addition, advice is given on how to look after your plants in severe weather conditions and when you are away on vacation. Also covered are general tips on container care – from cleaning and repairing pots to moving heavy containers.

The Plant Directory on pages 106–155 provides specific cultivation information for individual named plants.

LEFT: *The pale pink flowers of* Osteospermum jucundum *are borne in great abundance during late summer.*

ABOVE: *Two formal Versailles tubs are filled with topiary balls* (Ligustrum delavayanum), *petunias and variegated ivies.*

Buying and Setting Up

When you start gardening with containers, you need to have some idea of the plants' requirements as well as your own. Committed gardeners may know what is required and can skip this section, but if you are new to gardening, you will want to set yourself up with healthy, well-selected plants and some basic equipment to maintain them.

Although you can grow your own plants from cuttings or from seed, most people tend to buy them as young container-grown plants, from garden centers. When choosing plants, look for those that are suitable for the purpose you have in mind, rather than simply selecting the most obviously attractive. Take the time and trouble to check the conditions that each plant requires: for example, sun/shade, temperature and situation. The label should explain everything you need to know. You will not be able satisfactorily to plant up a container with plants from disparate regions that require diametrically different growing conditions: hot sun- and drought-loving silver-leaved helichrysums,

CHOOSING EQUIPMENT

The equipment below is a suggested range of basic tools for container gardening. As a general rule, go for the best quality equipment which will last longer and do the job better.

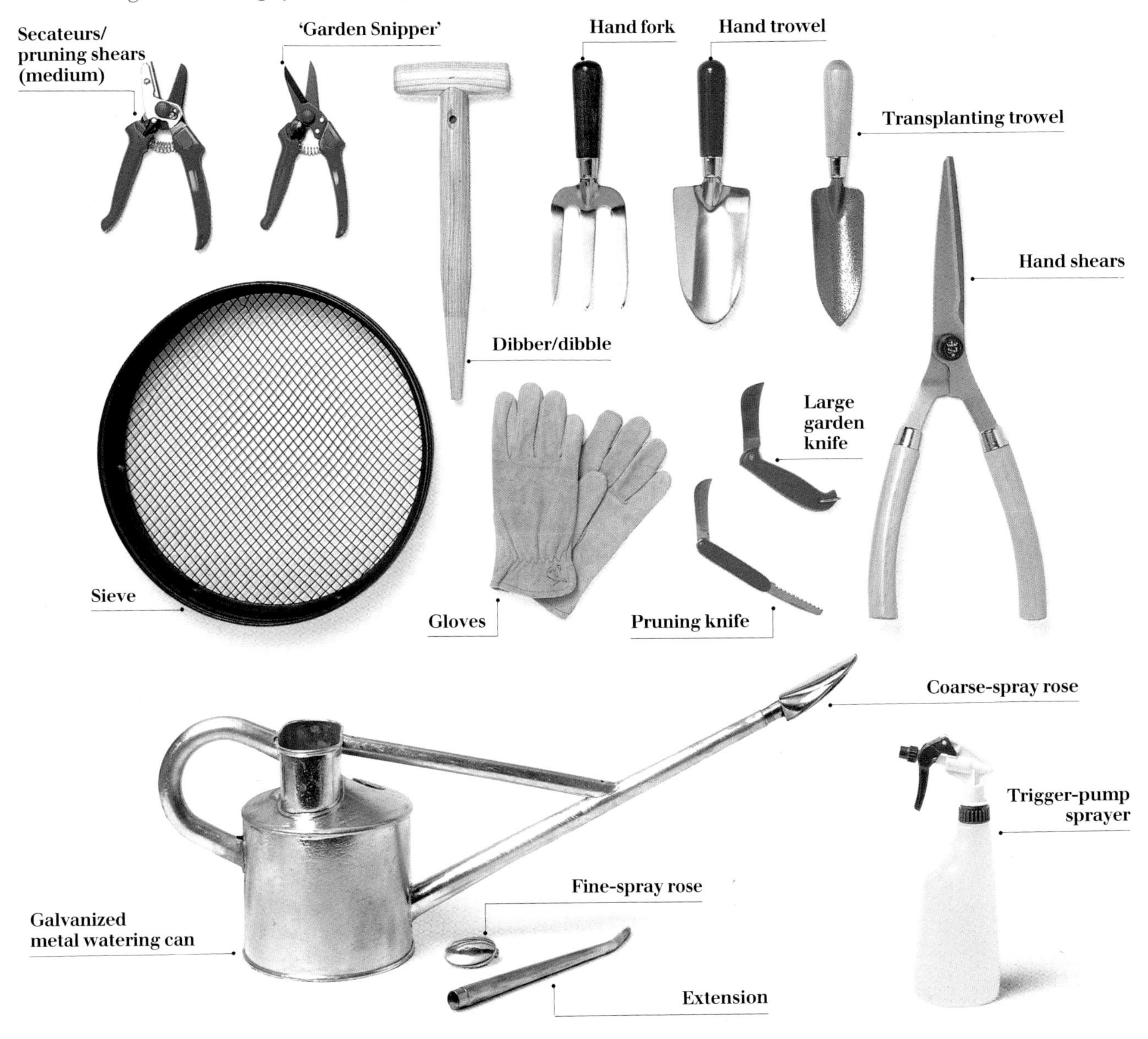

for example, with bog plants like *Zantedeschia aethiopica*, although you can grow both in separate containers in sun and shade respectively.

Setting up a workspace

One of the main problems for container gardeners is lack of space since container gardening is often the solution to high-rise gardening on roof terraces or balconies. This means that you have to be economical with tools and other equipment, because you will probably lack storage space for them. The illustration opposite shows a range of basic tools and if you have the space you can certainly add to these. It is a good idea to devote an area to garden equipment, with storage boxes, shelves or drawers. You can even build seating which doubles up as a storage box.

Depending on the kind of garden you wish to create, you may also need more specialized equipment. For example, a sunny roof terrace could well benefit from a permanently installed drip-watering system, which greatly reduces the chore of watering. Equally, if you want to propagate your own plants, you are probably going to need a propagating unit.

STAKING EQUIPMENT

It is important when growing plants in containers to display them to advantage and to this end you may have to stake them. The method you use depends on the habit of the plant, but it is worth keeping a few canes and ties handy. More details are given on page 95.

Split canes

Bamboo canes

Oiled garden string

Green garden string

Plastic ties

Aluminum labels

Garden wire

Plastic labels

CHOOSING A HEALTHY PLANT

Not all garden centers and nurseries look after their plants as well as they might. When buying plants, it is important to select healthy stock. The captions to the plant shown below illustrate the main points to look for.

Stems
The plant should have strong healthy stems, with no signs of disease or mildew. The overall shape of the plant is important and the leading stem should be strong and well formed. Check for signs of bruising or tearing.

Leaves
Check the leaves for signs of pest damage, and diseases such as mildew (see pages 104–105). The leaves should not be discolored or drooping. Turn the leaves over to check for pests.

Base
The roots of any container-grown plant are of necessity hidden in the container. If the plant looks at all unhealthy, turn the container upside down and check to see if the roots are growing out through the holes in the base. If they are, the plant is probably pot-bound – that is, grown too large for its pot (see Potting on, pages 82–84).

CLEANING LEAVES

In heavily polluted areas it pays to clean the leaves to get rid of grit and grime. Use a damp rag or sponge to wipe them, then spray with a proprietary leaf shine if you wish.

Growing in Containers

MANY OF THE PLANTS grown in containers originate in parts of the world where diverse growing conditions exist. In their natural environment some are exposed to intense light and heat, others are protected by neighboring plants or given shade by overhanging trees, while others live in cool swampy conditions. This diversity of habitat explains why the plants we grow have different cultivation requirements. Some plants are much more resilient than others and can adapt more readily to different conditions, but in order to thrive rather than simply survive all plants need adequate light, a suitable temperature range, the right levels of nutrients and moisture, and possibly a dormant period to permit rest.

Light

Light is essential to all plants. Without sufficient light, leaves become small and pale, and the overall growth of the plant is retarded. Healthy plants depend on the process known as photosynthesis, in which their green parts react to light energy to produce carbohydrates and oxygen from water and carbon dioxide.

HOW THE PLANT GROWS

Roots, leaves, stems and even the growing medium all play a part in the healthy growth of a plant, as do external factors such as light and temperature. Successful cultivation depends on a proper balance between these factors.

Stems
Nutrients taken in by the leaves and up by the roots are transported along the stems to other parts of the plant. The substance in which nutrients are carried, known as phloem, is made up mainly of water. During periods of drought this activity is disrupted, and since container plants dry out faster than garden plants, regular watering is especially important for plant health (see pages 88–89).

Leaves
During daylight hours, carbon dioxide is taken from the air through the pores (stomata) on the undersides of the leaves, and light energy absorbed through the green pigment (chlorophyll) in the leaves is used to split water molecules into hydrogen and oxygen. The hydrogen combines with the carbon dioxide taken in by the pores to form carbohydrates, providing the plant with food.

The growing medium
A good-quality planting medium is essential for the plant's well-being and, since nutrients and water leach rapidly from containers, it is important that you maintain the correct balance of nutrients and water at all times (see pages 85–89).

Roots
The functions of the roots are to anchor the plant in the growing medium, to absorb water and to supply mineral salts used by the plant in the manufacture of food. The thin thread-like roots in the rootball are the feeding roots, absorbing nutrients and water from the planting medium, while the thicker older roots transport these nutrients. Any restriction to the roots will limit the extent of the top growth.

STOMATA

Each leaf is made up of literally thousands of tiny pores (stomata), like human skin, through which light and water are passed, and via which the process of photosynthesis takes place.

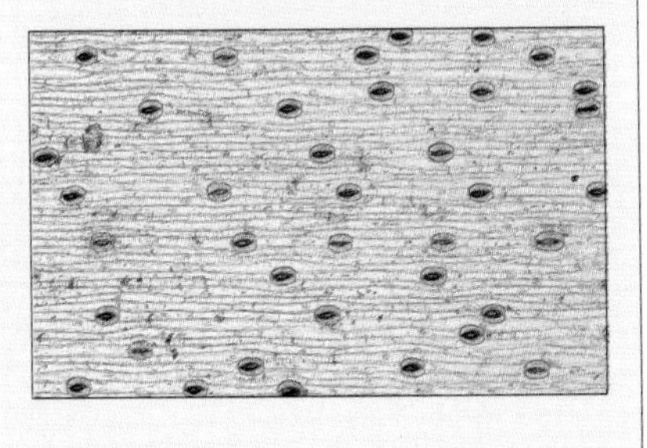

Light also influences flowering periods: many plants respond to the amount of daylight within a 24-hour daily cycle. Long-day plants flower only when daylight lasts for more than 12 out of 24 hours. Short-day plants flower only when daylight is limited to less than 12 hours. Nowadays, this reaction is manipulated to induce forced flowering in certain varieties of plant, such as chrysanthemums and poinsettias.

Light, or the lack of it, is the main stimulus for leaf fall in deciduous plants, the trigger being the shortening days of autumn. It also causes a response known as phototropism, in which the plant grows toward the light, and this is one of the reasons why plants adopt a one-sided shape when grown in front of a wall or fence.

Water

A plant is composed of over 90 per cent water, and loses a great deal of this through its leaves – a process known as transpiration. A mature lettuce, for example, loses half its own weight of water every summer's day. Failure to water a plant adequately reduces its growth potential, in part because water is the medium by which nutrients are passed from one part of the plant to another. The water appears as a continuous column, and if this column breaks through water shortage, the plant will simply stop growing until recovery occurs.

The container

Plants in containers, unlike plants in the garden, have their roots imprisoned by the pot. It is important therefore to ensure that the container is large enough to give the roots adequate space to breathe and grow, otherwise the plant will ail and eventually die. The illustration below gives some indication of the ideal container size in relation to the rootball of the plant.

Temperature

All plants have an optimum temperature range in which they thrive and a wider one that they will tolerate. The majority of popular plants grown outdoors in containers originate from temperate and sub-tropical areas where the temperature range is around 10 to 18°C (50 to 64°F). Although these plants dislike widely fluctuating temperatures, they can tolerate lower or higher temperatures for short periods of time. Frost damage and low temperatures affect the roots, and plants in containers are particularly vulnerable in this respect, mainly because the soil in a container chills more rapidly than that in open ground. Factors other than low temperatures will have a considerable effect on root death of plants grown in containers, such as the period of time that the temperature is low and the general health of the plant at the time.

Root-killing temperatures for container-grown ornamentals standing outdoors over winter:

0 to -5°C/32 to 23°F	-5 to -10°C/23 to 14°F	-10 to - 15°C/14 to 5°F
Buxus sempervirens	*Acer palmatum* 'Atropurpureum'	*Juniperus* spp.
Daphne cneorum	*Cornus florida*	*Leucothöe fontanesiana*
Hypericum spp.	*Magnolia stellata*	*Mahonia aquifolium*
Mahonia bealei	*Pieris japonica* cvs.	*Pieris floribunda* cvs.
Pyracantha coccinea	*Viburnum carlesii*	

POT-TO-ROOTBALL RATIOS

Plants need to be potted in a container roughly 5cm (2in) larger in diameter than the rootball. As the plants grow, they will need to be potted on into larger containers, unless you wish to prevent them increasing in size, in which case the roots will need pruning (see page 84).

A small plant like a young geranium (Pelargonium) *with a rootball roughly 10cm (4in) in diameter can be potted into a 15cm (6in) diameter pot.*

A medium-sized plant like a mature lavender with a rootball roughly 15cm (6in) in diameter can be potted into a 20cm (8in) diameter pot.

A large plant like a mature hosta with a rootball roughly 20cm (8in) in diameter can be potted into a 25cm (10in) diameter pot.

Potting and Repotting

THE WAY PLANTS ARE POTTED can affect their growth as much as feeding, watering and pruning. There are a number of closely related operations and the terminology can be confusing. The term planting is used for placing established plants in containers, as a permanent site. Planting up means planting young plants, bulbs or tubers in containers. Repotting means replenishing the growing medium in an existing container, and potting on is the term for transferring a growing plant to a larger container.

PLANTING UP BULBS

Plant spring-flowering bulbs six months before flowering and summer-flowering bulbs three months before flowering.

1 *Fill the pot with 50:50 peat (or peat substitute) and vermiculite. Place the bulb roots down inside.*

2 *Cover with growing medium to a depth of three times that of the bulb (see the table below).*

3 *Water at planting time, and keep moist until six weeks after flowering, then lift the bulbs and dry them for winter.*

Planting

A plant should be placed in a container that is at least 5cm (2in) larger than its rootball (see page 81). The container must have drainage holes in its base, and you should incorporate a layer of pebbles or shards of pots at the bottom to help drainage take place. Above that put in a layer of growing medium before inserting the plant (see left for planting up bulbs and right for planting container-grown shrubs, trees or perennials).

Repotting

Much depends on the plant's rate of growth, but most plants will need to be repotted in fresh growing medium every other year. Follow the instructions opposite for planting up, but return the plant to a container of the same size, filling it with planting medium to a level 1.5cm (½in) higher than previously to allow for settlement. As an interim measure you can replace only the top few centimeters (1in) of growing medium and add fertilizer, which will help to rejuvenate the plant. Annual repotting of vigorous trees and shrubs should maintain healthy growth without encouraging excessive vigor. However, if you notice the plant is ailing, check that it is not pot-bound (see page 84) and follow the measures suggested.

Potting on

Young plants that have still to reach maturity will need to be potted on into larger pots at regular intervals, the

PLANTING DEPTHS FOR BULBS

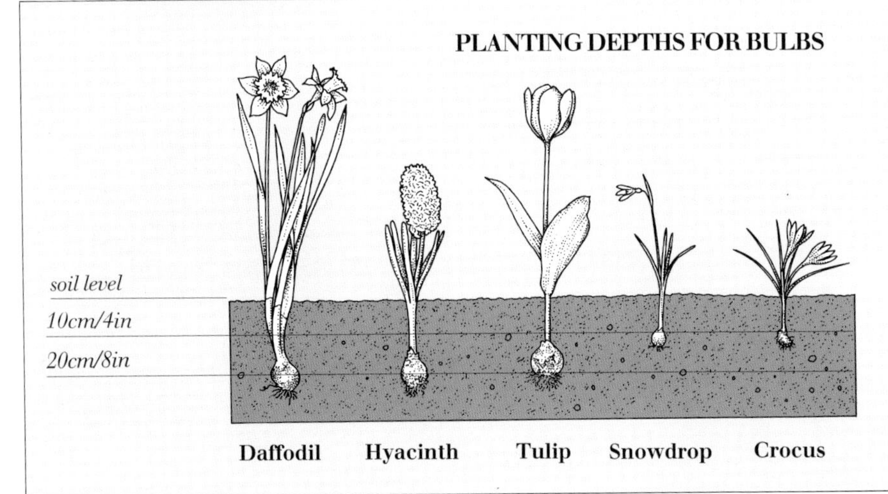

Bulbs vary in size from the tiniest snowdrops to large hyacinths, as do corms. For optimum growth they should be covered in three times their depth of soil. This means that their base is at a depth equal to four times their own depth. Most bulbs prefer free-draining soil, so incorporate a layer of pebbles in the base of the pot, and include a good proportion of grit or coarse sand.

frequency depending upon their natural rate of growth and the conditions in which they are kept. It is normally best to pot on a plant into a pot slightly – about 5cm (2in) – larger than the current one rather than suddenly moving it from constricted surroundings to a much larger container, which tends to cause an imbalance in the root-to-canopy (leaf spread) ratio. In slow-growing plants, such as dwarf conifers, the roots do not spread through the growing medium quickly enough and in vigorous plants you tend to get large quantities of soft leaf growth and delayed flowering.

A plant with a rootball 9cm (3½in) in diameter should be potted on into a container approximately 13cm (5in) in diameter. See the instructions given on page 84 for increasing plant growth.

Controlling plant growth

If a plant is growing too vigorously, you can curb its growth by root pruning. If you simply leave a vigorously growing plant in a pot, it will eventually become pot-bound as the roots run out of space to develop, become entangled, and fail to take up water and nutrients.

PLANTING UP A LARGE SHRUB OR TREE

Large shrubs and trees should be planted in a container roughly 5cm (2in) larger than the rootball of the plant. To support the plant and prevent wind rock, insert a strong stake or cane deep into the growing medium and secure it to the plant with ties (see staking a single-stemmed plant, page 95).

1 Cover the base of the container with a thin layer of pebbles or broken pottery. Place enough growing medium in the bottom to allow the top of the plant's rootball to lie about 2.5cm (1in) below the rim of the pot.

2 Place the plant carefully in the pot, taking care not to damage the rootball, and gently spread the roots out if necessary.

3 Add more growing medium around the sides of the rootball, making sure the plant is positioned centrally. Insert a support if necessary (see page 95) and tie it loosely to the plant.

4 Fill the pot with growing medium to within 2.5cm (1in) of the rim and firm in well with your hands. Water generously to ensure that the roots are able to take up moisture.

The planted apple tree

Unattended to, the plant will ail and eventually die. Obvious signs of a pot-bound plant are evident when the pot starts to dry out more and more rapidly, and when nutrient starvation produces hard stunted growth of the stems and leaves. In extreme cases, there will be premature leaf-drop, loss of vigor and die-back, followed by the death of the entire plant.

To curb the growth of an over-vigorous plant, you will have to prune the roots (see below). It should be remembered that if large sections of root are to be removed, the plant will need time to recover or some of the top growth may die back.

WHEN TO POT, REPOT OR POT ON

The ideal time for planting up, potting on or repotting should be related to the flowering period of the plant. Plants that grow quickly will need to be potted on more frequently than those that grow slowly, and if the plant is obviously under stress it should be repotted at the times indicated below.

Plant type	Best time for potting
Aquatic and marginal plants	late spring
Broad-leaved evergreen shrubs	
(spring-flowering)	autumn
(autumn- and winter-flowering)	spring
Bulbs	
(spring-flowering)	autumn
(summer-flowering)	winter/spring
Conifers	autumn/spring
Deciduous trees and shrubs	
(spring-/summer- or winter-flowering)	autumn
(autumn-flowering)	spring
Herbaceous perennials	autumn/spring
Hardy biennials	
(spring-/summer-flowering)	autumn
Summer bedding plants/annuals	spring (after late frosts)

INCREASING GROWTH

Young and pot-bound plants will need to be potted on to make sure that they reach their optimum size.

1 *Turn the pot on one side. Shake the pot to loosen the roots and trim any roots poking through the pot. Remove the plant from the container.*

2 *Using your fingers, gently tease out any enmeshed roots, taking care not to damage or tear the very fine roots which perform the feeding function.*

3 *Return the plant to a larger pot (normally about 5cm/2in wider in diameter than the rootball) and pot up in fresh growing medium, as shown on page 83.*

REDUCING GROWTH

Plants in containers occasionally need to be checked from growing too freely. The solution to this problem is to trim the roots, since it is the root growth that determines the size of the top growth of the plant.

1 *Remove the plant from the container. Shake off excess growing medium, then tease out the rootball to separate out the roots.*

2 *Start to cut away up to one-third of the main root growth using a pair of sharp secateurs/pruning shears.*

3 *When you have trimmed the roots, repot the plant in fresh growing medium in the same size container (see page 83.*

The Growing Medium

THE PLANTING MEDIUM used in containers must have certain desirable characteristics – it should crumble easily in the hand, retain moisture while permitting free drainage, and have the correct balance of nutrients and the appropriate level of lime or acidity for the chosen plants. For this reason most gardeners prefer to use a purpose-made growing medium for their containers. Remember that while soil looks inert, it is not. Growing plants will take nutrients from it, which must be replaced by you, the gardener.

Purpose-made growing mediums fall into two groups: loam-less (usually peat-based) and loam-based (soil-based). The most popular commercial general-purpose potting mediums are peat-based mixtures containing peat (or peat-substitute), grit or gravel, and slow-release fertilizer. Their main advantage for container-gardeners is that they are light and clean to use. However, they do dry out quickly and, as a result, lose their nutrients rapidly.

Loam-based growing mediums provide more stability and support for the plant than peat-based mixtures. However, they are heavier and not so clean to handle.

Growers who want to make their own loam-based mixtures can use 1 part rich, sterile loam; 1 part sterile garden compost, peat or peat substitute; and 1 part grit or gravel. To each gallon of this mixture, add 1tbsp (15ml) lime, 1tbsp (15ml) bonemeal, and ¼ cup dried manure. To sterilize the soil bake in the oven at 180°F. If you are growing acid-loving plants, use more peat and omit the lime.

Soil acidity is an important factor in plant health. It is measured on a scale of pH which ranges from 0 to 14. A neutral soil, suitable for most plants, measures about pH 6.5. You can take a reading, using a pH kit (available from garden centers), and can make the soil more alkaline (that is, with a higher pH) by adding lime or chalk-based fertilizers, or make it more acidic (that is, with a lower pH) by adding peat.

PEAT-BASED GROWING MEDIUMS

These are produced from three principal ingredients: peat (or peat substitute), base fertilizer and coarse grit or gravel. The three components, and the blended growing medium, are shown below.

SOIL pH TEST

Soil pH testing kits contain a tube in which you put a sample of soil, and some liquid which you add to the soil. You shake this up and allow it to settle and the liquid will change color – from green for alkaline to yellow for neutral and red for acid soils.

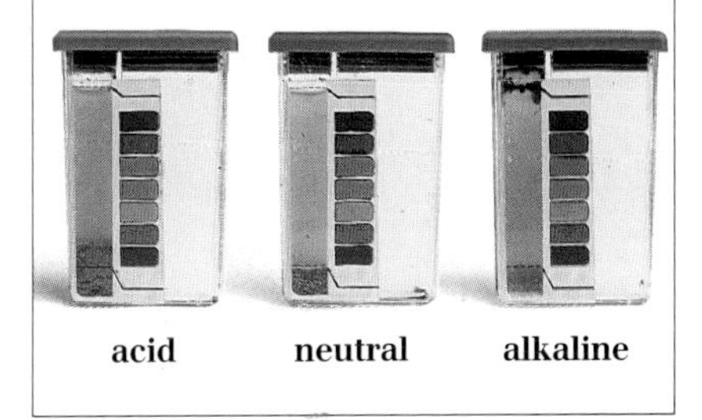

PEAT AND PEAT SUBSTITUTES

Although Canadian peat is still readily available, in other areas peat bogs have been so severely depleted that many gardeners prefer to use a more environmentally sensitive alternative. The illustrations below show other components of peat-based planting mixtures.

Shredded bark

Feeding Plants

IN THE NATURAL WORLD, plants obtain all the food they need from the soil, where the cycle of plant growth and decay creates all the minerals and nutrients required for survival. Gardening, especially gardening in containers, disrupts this natural balance, and it becomes imperative to add nutrients to the medium in which the plants are growing, since the plant remains are not returned to the planting medium in order to replenish it.

Types of fertilizer

Plant nutrients can be obtained in the form of proprietary fertilizers, which are grouped into two principal kinds: organic, from animal or plant substances; or inorganic, from mineral sources. Organic fertilizers are generally slower-acting but last longer – mixtures containing blood, fish and bone are typical examples of concentrated organic fertilizer. Fertilizers can be released into the soil rapidly (quick-release) or slowly (slow-release), depending on their solubility in water. The rate at which the nutrients are released also depends on soil-borne organisms, which are less active in cold weather, so nutrient release tends to slow down around then. Release rate is also linked to the particle size of the fertilizers, so the smaller they are, the more rapidly they will be released and absorbed.

To promote balanced growth in a plant, always start by using a potting medium of a known type and strength (see page 85); proprietary mixtures already contain measured amounts of slow-release fertilizers. Any supplementary feeding you do can be applied as quick-acting fertilizer in the form of top dressing. Be aware that as much as one-third of all nutrients are lost when water drains from the container.

Release rates for fertilizers

Fertilizer type	Plant response
Slow-release	14–21 days
Quick-acting	7–10 days
Liquid feed	5–7 days
Foliar feed	3–4 days

Most proprietary fertilizers contain a mixture of the main nutrients – nitrogen (N), phosphorus (P) and potassium (K) – and these are normally indicated on the bag, along with the relative proportions of each. Although these commercially mixed fertilizers are fine for most plants, bear in mind that container plants

FORMS OF DRY FEED

The illustrations below show the various forms of dry fertilizer. Most of these are highly concentrated and should be applied with care.

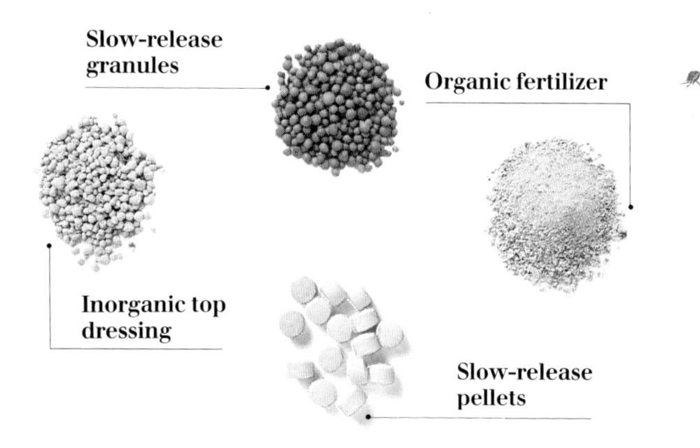

LIQUID FEEDS

When mixing liquid feeds, follow the manufacturer's guidelines and do not attempt to give extra 'for luck'; it may do active harm.

Mixing liquid feeds
Pour the concentrated liquid feed into a watering can, and dilute with water to the appropriate strength, as directed by the manufacturer's instructions. Stir with a clean stick before use.

Foliar feeds
Mix the feed with water, if necessary, as above.

Spray the leaves of the plant evenly, preferably on a cloudy day. Spraying in sunshine will encourage unsightly leaf scorch.

grown for foliage do better when given extra nitrogen, while flowering plants may need more potash and nitrogen. In general, choose a balanced fertilizer in the growing medium and top up the nutrient content with liquid feeds.

Loam/soil-based growing mediums contain sufficient residual trace elements to support plant growth, whereas loamless growing mediums, such as those based on peat and coir, have no residual trace elements, and these will have to be added.

Forms of fertilizer

Fertilizers are available in two main forms: dry and liquid. Dry fertilizers (shown on page 86, bottom) are used in the base dressing of the growing medium or they can be added during the growing season to top dress permanently established plants in large pots. Carefully measured top dressings of quick-acting inorganic fertilizers can be applied to shrubs, or to hungry feeders like chrysanthemums, in the growing season.

You can buy liquid feeds either as liquid concentrate or as granules or powder which are dissolved in water (see left). The latter are applied by watering can or hose, and are ideal for container plants.

Foliar feeds are used to give instant supplies of specific nutrients as and when needed, or to correct specific deficiencies. Foliar feeds containing magnesium are used to assist fruiting and feeds with iron are useful for acid-loving plants like azaleas, which are susceptible to iron deficiency when grown in containers.

HOW BULBS STORE FOOD

Bulbs, corms and tubers store food in the swollen part of the root or stem; other plants rely on drawing nutrients from the soil as needed. In order for bulbs to lay down these stored nutrients, on which the next year's flowers depend, they need to be fed after flowering. Failure to provide nutrients at this point will lead to poor or non-flowering the following year.

Flowering shoot
The main bud which forms the leaves and flower of the daffodil.

Bulb
This is an underground stem made up of food-storing tissue from which a flowering shoot will emerge, after the previous year's shoot has died.

Roots
These take in nutrients from the soil and pass them up to the underground stem (bulb) for storage. They also serve to anchor the bulb in the soil.

COMMON NUTRIENT DEFICIENCIES

The deficiencies described below are among those most commonly found in plants. First identify the cause, then apply the appropriate remedy.

Nitrogen
Symptoms Overall growth is reduced. Whole plant eventually becomes spindly. Results in pale-green leaves, which sometimes develop yellow or even pink tints.
Cause Planting in restricted surroundings, such as hanging baskets or window boxes, where soil is poor.
Control Add a high-nitrogen fertilizer, such as sulphate of ammonia or dried blood.

Phosphate
Symptoms Plant growth is slow and young foliage may appear yellow or dull.
Cause Leaching out of phosphates in acidic growing mediums, so acid-loving plants like azaleas are most likely to be affected.
Control Apply bone meal or superphosphate.

Potassium
Symptoms The foliage turns purple, yellow or blue, with a brown discoloration (necrosis) either in blotches or on the leaf tips and margins. Some leaves may curl downward, and they are soft and prone to attack by pests and diseases. Fruiting, flowering and general growth may be reduced.
Cause Growing plants in planting medium with either a light texture or a high alkalide or peat content.
Control Dress with sulphate of potash or a high-potash fertilizer, then make sure that the plant receives proper feeding thereafter.

Magnesium
Symptoms Distinct areas of discoloration, usually yellow or reddish brown, that develop between the veins on older leaves (known as intervinal chlorosis), which may then fall prematurely.
Cause Acid planting mediums and/or excessive watering can leach out magnesium; high potassium levels can lock up the magnesium and make it unavailable to the plant. Plants fed on high-potash feeds are susceptible.
Control Treat plants and growing medium with Epsom salts in autumn, either added undiluted to the soil or as a foliar feed, including a wetting agent such as soft soap.

Manganese and iron
Symptoms Yellowing or browning of leaves starting at the leaf margins and extending between the veins.
Cause Cultivation in unsuitable growing mediums; acid-loving plants are unable to absorb adequate trace elements from mixtures with a high pH (present in hard water).
Control Use rain water. Acidify the growing medium before or after planting and apply trace elements and sequestered iron.

Watering Plants

PLANTS OBTAIN THEIR WATER and nutrients mainly through their roots, but a small proportion is taken in through the leaves. When plants are grown in containers, their roots are restricted in their search for water and food, and the gardener has a role to play in overcoming this problem. To give you some idea of what is required, a container 1m (39in) in diameter, which is fully covered with growing plants, can lose up to 6 liters (1.3 gallons) of water on a hot, sunny, windy day, and the growing medium can rapidly dry out. The water is lost through transpiration (see page 81) and evaporation from the surface of the container. This rate of loss explains why efficient watering of plants in containers is so essential to their health.

CONTAINER SHAPE

The shape of some containers has an effect on how quickly the growing medium will dry out (see Container shape and construction, page 89).

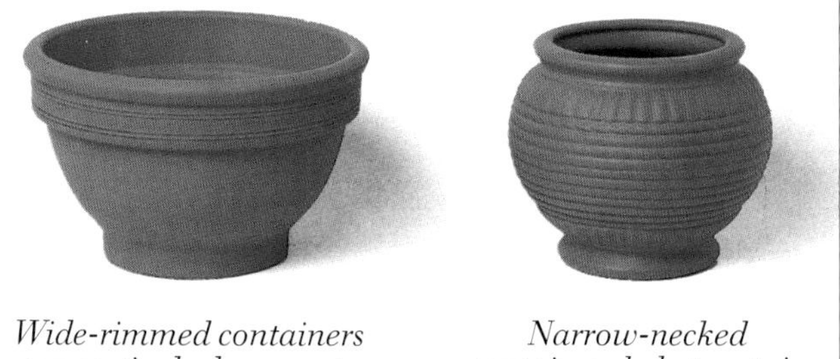

Wide-rimmed containers are particularly prone to drying out.

Narrow-necked containers help to retain moisture.

How to water

You must make sure that the water you supply the plant with penetrates right down to the roots, rather than simply wetting the surface of the growing medium. In most cases it is advisable to fill the container with water to the rim and let it slowly permeate down through the growing medium. Any surplus will simply drain away. A prolonged soaking will do more good than a light wetting, which simply encourages surface rooting and so makes the plant susceptible to drought. If sufficient water is not provided, the growing medium will shrink away from the sides of the container as it dries out. The next time you water the plant, the water will flow down the sides of the container and drain away without really soaking into the growing medium. The best remedy if this occurs is to insert canes into the surface of the soil, as illustrated below. Another useful tip when watering is to put a very small drop of washing-up liquid/detergent into a large container with about 9 liters (2 gallons) of water. This increases the ability of the water to stick to and penetrate the dry growing medium. With smaller containers, stand them in a tray of shallow water and the plant will gradually soak up the water.

Dangers of under- and overwatering

If you fail to water your plants on a regular basis, you subject them to stress which can damage the plant and

WATERING PROBLEMS

Over- and underwatering are common problems with container-grown plants. To compensate for over-or underwatering, follow the procedures shown below.

An overwatered plant should be tipped on its side to allow surplus water to drain out.

An underwatered plant will wilt badly before it dies. Revive by plunging the container in a bucket of water, and leaving it until air bubbles subside.

If a container is underwatered, the water will not penetrate the plant's roots easily. Use a cane to insert holes in the growing medium before watering.

KEEPING PLANTS MOIST

Various devices can be employed to help reduce water loss if you have to go away. Plants in small containers can be placed in a sealed polythene bag after a thorough watering or kept moist using the capillary method (see right). For larger plants, make an improvised drip system using a container at a higher level, filled with water, from which a wick (such as a piece of absorbent rope) extends to the plant container below.

Placing plants in a cool, shady area and spreading small pebbles on the surface of the growing medium will help to reduce moisture loss.

Standing the container in a shallow dish of water, filled with a fine layer of pebbles to prevent the growing medium at the base of the container becoming waterlogged, will also help prevent the growing medium from drying out.

lead to stunted growth. Overwatering does not compensate for drying out, and the plant simply has to suffer the additional stress of waterlogging.

Watering systems

The range of equipment for watering increases all the time and there are a number of highly sophisticated systems you can employ, rather than relying on the traditional watering can and hose of the past. The automatic watering systems include high-level ones, where the water is directed downward from overhead sprinklers, and low-level ones, where the water is delivered directly to the plant. Both systems are fairly expensive. However, if you have a lot of containers, particularly if they are grouped near to one another, these methods do save time. The low-level watering systems are based either on drip or trickle systems, which deliver water more or less accurately to the plants involved, or on capillary systems, which work as a form of sub-irrigation, the water being delivered to the base of the growing medium via a pipe with seep holes, although this is effective only for shallow containers.

Container shape and construction

The choice of container in terms of shape and material can make a considerable difference to the speed at which the growing medium dries out. Porous materials, such as terracotta, encourage the soil to dry out more rapidly than non-porous materials, such as polythene. Containers with a wide top offer a large surface area of growing medium to the elements and this too will encourage rapid drying; it may be wise, in a hot climate, to limit this shape of container to shady areas.

Frequency of watering

Much depends on the plant form and country of origin. Drought-loving plants from the Mediterranean, for example, will need far less moisture than bog-loving plants from cool temperate countries, so planting for the situation is a key element in keeping watering chores to a minimum.

In very hot sunny weather most container-grown plants will need watering on a daily basis, ideally late in the evening or early in the morning since less of any surface residue will be lost through evaporation at that time. Remember that windy conditions are also very drying for container plants.

HANGING BASKETS

Moisture leaches very rapidly from hanging baskets because so much of the growing medium is exposed. In addition, their position well above head height makes watering difficult. To ease watering, attach a bamboo cane to the end of the hose using plastic ties (see below) to make it more rigid.

Water retaining crystals
These can be incorporated in the growing medium of hanging baskets to reduce moisture loss.

Pruning Plants

Most container-grown plants need pruning and training regularly to keep them tidy and in good growing condition, and many will also require some sort of support, so that the operations of supporting (or staking) the plant and of pruning are often intertwined, particularly in the case of climbing plants and wall shrubs. Perennials, which may need staking (see page 95), do not need to be pruned.

The purpose of pruning

The main reason for pruning is to make the plant perform to your particular expectations and requirements. This may well involve the removal of unwanted growth in order to regulate and direct vigor, flowering or, in the case of fruiting plants, cropping. The range of plants grown will vary and their needs differ, but the principal techniques are roughly the same.

Pruning consists of removing sections of the plant with sharp secateurs/pruning shears to encourage buds lower down on the plant to develop in place of the removed shoots. To this end it is important to pay attention to where and how you prune. Normally you leave an outward-facing bud below the cut you have just made so that any new growth will grow away from the main stem, thus avoiding the problem of crossing or twisting stems.

PRUNING CUTS

It is important when pruning to use a sharp pair of secateurs/pruning shears so that the resulting cut is clean, rather than ragged, otherwise disease may set in. The shape of the cut is determined by whether the plant has alternate buds (left) or opposite buds (right).

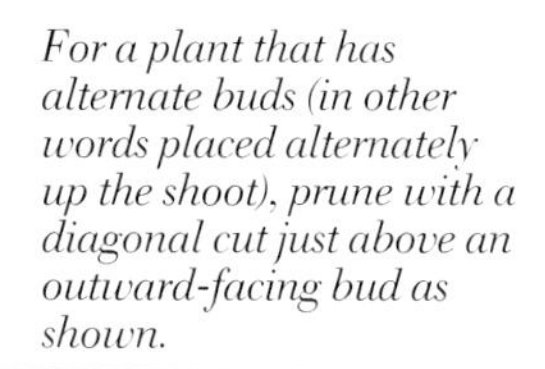

For a plant that has alternate buds (in other words placed alternately up the shoot), prune with a diagonal cut just above an outward-facing bud as shown.

For a plant that has opposite buds, prune with a straight cut across the stem directly above the buds.

BASIC PRUNING

There are three main reasons for pruning. The first is to train a young plant into an appropriate form, known as formative pruning. The second is to maintain shape, regulate vigor and control pests and diseases, known as containment pruning. The third is to correct existing problems, by removing dead and diseased wood or twisted branches; this is known as remedial pruning.

Formative pruning
Prune young plants after planting to a few strong shoots, leaving a couple of buds on each shoot. This will encourage the plant to grow in the direction required and help increase its bushiness and strength. The following season, climbers can be tied in to their supports and pruned and trained as required (see page 91).

Containment pruning
Shrubs tend to grow vertically, making most growth close to the stem tip. The growing tips of shrubs can be cut back to the appropriate size, provided at least a couple of pairs of buds are left on each shoot and you make the pruning cut just above an outward-facing bud (see Pruning cuts, left).

Remedial pruning
Dead, damaged or diseased wood should be cut out, as should any congested or mis-shapen shoots, or those where some of the branches have reverted from their variegated form to plain green shoots.

Plants have a control mechanism which encourages secondary buds to develop new shoots if the main (or apical) bud is removed. This ability is used by gardeners when pruning to create quite specific characteristics in the plant. For example, removing the apical bud in chrysanthemums will encourage the secondary buds to grow and the plants will develop several sprays of smaller flowers rather than one large flower.

The timing of pruning is clearly important since you are cutting off part of the developing plant. For instance, if you remove all the buds in spring from a plant that flowers in summer, it is obvious that the plant will not bear flowers that year. It is important, therefore, to establish whether your plants are spring- or summer-flowering, as this will determine how and when you prune them. An easy rule, if you are unsure, is to prune your plants just after flowering, when it is obvious which are the flowering shoots. In spring you need to know whether the plant flowers on current wood or last year's wood, to avoid cutting off all the flowering shoots. Further information on when to prune specific plants is given in the cultivation notes in the Container Plant Directory (pages 106–155).

Pruning therefore has several effects: it improves the shape of the plant, it stimulates growth and encourages secondary flowering shoots to develop (thereby creating more profuse flowering), and it also prevents some plants from growing out of control.

DISTINGUISHING SHOOT AGE

The color of the bark will help you to determine what is a first- or second-year shoot. First-year shoots are generally soft and pliable, and usually green or pale gray/brown in color; second-year shoots are usually gray or brown, and thicker in diameter; and shoots that are more than two years old have darker bark, and are tough and woody with a gnarled appearance.

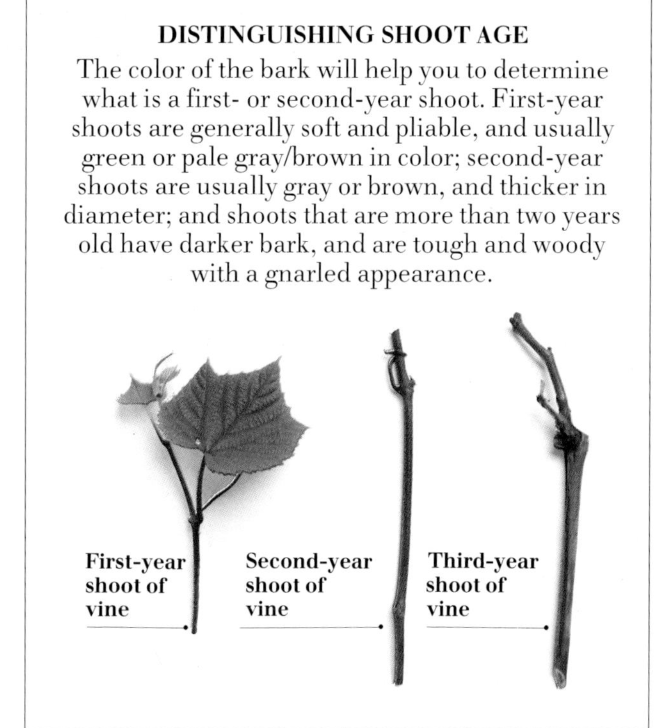

Climbers and wall shrubs

True climbers have their own method of support, and will satisfactorily support themselves if given a suitable host, such as a wall or trellis, to climb over. With most of these plants, training and pruning are combined operations, and depend to some extent on the plant's own means of support, which can be by twining stems, twining leaf stalks, aerial roots, tendrils, sucker pads or thorns (see page 17 for examples of the different support systems). As with any form of pruning, three categories apply.

Formative pruning and training

When you buy a young climbing plant, your aim is to encourage it to grow in the manner that you wish by

FORMATIVE PRUNING AND TRAINING OF CLIMBERS

Climbers should be trained as soon as possible after planting into the appropriate form for the situation. This consists of arranging and tying in the strongest-growing shoots to achieve a balanced framework. The young shoots will need to be loosely attached to the supports with plastic ties or garden string.

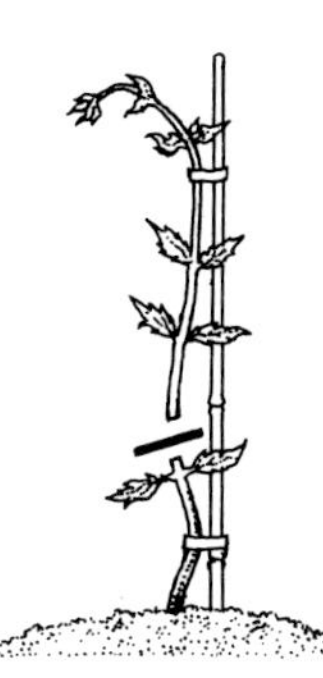

1 After planting, cut the main stem of the plant back to the lowest pair of strong buds. As they grow, these buds will provide the framework for the plant and should be tied in to the support with plastic ties or garden string.

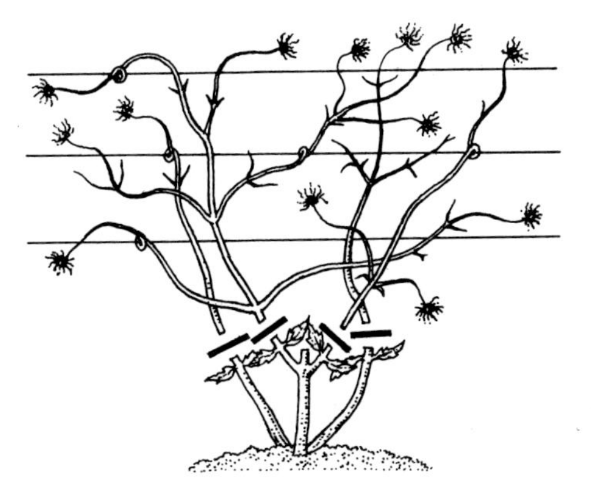

2 In the second year after planting, cut the stems back to a pair of buds on each main shoot, to leave an outward-facing bud from which new shoots will form. Cut back any other shoots to within two buds of the nearest stem. Again, use plastic ties or string to tie the stems in to the support.

PRUNING WISTERIA

When pruning vigorous climbers like wisteria, you will need to concentrate both on controlling the direction and overall growth of the plant as well as on encouraging the maximum number of flowering shoots to form.

Spring-flowering climbers, such as wisteria, are normally best pruned in summer, directly after flowering, as this encourages the new shoots to grow and produce the next season's flowers.

first providing an appropriate support and then encouraging the growing shoots to extend along it. Training should be carried out when the stems are still young or pliable enough to be bent into position without straining them.

Containment pruning

Once the climber is established, you will have to start containing its growth while at the same time making sure that you do not cut off the stems on which flowers form. This task is made more confusing by the fact that some plants flower on the current season's growth and some on the previous season's growth.

During the growing season, tie in the strongest new stems (before they become tough or woody), directing them across the support as required. The ties should be loose enough so as not to restrict growth, but tight enough to hold the stems in place in strong winds. Loosen the ties periodically as the stems thicken with age.

Remedial pruning

Climbers which have been left unpruned for a long time can become a tangled mass of stems. To rejuvenate them you will have to cut the stems back to soil level, removing dead or diseased wood, either in one season's operation or over two years (depending on the general health of the plant). Whichever option you adopt, you will probably lose the flowers for that season. Some climbers respond to being cut from their support and laid along the ground to encourage new lateral shoots to develop. When the lateral shoots are 15cm (6in) long, tie the main stems back into the support. Among the climbers that can be pruned to soil level are *Campsis*, most species of clematis, most honeysuckles (*Lonicera* species), *Passiflora* and *Vitis*.

Shrubs

How you prune a shrub will determine how it performs. To prune successfully, you need first to

REJUVENATING WALL SHRUBS OR CLIMBERS

When pruning shrubs or climbers that are trained against a wall, you need to develop flowering shoots and a strong framework of main stems. Timing will depend on whether the plants flower on current season's growth, in which case they are pruned in winter or early spring, or on previous season's growth, in which case they are best pruned after flowering. A plant that has been left untrained is best pruned hard to encourage new growth.

Cut the shoots back to within a couple of buds on each main stem and then tie them in to supports when the resulting shoots are about 15cm (6in) long. Remedial pruning of this type will mean the loss of one season's flowers but the gain will be a stronger prolific flowering plant the following year.

decide what kind it is. The categories are: slow-growing, broad-leaved evergreens; small-leaved evergreens; plants that flower in spring or early summer; and plants that flower in summer (see the box, below right).

For broad-leaved evergreens, the only pruning required is normally dead-heading (see page 94) and removing seed pods. You can cut away any straggly or mis-shapen shoots, but do not overdo it or you will lose the following season's flowers.

For small-leaved evergreens, where the flowers form immediately above the foliage, you will need to keep the bushes trimmed to stop them looking straggly. This also conserves their strength and prevents the plants from producing seed, which in turn encourages flowering. Lightly clip the bushes all over with secateurs/pruning shears after flowering has finished. For plants in this category used for topiary, see Training topiary, below.

For spring- and early summer-flowering plants, you should cut the shoots that have flowered back to a main framework of stems, leaving only young non-flowering growth. Thin out any weak, spindly, badly placed or crossing growth at the same time. If you do this immediately after flowering, you can see which are the flowering shoots.

For summer-flowering shrubs that flower on the new season's growth – for example, some of the large-flowered clematis – it is vital that you do not prune in late spring, thereby losing the flowers. Instead cut the plants back just after flowering to the main framework of branches. You can do this in early spring, if necessary, provided that the new season's shoots have not started to grow.

Deciduous shrubs that flower on short spurs of old wood need very little pruning other than the occasional remedial cutting back of dead or diseased wood or thinning out of the bush.

TRAINING TOPIARY

Evergreen shrubs with small leaves and a slow growth rate are ideal for training into formal clipped shapes, known as topiary (see page 67), but otherwise they need little or no pruning, except to remove damaged or diseased wood and mis-shapen stems.

1 *If you are training an evergreen into a formal shape, wait until the plant is about 30cm (12in) tall before starting to clip it. Make sure that the base is wider than the top or it will become straggly where the light fails to reach the leaves.*

2 *Once you have achieved a satisfactory shape, clip the plant only twice a year – at the start of the growing season and again in late summer – using secateurs/pruning shears to remove thick stems, and hand shears to clip the leaves.*

PLANT CATEGORIES FOR PRUNING

The following categories give an indication of which method of pruning to adopt. Broad- and small-leaved evergreens need occasional trimming, and some remedial pruning. Spring-flowering plants and those that flower on old wood require pruning in the previous summer; summer-flowering plants, flowering on the current season's growth, can be pruned in spring; and vigorous plants can be pruned back to the base if necessary.

Broad-leaved evergreens include:
Aucuba
Choisya
Berberis (evergreen forms)
Camellia
Escallonia
Euonymus
Photinia
Pieris
Rhododendron
Sarcococca
Viburnum tinus

Small-leaved evergreens include:
Artemisia
Buxus sempervirens
Calluna
Cistus
Erica
Lavandula
Helianthemum
Helichrysum
Hypericum
Santolina

Spring-flowering plants include:
Chaenomeles
Forsythia
Kerria
Magnolia
Philadelphus
Prunus triloba
Ribes
Skimmia
Syringa
Weigela

Summer-flowering plants include:
Bougainvillea
Caryopteris
Cornus alba
Deutzia
Fuchsia
Hydrangea
Hypericum
Kolkwitzia
Rosa
Sambucus

Plants that flower on old wood include:
Amelanchier
Buddleja globosa
Caragana
Cercis
Chimonanthus
Daphne
Enkianthus
Eucryphia
Hamamelis
Viburnum (deciduous forms)
Weigela

Plants that can be drastically pruned:
Choisya ternata
Cornus alba
Corylus avellana
Cotinus coggygria
Eucalyptus
Euonymus fortunei
Ilex aquifolium
Rhododendron ponticum
Santolina

REMEDIAL PRUNING OF SHRUBS

Occasionally disease or other unwanted problems strike your plants. Inspect leaves and stems on a regular basis for signs of damage. Prompt action – involving the removal of damaged stems and the treatment of pests and disease – will often help prevent them from spreading.

Here a cotoneaster has started to develop a shoot that has reverted to the plain green leaf of the species. Using sharp secateurs/pruning shears, remove any such shoots, or those that are diseased or damaged, perhaps with splits or tears in the bark or stems.

Remedial pruning

Sometimes old, tangled or overgrown shrubs need to be pruned drastically to rejuvenate them. However, not all shrubs can tolerate really drastic pruning (some of those that do are listed on page 93), so if in doubt phase the heavy pruning over a year or two, cutting part of the shrub down to the base.

To generate new growth, start by cutting out all the weak crossing stems and reduce the main stems to about 30–45cm (12–18in) above soil level. Masses of shoots should sprout from just below the cuts on the main stems during the following growing season. Keep the three or four strongest shoots and remove the others. For evergreens, remove the surplus shoots in spring after flowering, and for deciduous shrubs, remove them during the dormant season. In the second season some regrowth may occur where the shoots were cut, and these should be removed.

PRUNING ROSES AND CLEMATIS

Many gardeners are understandably confused when it comes to pruning roses and clematis, because within each genus there are plants that perform in different ways. For example, roses can take the form of climbers or of bush shrubs, and clematis includes a wide range of hybrids and species which flower at different times of the year. Provided you know whether the plant flowers in summer or in spring, you can organize your pruning accordingly. Prune any spring-flowering clematis after they have flowered, and summer-flowering ones in early spring.

Treat roses in the way described for either shrubs or climbers, as appropriate. Very vigorous climbing roses should have a number of the stems cut down to the base each year to prevent them becoming too thick and tangled. Dead-head roses when the blooms fade to promote further flowering, unless the hips are a feature of the plant.

DEAD-HEADING

To prolong the flowering season of your plant and improve its general appearance, remove fading or dead flowerheads promptly after flowering.

1 *As soon as the petals start to drop, pinch out individual dead or faded florets from the main flowerhead.*

2 *Once the whole flowerhead has died, snap off the flower stem at the base, taking care not to damage the other stems.*

Dead-heading

The main reason for a plant to flower is to produce seed. With many annuals and perennials, once the plant produces seed, flower production stops and a large proportion of the plant's energy will go into seed development. The best course of action is to remove the dead and faded flowers, a process called dead-heading, so that the plant concentrates its energies on repeated flushes of flowers. Dead-heading can extend the flowering season of many bedding plants and some shrubs, such as *Buddleja*, although the later blooms may be slightly smaller. Plants like lupins that produce flower spikes will bear a later flush of flowers on smaller secondary spikes if you remove the main flower spike once the flowers start to wither.

General tidying

If you are growing herbaceous perennials in containers, cut the dead stems and top growth by about half in late autumn, leaving some growth to provide frost protection for the crown of the plant during winter. In late spring cut away all the dead top growth to allow the new growth to develop unimpeded.

Staking Plants

When you grow plants in containers, you draw attention to their appearance by virtue of their isolation. As a consequence the shape each plant creates and its general vigor and health are all-important. If a naturally bushy perennial or annual grows in a less-than-tidy way, you can improve its appearance by supporting the stems with discreetly hidden stakes. Shrubs and climbers can be both improved in shape and encouraged to flower more profusely by pruning and dead-heading (see pages 90–94). Staking is therefore important if you want to keep your plant display in good order.

Another reason for staking a plant is to prevent damage from wind, which can harm or even snap the stems and can create wind rock in the rootball, effectively preventing the plant from taking up moisture and nutrients from the soil. Large plants, such as small trees, will need to be supported with a fairly stout stake, sunk deep into the container, to prevent this from occurring at the time of planting.

There are various methods of staking plants, and there is a wide range of staking equipment on the market (see page 79 for basic canes and ties, and below for more complex devices). Each sort is suitable for different kinds of display and different types of plant.

SINGLE-STEMMED PLANTS

A single-stemmed tree or a shrub grown as a standard will need to be supported to stabilize it.

After planting, water the growing medium to soften it. Insert the stake a few centimeters (1in) away from the main stem, making sure it is sunk deep into the container. Tie the support to the plant stem with ties.

The most commonly used staking device is a bamboo cane, with plastic ties for attaching the cane to the plant's stem. This simple stake will serve most purposes, particularly for smaller plants grown in containers, and it is worth keeping a small stock of them. A cane-type stake is normally about two thirds as tall as the plant's ultimate height.

The staking method you choose depends largely on whether the plant is single- or multi-stemmed.

MULTI-STEMMED PLANTS

The kind of support you choose is determined by the type of plant and the purpose of the display. The aim is for the stake to become invisible as the plant grows.

Link stakes
Link stakes, which connect together, are ideal for soft-stemmed perennials.

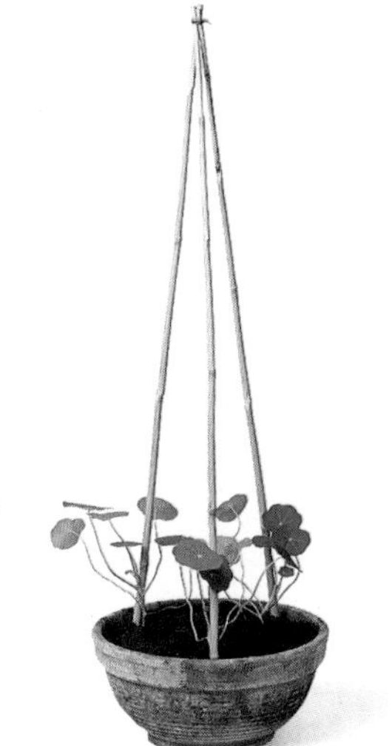

Wigwam of canes
Climbing annuals can be grown up a wigwam of canes and tied at the top.

Trellis
Trellis makes an ideal support for any climber that twines.

Brushwood
This can be used to provide a natural-looking support for delicate-stemmed perennials.

Propagation

If you raise your own plants, you save large sums of money. This also gives you the opportunity, if you particularly like a plant that you have bought or been given, to create more of the same for almost no extra expense. There are a number of ways to do this, referred to collectively as propagation. The different methods include taking cuttings (by using the shoots, leaves or roots of mature plants), growing from seed, layering the stems of plants, dividing the roots and grafting the stem of one plant onto the stem of another. Although this last technique is rather specialized and is principally used by professional growers to improve the strains of particular plants, the foregoing methods are all relatively easy to carry out.

Cuttings

Propagation from cuttings is the simplest method and the commonest one for increasing your stock of shrubs. Stem cuttings produce roots directly from the stem itself or from the mass of thin-walled wound-healing tissue (known as the callus) that develops at its base. The leaves of certain plants can also be used for cuttings and will develop roots from the point at which the vein in the leaf is severed. Vigorous young roots can be used for cuttings too.

The techniques for propagating from cuttings vary slightly according to the type of plant. The cultivation information in the Container Plant Directory (pages 106–155) explains which type of propagation is best suited to which plant.

CUTTINGS FOR PROPAGATION

The three parts of the plant used for propagation are the roots, leaves and stems. Some plants respond better to one form of propagation than another and this information is given in the cultivation notes on the specific plants in the Container Plant Directory.

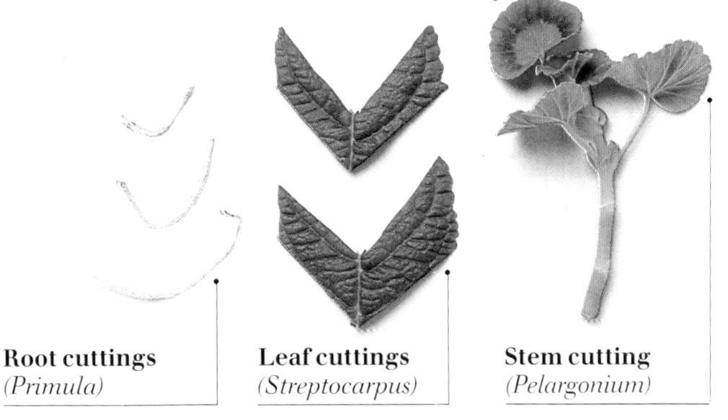

Root cuttings *(Primula)*

Leaf cuttings *(Streptocarpus)*

Stem cutting *(Pelargonium)*

Methods of taking cuttings vary slightly depending on the age of the stem and the time of year. The three main categories – softwood, semi-ripe and hardwood cuttings – are discussed below.

Softwood cuttings

These are taken from the new growth of the plant in spring. You use the fast-growing stem tips. As they have a soft base, they root more easily than cuttings taken from mature wood, and so this method is used for plants that can be difficult to propagate from more mature wood – fuchsias are a case in point. The method

SOFTWOOD CUTTINGS

This method is used for shrubs and herbaceous perennials in spring when the new growth has just formed. Here a geranium (Pelargonium *sp.) is being propagated. It roots extremely easily and you can increase your stock quickly.*

1 Select a side shoot from a healthy plant. Ideally the shoot should have two or three leaves and be about 8cm (3in) long.

2 Pull the side shoot away from the main stem, to leave a small heel of main stem tissue attached to the side shoot. This will help rooting take place.

3 Insert the side shoot in a small pot of growing medium formulated for cuttings. Keep well watered until the roots develop.

The rooted cutting

SEMI-RIPE CUTTINGS

These are taken in late summer from the current season's growth. Among the suitable plants are many of the herbs, such as rosemary, lavender and sage. The plant here is a camellia.

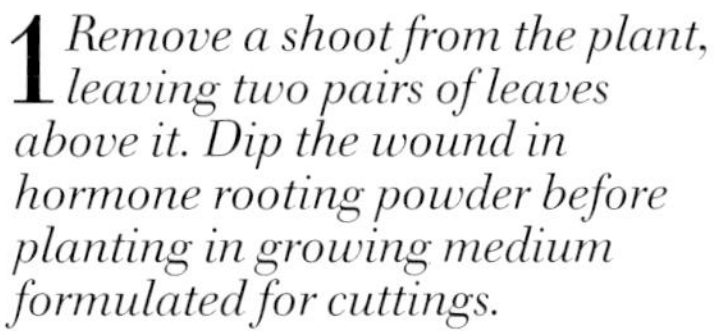

1 Remove a shoot from the plant, leaving two pairs of leaves above it. Dip the wound in hormone rooting powder before planting in growing medium formulated for cuttings.

2 Water in well and then cover the pot with a polythene bag to conserve moisture and provide a humid atmosphere for the growing plant.

The camellia cutting showing the callused wound (left) and the newly formed roots (right).

for taking softwood cuttings is shown below left. You have to pay particular attention to the aftercare of softwood cuttings, since they have a depressing tendency to wilt very quickly if not kept in humid conditions. Be sure to remove any fallen leaves before they start to rot, and apply a fungicidal spray once a week. When roots have formed, the cuttings can be transplanted into individual pots or containers and hardened off (gradually acclimatized to cooler conditions).

Semi-ripe cuttings

These are taken in late summer from the current season's growth and they are less prone to wilting than the softwood type. They root less readily than softwood cuttings, but their survival rate is better. You will find that some semi-ripe cuttings root more easily if a heel of the mature stem is taken with the cutting (see Semi-ripe cuttings, above). Once the cuttings have rooted, pot them on in standard growing medium and harden them off in a coldframe or on a window sill. Plant them into their permanent containers the following spring.

Hardwood cuttings

These are taken in autumn or early winter from one-year-old shoots. Because they do not root easily, you will need to apply hormone rooting powder to the wound to speed up the process. These cuttings can take a surprisingly long time to form roots, and the chief peril is that the propagator becomes bored with waiting and yanks the plant up to inspect it, killing it in the process. Most hardwood cuttings will root by the spring

HARDWOOD CUTTINGS

These are taken from growth that is one year old. You can take several cuttings from one shoot, but be sure to mark the top and bottom of each. Make a horizontal cut at the base and a diagonal cut at the tip. Cut just above a bud for the tip.

1 Using sharp secateurs/pruning shears, cut the stem into as many parts as you wish, making a horizontal cut at the base of each cutting and a diagonal cut at the tip, just above a bud.

2 Plant the cuttings, base down, in a container of growing medium formulated for cuttings, and water in well. These vine cuttings have rooted after approximately three months.

The hardwood vine cutting (right) and the rooted growing cutting (far right).

of the following year. You can then pot them into standard potting medium in a permanent container.

Division

This method of propagation is suitable for plants with a spreading crown or rootstock and which produce plenty of growth from the base. Not only does division increase the number of your plants, it also helps to keep them healthy and vigorous. Many herbaceous perennials die out at the center, spreading laterally: division is a means of keeping the clumps concentrated at the middle. You can easily divide many perennials by simply pulling the sections apart manually. Larger, more strongly rooted perennials, such as *Helianthus*, will require the use of a couple of hand forks, placed back to back, to pull the roots apart. Plants that have fleshy roots, such as *Rheum*, are best split using a trowel or sharp knife, to minimize damage caused by tearing.

Most plants should be divided when they are dormant, between late autumn and early spring, provided the weather is suitable. Do not divide plants when conditions are very wet or cold, or very dry, as re-establishing growth may be difficult.

Perennials with fleshy roots are best left until the end of the dormant season, in spring, before being divided. You will then be able to observe the new shoots forming, and see where vigorous growth is taking place, and can divide the plant accordingly.

When you are dividing plants, wash off the surplus soil and debris before you cut them. This will help to reduce wear and tear on your knife and secateurs/pruning shears, and will enable you to see more clearly what you are doing.

Follow the steps shown below to divide a plant. If you are not going to replant the divisions within a couple of hours, dip them in water and keep them in a sealed polythene bag in a cool place until you are ready to replant them. This will help prevent drying out.

Largish pieces of divided plants may flower in the same season, although the stems may be shorter than those of established plants. Smaller divisions are best grown on in small pots for a season until they are fully established. Remember that any new divisions should be planted at the same depth as the parent plant, but those that are prone to rotting at the crown – such as rosette-forming

DIVISION OF BULBETS

Many bulbs produce small bulbets around the parent bulb which you can use to increase your stock of plants. Lilies are a good example of this.

1 Break off individual scales from as near the base as possible, rotating the bulb so that the scales are taken evenly, and working toward the center. Up to 80 percent of the scales can be removed in this manner.

2 Place the scales in a bag of growing medium; store in a frost-free room until small bulbs develop.

ROOT DIVISION

Among the many plants that lend themselves to root division, the iris is one of the easiest to propagate. You will be able to produce many new plants from one largish tuberous root. Fleshy roots can be cut up after division to form new plants, provided each has a shoot attached.

The iris before division, showing lots of new shoots.

1 Dig up the rooted plant carefully and wash off most of the soil. Insert two handforks, back to back, through the root.

2 Exert pressure downward and outward to pull the roots apart.

3 Make sure that each new division bears a shoot. Cut the divisions off using a sharp knife. Replant in small pots until established.

alpines – may need to be planted slightly proud of the surrounding soil level.

Layering

You can propagate plants that have trailing or flexible stems by making a cut in the stem and then pegging it down into growing medium until the wound roots. This is rather more difficult to do when you are gardening in containers, and the similar technique of air layering (sometimes referred to as Chinese layering) is more useful. It works on the same principle, but you do not need to peg the shoot down to soil level.

In spring select a vigorous one-year-old stem from the plant, and trim away the side shoots and leaves to leave a clear length of stem behind the shoot tip. Carry out the layering procedure shown right.

The following spring you can open the bag to see if the layered stem has rooted. If very few or no roots have developed, reseal the bag and leave it in place for a couple more months. Once rooting has occurred you can sever the stem immediately below the point where the roots have developed and prune back the main stem to the nearest outward-facing bud. Pot up the resulting plant in standard potting medium, then place in a cool greenhouse until well-established.

Sowing seed

Most ornamental plant seeds are sown indoors or in a greenhouse in containers so that they develop in warmer conditions, under cover, and the young plants are planted out as soon as the conditions are favorable. Temperature is an important factor in seed germination: seeds simply remain dormant if they are too cold. Another important factor is moisture: seeds should be kept constantly moist as drought inhibits germination,

SELF-PROPAGATING PLANTS

A few plants propagate themselves by means of plantlets which grow on the plant, form roots and drop to the ground, where they root themselves.

1 *This* Tolmiea *is mature enough to start creating its own plantlets. Check the plant for signs of new growth within the leaf bases.*

2 *Transfer the plantlets to a tray of growing medium formulated for cuttings until established; water in well.*

AIR LAYERING

The technique of air layering can be carried out on most shrubs and some trees. The exclusion of light around the wound means that the stem becomes thin-walled and roots more readily, but even so rooting can take up to two years.

1 *In spring or summer, select a suitable stem and push a polythene bag over the stem past the point at which you wish to make the cut.*

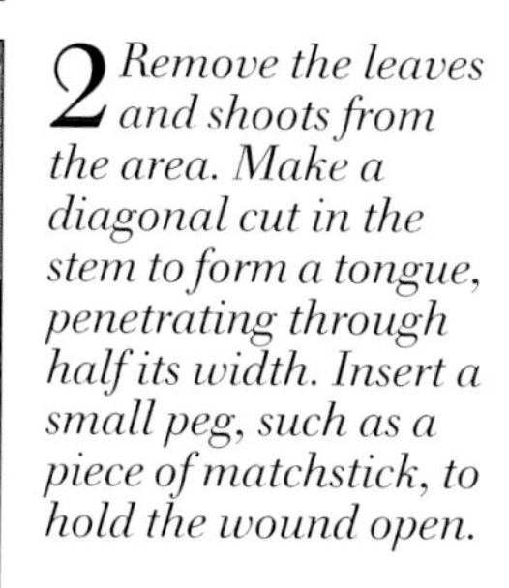

2 *Remove the leaves and shoots from the area. Make a diagonal cut in the stem to form a tongue, penetrating through half its width. Insert a small peg, such as a piece of matchstick, to hold the wound open.*

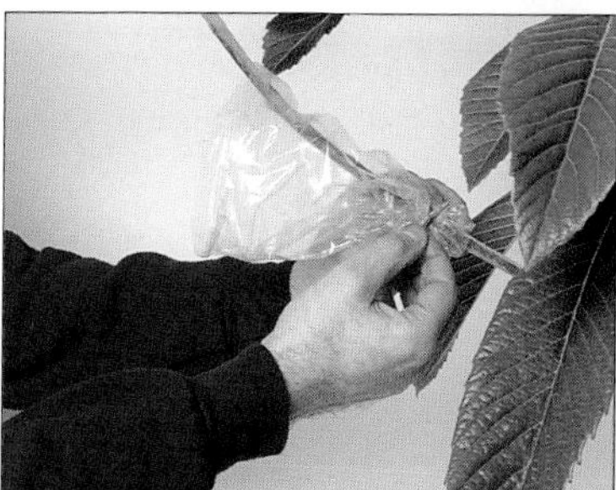

3 *Fasten the base of the polythene bag using a plastic tie. This should be tight enough to hold the bag in place but not so tight that it inhibits the stem's growth.*

4 *Fill the polythene bag with moist sphagnum moss, packing it around the wound. Fasten the top end with a second plastic tie.*

5 *Leave the sealed bag of moss on the plant for as long as it takes for roots to form. This can take up to two years. Open the bag carefully after six months to check on root development. Reseal if no roots have formed.*

though overwatering may well cause them to rot before they start to germinate.

The genetic package contained in seeds can last a surprisingly long time – poppies have been known to germinate from seed over 70 years old. That being said, it does not pay to keep seeds for long periods of time. They have the greatest chance of germination when they are fresh, and for each year they are kept the ratio of germinating to non-germinating seeds declines. If you want to keep seeds, it is best to store them at a temperature of 3–5°C (37–41°F) in an airtight container along with a packet of silica crystals (available from florists) or some grains of rice to absorb any moisture in the air.

To maintain an even amount of humidity, cover seeds after sowing with glass or plastic to provide warmth, watering as necessary. Do not put the seed trays in sunlight as the growing medium and seeds will dry out. You can cover the plastic or glass with paper to keep sunlight out until germination occurs, but once it does, the seedlings will need an adequate supply of light or they will become etiolated (tall and straggly). Another cause of etiolation is overcrowding. It is important, therefore, to try to sow seed as thinly as possible (see Sowing fine seed, step 4).

Very fine seed is normally broadcast (sprinkled at random) on the surface of the growing medium and watered from below to prevent it from being washed out of place. Watering from below encourages the embryonic seedlings to develop stronger roots, as the roots naturally grow downward to seek the water. Watering from the top can create problems as the water sometimes fails to penetrate to the lower part of the growing medium.

Coarser seed can be sown individually in pots – Jiffy pots are ideal – making the seedlings easier to handle when they are planted out.

Take care not to sow seed too deeply, as this may prevent germination from taking place. Each seed contains a reserve of food to enable the emerging shoot to cope until it gets to the light and starts making its own supplies. If you sow seed too deeply, the supplies of nutrients will run out before the shoot reaches the light and the shoot will die prematurely. The smaller the seed, the fewer the reserves, and therefore the more shallowly it should be sown.

SOWING FINE SEED

Fine seed is best sown broadcast (sprinkled on the surface) in a shallow tray of growing medium formulated for sowing at temperatures of around 15 to 18°C (60 to 65°F). The seed must be kept moist, but not waterlogged, after sowing. If the seed is coarse, dispense with step 3.

1 *Fill the seed tray with growing medium. Drag a straight edge along the top to remove any surplus and ensure a flat surface.*

2 *Firm the growing medium by tamping it down with a firming board – a piece of hardboard with a rough handle will serve the purpose.*

3 *Cover the tray with a fine layer of growing medium using a fine-gauge sieve, then firm again.*

4 *Sow the seed from the palm of one hand onto the firmed planting medium, sprinkling it as thinly as possible over the surface.*

5 *Water from below by placing the seed tray in another tray of shallow water.*

PRICKING OUT

Seed that has been sown broadcast or in shallow drills will need to be thinned and transplanted (known as pricking out) to give the seedlings more space to grow on.

Once the seedlings are large enough to handle prepare a tray of growing medium, moisten it thoroughly, and make small holes to receive the seedlings. Allow 4 or 5cm (1½–2in) of space around each one.

SEED SIZES

The seeds below vary in size. Sow the finest broadcast, the medium ones in shallow drills and the largest individually in pots. Seed packets will give methods and depths.

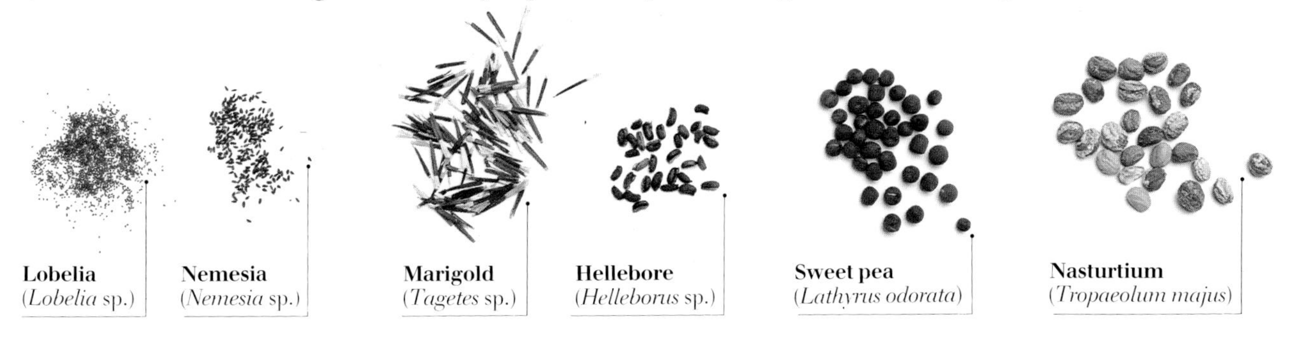

Lobelia (*Lobelia* sp.)
Nemesia (*Nemesia* sp.)
Marigold (*Tagetes* sp.)
Hellebore (*Helleborus* sp.)
Sweet pea (*Lathyrus odorata*)
Nasturtium (*Tropaeolum majus*)

Seed trays and growing mediums

There are now a wide variety of purpose-made containers for seed sowing, from shallow trays to small individual pots. You can make your own seed trays by recycling plastic ice-cream cartons or yogurt pots, for example, but you must puncture the base with enough holes for water to percolate through. For individual seeds, Jiffy pots are useful (see below). Made from biodegradable fiber or plastic, they allow you to transplant the seedlings without removing them from the pots, ensuring less damage to the fine root hairs.

The planting medium for seeds must be light and full of nutrients. Specially prepared formulations are available which are ideal, and they are based on a mixture of 50:50 peat (or peat substitute) and vermiculite. Because this type of growing medium dries out rapidly, it must be watered regularly with a very fine rose on the watering can (unless the seed is very fine, when it should be watered only from below – see Sowing fine seed, step 5 (left).

Pricking and planting out

Once the seedlings are large enough to handle, they can be thinned out (if sown broadcast) and pricked out (separated out and planted in deeper growing medium to give them more space to grow). Larger seeds that have been sown in individual containers or Jiffy pots can simply be planted in their final positions once the seedlings have grown to 10cm (4in) or so. Seedlings grown indoors in warmth may need to be hardened off in a cold frame or similar (gradually acclimatized to cooler, outdoor temperatures) before being planted out in their final positions.

JIFFY POTS

There is a range of biodegradable pots, in different forms, which allow you to sow in a container which is then itself planted in the growing medium. Jiffy 7s are compact fiber pots that swell when water is added to create a complete planting unit. They are ideal for growing plants for hanging baskets.

Dry Jiffy
Wet Jiffy
Established seedling

SOWING LARGE SEED

Large seed which is easily handled can be sown individually in pots, thereby reducing damage to the young plants, which can be transplanted to their final positions.

1 *If the seed coating is hard, chip or score it first with a sharp knife. This process aids germination.*

2 *Fill the pots with growing medium formulated for sowing, firm it and make a small hole 2.5cm (1in) deep. Insert the seed and cover with growing medium.*

3 *When the seedlings are large enough to handle, plant them out in their final positions, hardening them off beforehand, if necessary.*

Maintaining Containers

CONTAINERS MUST BE fit for the purpose you have in mind – in other words, they must be strong enough to hold the plant and the growing medium, and they must provide adequate drainage. Containers come in a variety of materials (see pages 30–35), some of which are more durable than others. If you are buying terracotta ones make sure that they are frostproof, otherwise you will have to bring them indoors during the winter to prevent them from cracking. Wooden containers need to be solidly constructed and, if made from softwood, treated with a horticulturally safe preservative.

Drainage

All containers, apart from water barrels used for aquatic plants (see page 54), must have drainage holes inserted in the base to allow the water to flow out. If this is not the case, the growing medium will become waterlogged and sour, and the plants will sicken and eventually die. Most purpose-made containers have holes already drilled in or near the base, but if you make your own or recycle an *objet trouvé* as a container, you will have to drill holes in the base yourself.

Aeration

The base of any container should be filled with a shallow layer of broken pots or pebbles to prevent the growing medium being washed out after watering, and to improve drainage and aeration. Be careful with acid-loving plants, such as rhododendrons, azaleas and blue hydrangeas, not to use limestone chippings as they are liable to neutralize the planting medium.

MOVING CONTAINERS

If you have to move a heavy container, there are several methods open to you, depending on its size, its position and what you have available.

Here, small sections of metal piping are used with a board, as rollers. Alternatively, you can put the container on a large sack and use one corner of the sack to drag the container along a flat surface.

CLEANING A USED POT

If you are reusing a pot that has previously contained plant material, you must clean it out well first, otherwise you risk spreading pests and diseases.

1 *Remove any existing growing medium and plant debris by scrubbing the pot, inside and out, with a stiff brush.*

2 *Wash out the container thoroughly with detergent and warm water, then rinse well before refilling with fresh growing medium.*

Cleaning

Any container that has been used already for planting must be cleaned before reuse, as should stakes and canes, to prevent pests and diseases being transferred to the new plant or plants (see above). If the container is particularly dirty, soak it overnight in a bucket of water to loosen the dirt before giving it a good scrubbing. Check that wooden containers are sound before reusing them, particularly if they are to be positioned on window sills where they might damage the paintwork. If you want to avoid having to clean out containers each time you reuse them, it is a good idea to line them with a plastic bin liner before adding the growing medium, remembering to create some drainage holes in the base of the liner first.

Extending the life of your containers

Containers will last longer if they are slightly raised off the ground. This will also encourage free drainage. You can buy small chocks on which to raise terracotta containers (see page 51) and you can make small wooden wedges for window boxes. Alternatively, place a brick under each corner of the container. Drip trays are useful for containers on window sills to prevent excess water from spilling over the sill. These are available to match terracotta or plastic pots.

Wooden containers are likely to rot unless they are raised so that the air can circulate underneath. You should treat any new wooden container that has not been previously treated with a horticulturally safe preservative or with two coats of oil-based paint in your chosen color. This should remain in good condition for a couple of years before it needs repainting. If you are using copper, iron or lead containers, it is a good idea to line them first with plastic before planting up to prevent the minerals in the container from leaching into the growing medium.

Moving containers

Containers that are full of wet soil are extremely heavy and it is better, if you can, to avoid moving them around. For this reason, heavy containers should be placed into position before planting. You can then take the planting medium and the plants to the container. To save dragging a heavy bag of growing medium, decant some of it into a bucket beforehand. Guidelines for moving heavy containers using metal pipes as rollers are given below, left.

To move a container down a flight of steps, place it first on a couple of wooden planks and tie a rope around it. Slide the container along the planks, using the rope to control its speed of descent; you might require a helping hand if the pot is really heavy. You can also buy a small pot trolley with wheels from garden centers, but you will still have to lift the container onto it.

Be extremely careful when lifting heavy weights, and do so with a straight back; bend at the knees, not at the waist. If you have to move containers full of plants any distance, first wrap them in sacking to prevent damage to the leaves and stems.

WOODEN BARRELS

If you are planning to use a wooden barrel for aquatic plants, you need to soak it first to make it watertight (see steps 1 & 2). If you intend to fill it with growing medium, line it with a plastic liner, inserting a few holes in the base to permit drainage.

1 *Before soaking, the joints on the barrel will be open, and it will not be watertight.*

2 *Fill the barrel with water and leave for a couple of hours to allow the wood to swell. Drain off any surplus water.*

3 *Insert the plastic liner, and use a knife to make a few holes in the base to permit drainage.*

4 *Fill with growing medium, then cut off any surplus liner level with the top of the soil.*

WINTER PROTECTION

Containers of frost-tender plants will have to be protected in winter. Normally sacking or burlap is used to tie around the base of the plant, but the bubble plastic that comes with many packed items is equally effective provided it is tied securely.

The reinforced terracotta pot, ready for use

Reinforcing a terracotta pot
Terracotta containers which are prone to crack in cold weather can be strengthened with reel wire. Take the wire around the neck of the pot and twist the ends to secure.

Lagging a pot
A tender or half-hardy plant may need to be protected against frost in hard climates. Lag the pot with burlap or sacking.

The lagged pot

Diseases and Pests

However well you care for your plants, some will fall prey to pests or disease. You can reduce the risk by taking the best possible care of your plants: using sterile pots and potting mixtures will help prevent disease, and checking regularly for signs of infestation so that you can treat pests and diseases at an early stage when they are easily remedied will help control. Keeping your stock of plants as healthy as possible through regular care and maintenance is another important factor – constantly allowing them to dry out and then overwatering them in an effort to compensate leads to stress and makes them more susceptible to attack.

Whether you opt for chemical treatments or organic alternatives when pests or diseases strike is a matter of personal choice, but bear in mind that, by and large, you do not have the problems facing commercial growers – it is not imperative to your livelihood to save the plants, although you may well feel that a particular plant deserves chemical treatment.

If you do use chemicals, make sure that you spray on a windless day after sunset, when beneficial insects, such as foraging bees, have retired for the night. Always follow the manufacturer's instructions as to precautions, wearing gloves, mask or whatever is required, and keep children and pets away during application.

DISEASES

Plant diseases can be classified into three main groups: bacterial, fungal and viral. Some of the symptoms appear confusingly similar. Plant hygiene is the best preventor of bacterial diseases, which can be hard to control, though they are less common than fungal or viral disorders. The main bacterial diseases are mildew, root rot and canker.

Bacterial leaf spot Spots or patches, often with a yellow edge, appear on leaves. Causes death of affected leaves.
Remedy Avoid wetting leaves of susceptible plants. Remove affected areas and spray with benomyl.

Black spot A fungal disease confined to roses, this causes small, circular, black spots on leaves, which enlarge and result in premature leaf drop.
Remedy Spray regularly with benomyl or other fungicide.

Canker Shrunken areas appear on the bark of trees with flaking and blistering. The wounds grow and enlarge, and when all the bark is damaged, the affected branch dies.
Remedy Remove the affected area of the plant immediately.

Damping off Causes seedlings to topple over, wither and die. The warm moist conditions preferred by seedlings make them particularly prone.
Remedy None. Use clean containers as a preventative measure.

Downy mildew White, downy growth develops on undersides of leaves; upper surface blotched with yellow or brown. Causes stunted growth and results in secondary infection.
Remedy Remove affected leaves, and improve overall ventilation and drainage. If serious, spray with mancozeb.

Fireblight Leaves turn blackish-brown, shrivel and die, but usually remain clinging to the stem. May result in death of entire stem or plant.
Remedy Prune back affected stems.

Fungal diseases The spread of fungal spores in humid conditions will cause a variety of molds, mildews, rots and wilts, showing themselves in a range of dusty coverings on leaves, flowers or fruits, or in brown, yellow or orange blisters on leaves.
Remedy Remove affected areas and spray with benomyl.

Fungal leaf spot Discrete, concentrically arranged spots appear on leaves. These vary in color, but are usually brown or dark gray. Results in premature leaf fall.
Remedy Remove affected areas and burn plant debris. Spray with copper-based fungicide or benomyl.

Powdery mildew White, powdery growth appears on leaves – most commonly on the upper surface, but also on the undersides. Causes yellow leaves which fall prematurely.
Remedy Remove affected areas immediately. If problem persists, spray with fungicide – for example, benomyl.

Root rot The bacteria that cause this disease tend to invade through damaged or bruised tissue; rough handling of roots should therefore be avoided. Contamination of soil, containers or tools is another cause, stressing the need to clean containers and equipment before re-use. The root rots, causing discoloration and, finally, wilting of the entire plant.
Remedy None. Burn the affected plant immediately.

Rust forms Red-brown pustules, principally on leaves. Releases powdery orange and yellow spores into the air, which carry the disease.
Remedy Spray with a copper-based proprietary spray.

Smuts These start as pale green or cream swellings on leaves and/or leaf stalks, which eventually rupture to reveal black, powdery spores. Can result in premature leaf drop or death of the entire plant.
Remedy Burn affected plants immediately since infection may recur or spread to neighboring plants.

Viral diseases These are untreatable, so prevention is the only cure. The usual signs are stunted growth and mottling of leaves and flowers, sometimes with severe distortion. The disease can be carried from one generation to the next via cuttings.
Remedy None. Destroy suspect plants taking care to burn plant debris.

PESTS

The physical damage caused by pests will help you identify the type of insect attacking your plants. Some feed by biting and chewing, leaving the foliage ragged and full of holes; others distribute clues or trade marks, such as the slime trails left by slugs and snails. Other insect pests suck the plant's sap and secrete a sticky deposit (honeydew) on which sooty molds develop, making the plant weak and stunted and at greater risk of virus infection.

Ants These feed on the honeydew produced by aphids, and farm the aphids, moving them to fresh areas and plants, just as a farmer shifts his cattle to new pastures. This makes control of aphids difficult and therefore you need to combine your control of both pests. Equally, ants burrowing around the roots of plants can cause considerable damage.
Remedy Dust with ant killer.

Aphids Certain plants are more vulnerable to aphid attack than others. If left unchecked, aphids will form huge colonies on the soft, sappy, young growth of particular plants. The sooty mold that develops on the sticky honeydew they leave behind blocks light from the leaves and so weakens the plant. Greenfly and blackfly are the most common forms of aphid, causing deformed buds and stunted growth, and spreading viral diseases. High-nitrogen fertilizer increases the soft sappy growth which attracts them.
Remedy Spray with greenfly killer (pirimicarb).

Capsid bugs The young leaves and shoot tips of plants are particularly prone to attack, which causes the development of brown spots that enlarge to form ragged holes.
Remedy Spray with permethrin.

SLUG TRAP

Slugs can be trapped by sinking a shallow pot containing drinks with a high sugar content, such as cola, into the growing medium. The pests will be attracted to the liquid, fall in and drown.

A saucer filled with cola to attract slugs.

Caterpillars Can be extremely voracious, attacking buds, leaves and shoots and sometimes completely defoliating a plant.
Remedy Spray with derris.

Chafers Winged beetles that feed mainly on flowers in spring and summer, while the grubs feed on plant roots leaving large cavities.
Remedy Dust with pirimiphos-methyl. Remove beetles manually.

Earwigs Particularly partial to chrysanthemums, clematis and dahlias, these fast-moving insects with characteristic pincers feed at night, and are therefore hard to spot.
Remedy Trap with an earwig trap (see right) or dust with permethrin.

Eelworms These are tiny pests which enter the leaves, stems and roots of many plants, particularly in humid conditions, causing stunted, distorted and shrivelled growth.
Remedy Infected plants should be burnt immediately and the container sterilized. Discard any growing medium in the container.

Slugs and snails These are a common problem in many gardens, and container-grown plants are not immune from their predations. Hostas are particularly prone to attack, as are the young tender shoots and leaves of most other plants.
Remedy One of the simplest remedies is to scatter grit around the base of the containers as this makes it difficult for slugs and snails to move up to the plants. Another is to bait them using the method shown left.

Vine weevil These nocturnal insects have a pronounced snout. They are most active in autumn and early spring, feeding on the leaves of plants and leaving them holed and eaten around the margins. The young weevils feed on the roots, resulting in stunted growth, severe wilting and even the death of the entire plant.
Remedy Maintain good hygiene and remove plant debris to reduce attack. If problem is serious, spray with HCH.

Whitefly The nymph stage of this small insect secretes honeydew (whitefly excrement) on leaves, stems and fruit, and this is followed by gray, sooty molds. Whitefly are visible to the naked eye. Severely attacked plants become weak and stunted.
Remedy Dust or spray with permethrin at intervals.

EARWIG TRAP

Earwigs can be trapped using a simple device whereby a small flower pot is stuffed with straw and inverted over a cane placed close to the plant. The earwigs will crawl inside the pot and can be removed each morning.

A terracotta pot is filled with straw to attract earwigs.

CONTAINER PLANT DIRECTORY

THIS SECTION LISTS, in alphabetical order of the Latin (genus) name, a selection of the most valuable plants for growing in containers. The accompanying descriptions and cultivation information provides you with all you need in order to grow and maintain the plant successfully, including preferred situation and growing medium.

In addition, each plant is given a hardiness rating in four categories. In this book, **very hardy** refers to plants that can survive temperatures of -18°C (0°F) for long periods; **hardy** plants are those that can tolerate temperatures of 0°C (32°F) down to -18°C (0°F) during winter periods; **half-hardy** plants can survive temperatures of 1°C (34°F), although growth rate is slow at the lower temperatures; and **tender** plants will suffer cold damage if temperatures fall below 10°C (50°F).

The watering and feeding information is given as a general guide: **water sparingly** refers to plants that should be watered thoroughly every 7–10 days; **water normally** refers to plants that should be watered every third or fourth day; and **water frequently** refers to plants that should be watered daily.

LEFT: *The magenta and white daisy-like flowers of* Pericallis *cv.*
ABOVE: *A cottage-style display of summer-flowering annuals.*

MALVACEAE

ABUTILON

H Up to 2m (6ft) **S** 3–3.5m (9–11ft)

Abutilon megapotamicum or the **flowering maple**, as it is commonly known, originates from South America (mainly Brazil). In a protected situation it may be grown as a wall shrub and forms a thin branched plant of semi-weeping habit; it will survive in all but the coldest areas. Other more tender species, such as *A. pictum* 'Thomsonii' with its green leaves mottled yellow, are popular in summer container displays. The leaves often have three points, are usually bright green and resemble miniature grape vine leaves. The bell-like flowers, which are 4cm (1½in) long, are seen from early summer to mid-autumn.

Cultivation Tender; full sun/partial shade; water frequently; susceptible to attack by scale insect, whitefly and mealy bug; propagate from semi-ripe cuttings from early to late summer.

Special needs These plants thrive in a free-draining growing medium and are at their best in a sheltered spot. If sheltered, *A. megapotamicum* may overwinter outside in mild areas; *A. pictum* types must be overwintered indoors. No special feeding is required; high levels of nitrogen reduce flowering and may cause the leaves to lose their variegation. If necessary, rooted cuttings can be overwintered indoors and grown on the following spring.

Display ideas *Abutilon pictum* cultivars make an excellent centerpiece, but the lower leaves often die off, exposing the stem. See page 41.

ACERACEAE

ACER

Acer palmatum

H 5m (15ft) **S** Up to 2.4m (8ft)

This slow-growing large shrub or small tree, commonly known as **Japanese maple**, originates from China and Japan. It is grown for its attractive rounded crown and spreading habit, as well as for its delicate palm-like leaves which range in color from pale green through golden-yellow or deep reddish-purple. The leaves of most cultivars turn vivid shades of red, yellow and orange in autumn, and some are prized for their brightly colored bark, such as *A. p.* 'Senkaki'. In addition, there are some cultivars with finely cut leaves, such as *A. p.* 'Dissectum Atropurpureum'. An outstanding cultivar is *A. p.* 'Beni-tsukasa' which has remarkable leaf coloration, the young leaves turning from peachy-yellow to pinky-red, and darkening during late summer to reveal faint variegation and small flecks of lighter colors.

Acer palmatum 'Beni-tsukasa'

Cultivation Very hardy; full sun/partial shade; water sparingly, but do not allow roots to dry out; top dress with organic matter in spring; remove any dead or diseased wood in spring; susceptible to coral spot; propagate from seed sown outdoors in early spring or, for named varieties, from softwood cuttings in mid-summer.

Special needs Grow in an open, free-draining, lime-free growing medium. Water with rain water in hard-water areas. Keep plants sheltered from severe winds.

Display ideas The thin leaves produce only light shade, which makes them ideal for growing other plants beneath or for use as a light canopy over seating areas on a deck or patio.

ACTINIDIACEAE

ACTINIDIA

H Up to 10m (30ft) **S** 2m (6ft)

Actinidia is a genus of hardy, deciduous, climbing shrubs with twining stems originating in Asia. Commonly known as **kiwi fruit**, *A. deliciosa* (syn. *A. chinensis*) is a vigorous ornamental climber with dark green, heart-shaped leaves and hairy stems. The male and female flowers, borne in clusters 4cm (1½in) wide, are creamy-white deepening to buff-yellow; they appear in mid- to late summer. Male and female flowers are usually carried on separate plants, and specimens of each sex should be grown to ensure pollination. *Actinidia kolomikta* is grown for its attractive foliage – its heart-shaped, dark green leaves are marked with pink and white at the tip – and will reach a height of 4m (12ft) in a sheltered position.

Acer palmatum 'Dissectum Atropurpureum'

Cultivation Very hardy; full sun; water frequently; top dress in spring and liquid feed in mid- to late summer; thin out overcrowded growth and shorten vigorous shoots in early spring; propagate from semi-ripe cuttings in mid- to late summer.

Special needs Prefer a neutral, free-draining growing medium. Provide support for young, twining stems.

Display ideas This plant is vigorous and ideal for growing on walls, and over supports to provide summer shade.

ADIANTACEAE

ADIANTUM

Adiantum capillus-veneris

H 25cm (10in) S 25cm (10in)
Commonly called the **maidenhair fern**, this is one of 200 fern species in its genus. It has light green, triangular leaves on tough black stalks. Often treated as a pot plant because it is not fully hardy, it can be grown all year round in the open in warmer areas, but only if in a sheltered position.
Cultivation Half-hardy; shade; water frequently; top dress lightly with bone meal in spring; susceptible to attack by nematode eelworms (see page 105) which enter the leaves – their feeding causes large dead patches, black in color; woodlice can cause severe wilting and stunted growth by feeding on the roots; propagate by division of the rhizomes or by spores sown in spring.
Special needs Prefers a peaty growing medium with a high fiber content. Do not allow to dry out – but do not let it become waterlogged either. Avoid exposure to direct sunlight which will quickly dry out the plant and shrivel the leaves. Overwinter indoors; place outdoors from late spring, usually kept in its pot and plunged into a display container.
Display ideas Very good for giving life to a damp shady corner.

LILIACEAE

AGAPANTHUS

Agapanthus campanulatus

H Up to 60cm (2ft) S 75cm (2½ft)
This hardy, clump-forming perennial plant originating from South Africa is commonly known as the **African lily**. Deep green, strap-like leaves up to 60cm (2ft) in length, with large clusters of light blue flowers, each individual bloom being up to 7.5cm (3in) in size, are produced from mid-summer on stalks 75cm (2½ft) above the ground. Selections such as the well-known **Headbourne Hybrids**, with their deep blue flowers, are usually hardier than many of the species.
Cultivation Half-hardy; full sun/partial shade; water frequently; feed sparingly so as to encourage maximum flowering; propagate by division in late spring/early summer or by seed in late spring (though in the latter case it will take at least three years for the plants to start flowering).

Agave americana 'Marginata'

Special needs A heavy feeder that requires a fertile growing medium rich in organic matter. Very good for coastal regions as this plant is tolerant of salt-laden air, but needs shelter from wind, especially in cold areas. Dead-head as soon as flowering has finished to prevent seed forming and promote better flowering next year. Divide the clumps every second or third year: this prevents overcrowding which in turn increases flowering.
Display ideas This showy plant is ideal as a single specimen. The attractive dried seed heads may be used for winter decoration. See page 20.

AGAVACEAE

AGAVE

Agave americana

H Rosette up to 1m (3ft) S 1.2m (4ft)
The **century plant**, as this is commonly known, is a succulent which originates from Mexico. The gray-green leaves are narrow, thick and fleshy, tipped with very tough, sharp spines. In flower, *A. americana* is most dramatic with creamy-yellow bell-shaped blooms on spikes of 3–5m (10–15ft) in height. Usually the rosette bearing the flowers will die along with the flower stalk as it dies down. There is a variegated type, *A. a.* **'Marginata'**, with a yellow leaf margin.
Cultivation Tender; full sun; water sparingly; susceptible to attack by mealy bug; propagate from offsets (small rosettes that emerge through the growing medium around the base of the plant) in spring or summer.

Alchemilla mollis

Special needs Requires a well-drained growing medium containing a small amount of loam. Tolerant of irregular watering, but do not overwater. Must be provided with frost protection if temperatures fall below 5°C (41°F).
Display ideas Ideal as a single specimen in the hottest place on the patio.

ROSACEAE

ALCHEMILLA

Alchemilla mollis

H 45cm (18in) S 45–60cm (18–24in)
This herbaceous perennial, commonly known as **lady's mantle**, originates from eastern Europe and Asia. The furry-textured, light green leaves are shaped like the open palm of a child's hand and they often have a crinkled appearance to their margins. The plant develops an open center and has a tendency towards a lax, floppy, growth habit. The small, creamy-yellow, flower-like structures are in fact bracts (modified leaves) which appear in large clusters from mid- to late summer.
Cultivation Very hardy; full sun/partial shade; water frequently; mulch with organic matter in spring and apply a liquid feed every two weeks from early to late summer; cut back stems to 2.5cm (1in) after flowering; propagate from seed sown outdoors in mid-spring or by division in winter.
Special needs Must have a free-draining growing medium.
Display ideas Very useful for ground cover around the base of container-grown specimen trees and shrubs.

LILIACEAE/ALLIACEAE

ALLIUM

H Leaves up to 45cm (18in) S 10–12cm (4–5in)
This is a large familiar genus of bulbous plants with 280 hardy, easy-to-grow, ornamental and edible species, almost all of which have the typical onion smell. Those selected for ornamental garden use bear a globe or dome of brightly colored flowers at the tip of the main shoot. The most popular species is *A. albopilosum* which originates from the Middle East. Its strap-shaped leaves are a dull bluish-green and slightly hairy on the underside. The globe-like flowers, up to 15cm (6in) across, are made up of as many as 80 small, lilac-pink, star-shaped blooms produced in mid-summer. Another favorite species is the smaller *A. moly* from southern Europe, whose 5cm (2in), bright yellow, star-shaped flowers appear during mid-summer on 30cm (12in) stems; its thin, strap-like leaves are about 20cm (8in) long and a gray-green color.
Cultivation Hardy; full sun; water sparingly; top dress with slow-release fertilizer in spring; young plants susceptible to slug attack; white rot may attack the roots and the base of the bulb; propagate from seed sown between mid-autumn and early spring, or by division after flowering.
Special needs Require a free-draining, loam-based growing medium. These plants are hardy provided they are deeply planted. Lift and divide the clumps every three years to maintain a good flowering performance.
Display ideas Leave dead flowers on the plants over winter to provide an extra display. See page 27.

BORAGINACEAE

ANCHUSA

Anchusa azurea

H Up to 1.5m (5ft) S 60cm (2ft)
This is a genus of hardy biennials and herbaceous perennials with a deep fang-like root system, originating mainly from Europe and Africa. Mid-green, thickish, strap-like leaves are covered with fine bristly hairs which help to reduce moisture loss in warm conditions. The most widely grown species is *A. azurea*, and its cultivars are most popular because they provide a vivid display of color from mid- to late summer. The 1cm (½in) dish-shaped flowers are produced in long blunt spikes carried high above the foliage. Among the most popular varieties used for container growing are: **'Loddon Royalist'**, 1m (3ft) high, with rich gentian-blue flowers; the more vigorous **'Dropmore'**, 1.5–2m (5–6ft), with rich deep blue flowers; **'Morning Glory'**, 1.2–1.5m (4–5ft), and deep blue; and the sky-blue variety **'Opal'**.
Cultivation Hardy; full sun/partial shade; water frequently; mulch with organic matter in spring; propagate from root cuttings 5cm (2in) long in early spring.
Special needs This plant has very long, drought-resistant roots, which require a deep container and a well-drained growing medium. Use brushwood or canes to provide support for the more vigorous varieties.
Display ideas Plant with other sun-loving plants on a hot, dry patio.

Anchusa azurea 'Loddon Royalist'

Anemone × hybrida

RANUNCULACEAE

ANEMONE

H Up to 30cm (12in) S 45cm (18in)
This is a large genus of some 150 species of both fibrous- and tuberous-rooted plants, originating mainly from alpine and Mediterranean regions. The small blooms of the **windflower**, as it is commonly known, are up to 5cm (2in) across and cup-shaped with as many as 20 colored sepals; they open out almost flat as the flower ages, just like the majority of clematis (which is related). The small leaves of mid- to dark green, finely cut or divided into three to fives lobes or sections, are arranged in whorls on short hollow stems, which are often reddish-green. The most popular forms are hybrids derived from *A. coronaria*. There are also herbaceous species such as *A. hupehensis* and *A. × hybrida*, which flower from late summer to autumn; these are much taller, achieving up to 1m (3ft) in height, and offer colors ranging from deep pink to pure white.
Cultivation Hardy; full sun/partial shade; water frequently; liquid feed from early to late summer, depending on flowering time; young shoots susceptible to slug attack (see page 105); anemone rust causes distorted leaves and may prevent flowering; viruses may cause yellow foliage, distorted growth and poor flowering (see page 104); propagate by division: spring-flowering species in late summer, summer- and autumn-flowering species in mid- to late autumn.
Special needs Require a free-draining growing medium.
Display ideas The spring-flowering species are ideal for early displays and can be grown in pots and plunged into display containers. See pages 52 and 58.

ARISTOLOCHIACEAE

ARISTOLOCHIA

Aristolochia macrophylla

H Up to 8m (23–26ft) S 3m (10ft)
Originating from North America, and commonly known as **Dutchman's pipe**, *A. macrophylla* (also known as *A. Durior*) is a hardy vigorous species of deciduous climber with twining stems. The large mid-green leaves are heart-shaped. The small 2.5–4cm (1–1½in) flowers, curiously shaped and curved like an old Dutch pipe, are yellow, brown and green in color and appear in mid-summer.
Cultivation Hardy; full sun/partial shade; water frequently; top dress with slow-release fertilizer in spring; thin out

Artemisia pontica

spindly growths and reduce strong vigorous growths by one third in early spring; susceptible to attack by aphids (see page 105), which can severely restrict growth, and in warm summers by red spider mite; propagate in midsummer from 7.5cm (3in) softwood tip cuttings in a propagating case, or in late summer by layering, or in mid-spring from seed.
Special needs This is a vigorous plant which needs a deep container and a well-drained fertile growing medium. Requires plenty of growing room. Young shoots will need to be tied in until they are established.
Display ideas Very good for a wall, arbor or pergola, or for training up a vertical pole.

COMPOSITAE/ASTERACEAE

ARTEMISIA

Artemisia pontica

H 60cm (2ft) S 20cm (8in)
One of a large group of hardy and tender perennials, evergreen and semi-evergreen shrubs, originating from Europe, Asia, and America. This hardy deciduous sub-shrub, commonly known as **wormwood**, is grown for its attractive, silver-gray, slightly aromatic leaves rather than its small, grayish-yellow flowers borne in late summer/early autumn.
Cultivation Hardy; full sun/partial shade; water sparingly; cut down to 15cm (6in) in late spring; lower leaves and roots susceptible to aphids and rust (see pages 104–105); propagate from semi-ripe heel cuttings, 7.5–10cm (3–4in) long, in late summer.
Special needs Prefers an open free-draining growing medium and a very warm sunny position. Pot on or repot in spring every other year. Normally hardy but you may need to protect the roots in a hard winter.
Display ideas Perfect for mixing with purple-foliaged plants.

GRAMINEAE/POACEAE-BAMBUSOIDEAE

ARUNDINARIA

Arundinaria nitida

H Up to 3.5m (11ft) S 1.5m (5ft)
Originating from across China, this plant is a member of the genus which consists of many species of hardy evergreen **bamboos**. Some of these have been reclassified and renamed, and *A. nitida* is also known as *Sinarundinaria nitida* and *Fargesia nitida*. The main attraction of this particular plant is the new growth, which is coated with a waxy bloom giving the young shoots a purple color. The narrow strap-like leaves are bright green on the upper surface and bluish-green on the underside. The natural habit of this plant is to form a dense clump or thicket.
Cultivation Very hardy; full sun/partial shade; water normally; mulch with bulky organic matter in spring; cut out dead shoots at soil level; propagate by division in late spring.
Special needs Thrives in a moist, well-drained growing medium; never allow the roots to dry out. Do not expose to cold winds.
Display ideas Ideal for growing as a decorative screen.

SCROPHULARIACEAE

BACOPA

H Up to 60cm (2ft) S 1m (3ft)
This is a genus of about 60 species of summer-flowering water or moisture-loving plants, mostly originating from the American continent. The common name is **water hyssop**. The succulent fleshy stems are usually creeping and the leaves are arranged in opposite pairs, but the flower-bearing shoots tend to be more erect. The blue or white, bell-shaped flowers, borne in clusters of three to five appear in summer. The two most common species are: *B. caroliniana*, which forms a creeping plant, with 2.5cm (1in) long, oval-shaped, lemon-scented leaves, and blue flowers, 1cm (½in) in diameter, produced in the summer; and *B. monnieri*, with pale blue or white flowers, also borne in summer.
Cultivation Slightly tender; full sun/partial shade; water frequently; apply a base dressing of slow-release fertilizer when planting out; propagate either from 10cm (4in) softwood cuttings in spring or by division in spring or autumn (see pages 96–98).
Special needs These are marginal plants and need waterlogged conditions. Take care to provide them with frost protection throughout winter.
Display ideas Very good for containerized water features.

Begonia × tuberhybrida cvs.

Begonia × tuberhybrida

BEGONIACEAE

BEGONIA

H 7.5cm–1m (3in–3ft) S 30–75cm (1–2½ft)

This is a very large genus of over 900 species of perennial plants grown for their colorful leaves and attractive flowers, which are characteristic of all begonias but vary considerably in size and particularly in the number of petals. Because of the large numbers of species and hybrids, begonias are grouped into three basic categories: fibrous-rooted, rhizomatous and tuberous.

Fibrous-rooted begonias are generally evergreen and have an erect shrubby appearance. These plants are the most popular for container growing outdoors and are available with both green and bronze-colored leaves. The most important species within this group is *B. semperflorens*, a slow-growing evergreen begonia which bears masses of red, pink or white flowers from mid-summer to mid-autumn. There are mixed color selections such as **New Generation** with shades of salmon, rose, pink, scarlet and white; and individual named cultivars, such as **'Volcano'** (orange-scarlet flowers on bronze foliage) and **Thousand Wonders**, available in separate red, pink or white strains.

Rhizomatous begonias have a creeping horizontal stem and generally hairy evergreen leaves. They are grown mainly for their attractive foliage and prefer partial shade. Well-known species from this group include *B. masoniana*, which has mid-green leaves with a blistered appearance and a large chocolate-brown cross in the center which gives it the common name of **iron cross begonia**. Even better known is the red-leaved *B. rex*, and the silver-leaved hybrids originating from it.

Tuberous begonias are mostly deciduous, have a tendency to produce double male flowers and single female flowers, and like full sun. They are often grown as greenhouse or house plants, but do well outdoors in a hot summer, with the *B. × tuberhybrida* types producing huge, brightly colored flower heads up to 15–20cm (6–8in) across. Named varieties include **'Harlequin'**, which bears white flowers edged with bright pink; **'Jamboree'** which is orange-red; and the bright crimson **'Olympia'**.

Cultivation Tender; full sun/partial shade; water frequently; susceptible to damping off and powdery mildew (see page 104), and to attack by vine weevil (see page 105) and tarsonemid mites; propagate fibrous-rooted begonias from seed sown in mid-spring or cuttings taken from early to late summer; propagate rhizomatous begonias by division in late spring; propagate tuberous begonias by 7.5cm (3in) softwood cuttings or division in late spring.
Special needs Will thrive in a free-draining growing medium containing plenty of bulky organic matter. Those grown for their leaf color produce brighter colors in partial shade. The large-flowered tuberous types will need some means of support.
Display ideas Superb for tubs, troughs, hanging baskets and window boxes. See pages 24 and 46.

BERBERIDACEAE

BERBERIS

H Up to 1.5m (5ft) S Up to 2m (6ft)

Commonly known as **barberry**, and originating mainly from Asia and South America, this is a genus of about 450 species and numerous varieties and cultivars of hardy deciduous and evergreen shrubs. Evergreen species are grown for their attractive, glossy-green foliage, while deciduous species are grown for their vivid foliage color in autumn. The leaves are arranged in rosette-like clusters, and protected by tough spines. All produce large quantities of yellow-orange flowers in spring

Berberis prattii var. *laxipendula*

and some have clusters of brightly colored berries, which persist well into winter. ***Berberis prattii*** **var.** ***laxipendula*** is a deciduous plant with long, gently arching stems, and attractive finely toothed, bright-green leaves and deep coral berries.

Cultivation Hardy; full sun or partial shade; water normally; top dress with a slow-release fertilizer in spring; no regular pruning required; generally free of pests and diseases; propagate from 10-cm (4-in) long semi-ripe cuttings taken in autumn.

Special needs Prefer a well-drained growing medium and dry conditions.

Display ideas Evergreen varieties are very good for providing a glossy-green, year-round framework; deciduous varieties provide good autumn color of foliage and berries.

COMPOSITAE/ASTERACEAE

BRACHYCOME

Brachycome iberidifolia

H Up to 45cm (18in) S 45cm (18in)

Commonly known as the **Swan River daisy**, this is one of 66 species of annual and hardy perennial plants, closely related to the daisy and chrysanthemum. A half-hardy annual, it produces strongly scented, daisy-like flowers 2.5–4cm (1–1½in) across on 45cm (18in) tall slender stems from mid-summer to early autumn. The flowers vary in color from white and pink, through to mauve and even a bluish-purple. The pale green leaves are so deeply cut that they resemble fern fronds.

Cultivation Tender; full sun; water sparingly; feed with a base dressing of fertilizer when planting out; propagate from seed sown under protection in mid-spring.

Brachycome iberidifolia

Special needs Must be planted in a free-draining growing medium. Will tolerate dry conditions. Needs a sheltered position. Requires some means of support because of its loose, lax habit; may be flattened by heavy rain.

Display ideas Very good for low or shallow containers.

CRUCIFERAE/BRASSICACEAE

BRASSICA

Brassica oleracea

H 30–45cm (1ft–18in) S 25cm (10in)

There are four basic color variants of the **ornamental cabbage**, as it is commonly known: smooth-leaved pink, smooth-leaved white, and the Savoy type with densely crinkled, white or pink-splashed leaves. Usually these plants are grown from mixed seed selections providing a combination of colors with the central leaves either bright rose-pink or creamy-white, while the outer leaves are curled and fringed or veined with rose and white. One popular named selection is ***B. o.*** **'Sekito'**, which is bright green flushed with red. It is worth noting that these variegated cabbage selections are not as hardy as the green edible types.

Cultivation Hardy; full sun/partial shade; water normally in autumn and spring, sparingly in winter; feed with a base dressing of fertilizer when planting out; susceptible to club root (which causes wart-like swellings on the roots resulting in stunted growth), downy mildew (see page 104), and white blister fungus (which produces powdery white spores on the leaves and stems); also prone to attack by cabbage aphids and caterpillars; propagate by seed sown outdoors in mid-summer or under glass in early spring.

Browallia speciosa 'Heavenly Bells'

Special needs Prefers a free-draining, loamless (peat-based) growing medium. Protect young plants from feeding birds, particularly pigeons.

Display ideas Grow amongst spring bedding plants and bulbs. See page 41.

SOLANACEAE

BROWALLIA

Browallia speciosa

H 1.2m (4ft) S 1.2m (4ft)

Browallia, a genus of six species of half-hardy annuals from South America, is a distant relative of the humble potato. Commonly known as **bush violet** or **sapphire flower** (USA), *B. speciosa* makes an ideal container plant, mainly because, when it starts to flower, it will quite literally flower itself to death. The leaves are deeply veined, bright green, oval-shaped with a pointed tip, and feel slightly furry to the touch. The five-petalled flowers, up to 5cm (2in) across and a strong violet-blue, with darker veins leading down into a white throat, are produced from mid-summer to early autumn. The newer dwarf hybrids are particularly recommended for container gardening, the **Bells** series being the most popular, with tubular, violet-shaped flowers in a wide range of colors – from white, through both light and dark shades of blue.

Cultivation Tender; full sun; water frequently; mulch in spring; pinch out growing tips to produce a strong bushy plant; propagate from seed sown in mid-spring.

Special needs Prefers a fertile growing medium, though can grow reasonably well in poor conditions. Keep moist or the leaves will quickly shrivel.

Display ideas Dwarf hybrids are ideal for hanging baskets.

Brugmansia × candida 'Knightii'

Calendula officinalis 'Lemon Beauty'

SOLANACEAE

BRUGMANSIA

Brugmansia × candida 'Knightii'

H 1.2–1.8m (4–6ft) **S** 1.8–2.4m (6–8ft)
Brugmansia is a genus of 10 species of half-hardy annuals and tender perennial shrubs or trees, originating from places as far apart as central South America and India, with the annual species often listed separately as *Datura Brugmansia × candida* 'Knightii'. It is grown for its exceptional display of 30cm (12in) long, white, heavily scented, pendulous, trumpet-shaped flowers, produced from mid- to late summer. The large evergreen leaves, up to 25–30cm (10–12in) in length, are oblong with an elongated point. They are mid-green in color, with wavy margins, and are covered with a coating of soft felt-like hairs.
Cultivation Tender; full sun/partial shade; water frequently; liquid feed every two weeks from early summer to early autumn; cut all growths back to within 15cm (6in) of the main stem in early spring; susceptible to viruses (which cause yellow markings on the leaves, see page 104) and attack by aphids and whitefly (see page 105); propagate by 13cm (5in) semi-ripe heel cuttings grown under protection in early summer.
Special needs Prefers a free-draining growing medium. These plants are poisonous and the seeds in particular should be kept well away from young children and pets.
Display ideas Very good for use with garden lighting.

BUXACEAE

BUXUS

Buxus sempervirens

H Up to 3m (10ft) **S** 1.2–2m (4–6ft)
The **common box** (UK) or **boxwood** (USA) is the most popular species of *Buxus* in cultivation. It is a slow-growing bushy plant, originating from Europe, north Africa and western Asia. The glossy, dark green leaves are slightly oval with a small ovate notch at the tip; they are arranged close together in pairs on the unusually shaped (square) stems. In addition to the standard green species there are a number of attractive variegated cultivars, including *B. s.* **'Elegantissima'**, with its gray-green leaves each with a silver margin, and the very slow-growing *B. s.* **'Latifolia Maculata'**, with a splash of yellow in the center of each leaf.
Cultivation Very hardy; full sun/partial shade; water normally; mulch with organic matter in spring and liquid feed at half the recommended strength from early summer to early autumn; clip topiary specimens in late summer or early autumn; susceptible to leaf spot (which shows as pale brown spots on the leaves) and leaf rust (see page 104); propagate from 4–10cm (1½–10in) tip cuttings in a coldframe in autumn, allowing six to eight months to root.
Special needs Prefers a free-draining growing medium containing plenty of organic matter.
Display ideas Exceptionally good as a clipped specimen plant (topiary) and very useful as a focal point in a display. See pages 2, 8, 12, 36 and 66.

COMPOSITAE/ASTERACEAE

CALENDULA

Calendula officinalis

H Up to 60cm (2ft) **S** 35cm (14in)
One of a number of hardy annual species originating from southern Europe, *C. officinalis* is commonly known as **pot marigold**. The most popular varieties are grown for their 10cm (4in), yellow-orange flowers which are produced in flushes from early summer until the first frosts. This plant has a dense bushy habit, with long, narrow light green leaves that produce a pungent aroma when crushed. There are many varieties, including **'Lemon Beauty'**, with masses of double or semi-double, pale yellow flowers, and **Pacific Beauty**, which has orange or yellow double flowers.
Cultivation Hardy; full sun; water normally; feed with a base dressing of fertilizer when planting out; susceptible to rust and smut (see page 104); propagate from seed sown direct into the intended flowering site in mid-spring.
Special needs This plant will grow in very difficult conditions, but prefers a well-drained growing medium.
Display ideas Very good for planting close to the edge of pots and tubs, and very colorful in window boxes.

THEACEAE

CAMELLIA

Camellia japonica

H 2.4–5m (8–15ft) **S** Up to 4m (12ft)
This very popular, hardy, evergreen shrub, a close relative of the tea plant, originates from Asia. Known as the **common camellia**, it is ideal for growing in a container, especially if your garden soil is limy. The dark glossy green leaves, up to 13cm (5in) long, are oval-shaped, ending in a pointed tip, with a very finely toothed margin. The main attraction is the flower display produced in early spring, with color shades ranging from pure white, through every shade of pink, to deep red. The number of petals can vary from one cultivar to another; flowers of up to eight petals are described as single, those with 12–14 petals as semi-double, and those with over 16 petals as double, but the doubles are the most popular. The flower size, ranging from 2.5 to 15cm (1 to 6in), differs greatly from variety to variety, although it can

be influenced by variations in soil type, location, and general maintenance. The most popular varieties are found in the *C.* × *williamsii* hybrids, as these plants tend to flower freely even when young.
Cultivation Slightly tender; full sun/ partial shade; water normally; liquid feed every two weeks from early summer to early autumn and mulch with organic matter in spring after flowering; always dead-head flowers; may be attacked by vine weevil (see page 105); propagate from 10cm (4in) semi-ripe cuttings in mid-summer.
Special needs Must have a lime-free growing medium. Water with rain water in hard-water areas. Never allow to dry out as this can cause flower bud drop. Requires a sheltered spot away from morning sun and cold winds.
Display ideas Grow as a single specimen against a south-facing wall.

CAMPANULACEAE

CAMPANULA

H 5cm–1.2m (2in–4ft) S 30–90cm (1–3ft)
This is a large genus of annuals, both hardy and half-hardy, biennials and perennials, commonly called **bellflowers**. The herbaceous perennial *C. carpatica* is a hardy clump-forming plant with small, rounded, mid-green leaves, slightly toothed along the edges. The 2.5cm (1in) bell-shaped flowers, which form cups as they open out and range from purple through to pure white, are produced from mid- to late summer. Popular named cultivars include **'Isabel'** with purple flowers; **'Jewel'** with deep violet flowers; and the pure white **'White Star'**. The low-growing *C. isophylla* is also a perennial, producing light-blue star-shaped flowers from mid-summer to early autumn.
Cultivation Hardy/half-hardy; full sun/ partial shade; water normally; top dress with slow-release fertilizer in spring; susceptible to rust and to attack by slugs (see pages 104–105); propagate from seed sown in mid- to late spring, from cuttings in late spring or early summer, or by division in mid-autumn.
Special needs All species like their roots to be moist but well drained. *Campanula alpestris* (syn. *allionii*) prefers a lime-free growing medium.
Display ideas Very good for providing carpets of color over the surface of the growing medium.

BIGNONIACEAE

CAMPSIS

H Up to 10m (30ft) S Up to 10m (30ft)
Commonly known as **trumpet vine** (UK) and **trumpet creeper** (USA), this is a genus of three species of deciduous self-clinging climbing shrubs of vigorous habit. The trumpet-shaped, yellow, orange and pinkish flowers are borne in clusters of three or five on small flower stalks. The mid-green leaves consist of a central stalk with seven or nine leaflets arranged in pairs with a single leaflet at the tip. In autumn the stems change from green to a pale golden-yellow, and small clusters of tufted hair-like roots appear which anchor the plant to its support. The most popular hybrid is *C.* × *tagliabuana* which has 7.5cm (3in) long salmon-red flowers.
Cultivation Hardy; full sun; water normally; mulch with organic matter in spring; hard prune in spring, cutting back one-year-old shoots to four or five buds; propagate by hardwood cuttings in late autumn.
Special needs Flower buds may abort if growing medium dries out.
Display ideas To form a weeping standard, tie the stems together up to 2m (6ft) high and restrict the overall spread.

CANNACEAE

CANNA

H Up to 2.1m (7ft) S 75cm (2½ft)
This herbaceous perennial, commonly known as **Queensland arrowroot**, originates from South America. It has large broad leaves, up to 75cm (2½ft) long and 30cm (1ft) wide, and bright flowers, often carried 2–3m (6–9ft) high. A popular plant is *C.* × *generalis* **'Wyoming'** which has brownish-purple, spear-shaped leaves 60cm (2ft) long and 30cm (1ft) wide. The orange-red or golden-yellow flowers are borne from mid-summer to early-autumn. Another interesting cultivar is the purple-leaved **'Lucifer'** which has vivid red flowers with a wavy yellow margin to each petal.
Cultivation Tender; full sun; water normally; mulch with organic matter in spring; liquid feed every two weeks until flowers emerge; propagate from seed in early spring or by division in spring.
Special needs Prefer a peaty growing medium, rich in organic matter.
Display ideas Excellent for dramatic single specimen displays.

Camellia × *williamsii* 'Anticipation'

Campsis × *tagliabuana* 'Madame Galen'

Canna × *generalis* 'Wyoming'

CUPRESSACEAE

CHAMAECYPARIS

H Up to 5m (15ft) s 2m (6ft)
Chamaecyparis, commonly known as **cypress**, is a small genus of seven species of evergreen trees and shrubs native to North America, Japan and Taiwan. They have flattened frond-like branchlets and smaller cones than *Cupressus*, their leaves opposite and densely arranged, often with drooping tips and small cones.

For container culture the slower-growing cultivars of *C. lawsoniana* are very popular; **'Pygmaea Argentea'** has young leaf sprays splashed creamy-white, turning deep green with age; **'Grayswood Pillar'** has a narrow upright habit and tightly packed, bluish-green growth. The *C. obtusa* cultivars provide some interesting plants, among them *C. o.* **'Nana Gracilis'** which produces dark green leaf sprays, held in angled fan-like groups, and takes 10 years to reach 1m (3ft) high and 40cm (16in) across. For shape and grace there is no better weeping conifer than *C. nootkatensis* **'Pendula'**, with its slender branchlets hanging vertically in long streamers.

Cultivation Very hardy; full sun/partial shade; water frequently and do not allow the roots to dry out; top dress with slow-release fertilizer in spring; propagate from semi-ripe heel cuttings up to 15cm (6in) long in early summer.

Special needs Prefer a moist, well-drained growing medium. The golden and variegated cultivars should be grown in full sun to keep their color. The white variegations on some cultivars may be damaged by winter frost.

Display ideas The slow-growing and dwarf varieties are very good for developing miniature gardens in tubs or troughs.

RUTACEAE

CHOISYA

Choisya ternata

H 1.5–2m (5–6ft) s 2–2.4m (6–8ft)
Commonly known as **Mexican orange blossom**, *C. ternata* hails from Mexico and the south-western USA. A medium-sized shrub, it has a wide-spreading bushy habit often establishing a flat dome-shaped appearance. The glossy green, deeply lobed leaves are divided into three segments and are highly aromatic when crushed. The white flowers are sweetly scented and produced in clusters of up to six from the leaf joints (nodes) throughout late spring and early summer, with a second smaller flush in autumn. *Choisya t.* **'Sundance'** is a striking cultivar with yellow coloring on the young foliage.

Chrysanthemum (Leucanthemum) maximum

Cultivation Hardy; full sun/partial shade; water normally; top dress with slow-release fertilizer in spring; remove frost-damaged shoots at soil level and trim into shape after flowering; propagate from 7.5cm (3in) semi-ripe cuttings in late summer.

Special needs Prefers a free-draining growing medium. Likes a sheltered position. In colder areas it does best against a south-facing wall as the foliage is prone to frost damage; and the container may need to be lagged (insulated) in winter to protect the roots.

Display ideas Makes a good but bushy screening plant.

COMPOSITAE/ASTERACEAE

CHRYSANTHEMUM

H 15cm–1m (6in–3ft) s 30–60cm (1–2ft)
Chrysanthemum, commonly known as **daisy flower**, is a genus of over 200 species of hardy and half-hardy annuals, herbaceous perennials and sub-shrubs, which are suitable for growing as pot plants both indoors and outside. All have broadly oval, deeply lobed, dark or mid-green leaves, and many are free-flowering with single daisy-like flowers or larger mop-headed blooms with no center visible. Recently many of the plants we know as *Chrysanthemum* have been reclassified into different groupings: *Argyranthemum, Rhodanthemum Leucanthemum, Leucanthemopsis* and *Tanacetum*.

Cistus × aguilari 'Maculatus'

The species *Argyranthemum*, commonly known as **Marguerite**, has pretty daisy-like flowers produced from mid-summer up to the first hard frosts. They grow to 2m (6ft).

Leucanthemum (formerly *Chrysanthemum*) ***maximum*** has single white flowers with a golden eye in the center. It has many named cultivars, such as *L. m.* **'Esther Read'** (double white flowers) and the closely related *L. m.* **'Horace Read'** (double creamy-white flowers). These plants produce their blooms in summer on stiff stocky stems up to 45cm (18in) long.

Tanacetum (formerly *Chrysanthemum*) ***parthenium*** is a hardy but short-lived perennial that is usually treated as an annual. It is low-growing, reaching only 30cm (1ft), and produces large quantities of 4cm (1½in) flowers held above aromatic, light green leaves.

There are also alpine species, such as *Rhodanthemum* (formerly *Chrysanthemum*) ***hosmariense***, a low, spreading dwarf, shrubby perennial bearing silver-gray leaves and single, white daisy-like flowers. This is hardy, flowering from mid-spring through late autumn.

Cultivation Hardy to tender; full sun/partial shade; water normally; liquid feed every two weeks from mid- to late summer; susceptible to downy mildew, rust and viruses, and to attack by eelworm, leaf miner, aphids and thrips (see pages 104–105); propagate perennials from 5–8cm (2–3in) cuttings from late spring to mid-summer and annuals from seed sown in early spring.

Special needs Prefer a free-draining growing medium. Most plants need some support to keep them upright.

Display ideas Grow in pots indoors to achieve early flowering before plunging into display containers outdoors. See pages 10, 48 and 75.

Clematis florida 'Sieboldii'

Citrus sp.

CISTACEAE

CISTUS

H Up to 2m (6ft) S Up to 1.5m (5ft)
A small genus of about 20 species and many hybrids, these evergreen shrubs originate from the Mediterranean. They are commonly called the **rock rose** because of their single rose-like flowers, produced in large quantities from late spring through early summer. The saucer-shaped flowers, with thin, papery petals and a conspicuous cluster of stamens, are short-lived, opening in the morning and shedding their petals by late afternoon. These plants vary in habit from low-growing shrubs to much larger plants. Most have dull green leathery leaves, often sticky to the touch, carried on tough woody stems. The cultivar *C.* × *aguilari* **'Maculatus'** has purple-crimson blotches at the base of each petal. The most commonly seen plant is *C.* **'Silver Pink'**, which has gray foliage and clear pink flowers with a pale yellow eye in the center.
Cultivation Hardy to half-hardy; full sun; water normally; mulch with organic matter in spring; lightly trim straggly growth, but do not prune into old wood; propagate by semi-ripe cuttings, 7.5cm (3in) long, in late summer.
Special needs Plants may require overwintering indoors.
Display ideas An excellent choice for coastal gardens or hot dry patios.

RUTACEAE

CITRUS

× *Citrofortunella microcarpa*

H 45–60cm (18in–2ft) S 45–50cm (18–20in)
A genus of 15 partially spiny, semi-evergreen trees and shrubs, commonly known as **orange**, **lemon** and **grapefruit**, originating from South-east Asia. They produce glossy, dark green, oval leaves up to 10–15cm (4–6in) long which are often scattered sparsely over the gray-green spiny branches. The 2.5cm (1in) flowers are heavily scented, usually pure white with prominent yellow stamens, and borne in clusters of three to eight. The hardiest species is *C. ichangensis* **(Ichang lemon)**, its white flowers producing lemon-like fruits in temperate climates. Certainly the most commonly grown ornamental citrus is the hybrid × *Citrofortunella microcarpa* (formerly known as *Citrus mitis*). This 'orange bush', reaching only 45–60cm (18in–2ft), is the best-known species for pot culture, with its healthy-looking, mid- to deep green, oval leaves. It has the added advantage of fruiting early in its life, producing clusters of three or four fragrant white flowers 1cm (½in) across; these are in turn followed by round orange-yellow fruits up to 5cm (2in) in diameter.
Cultivation Slightly tender; full sun; water normally; liquid feed every two weeks from late spring to early autumn; prune only to remove dead or diseased growth; susceptible to attack by scale insect and mealy bug; propagate from seed sown in mid-spring or by semi-ripe cuttings in summer.
Special needs Prefer a sheltered spot and free-draining growing medium; overwinter indoors.
Display ideas Ideal as a single specimen on a sunny patio.

RANUNCULACEAE

CLEMATIS

H Up to 5m (15ft) S Up to 7m (23ft)
Clematis is a large genus of some 250 species of hardy and tender, woody, climbing plants, including several evergreen and herbaceous species, all grown for their attractive flowers and silken seed heads. The climbing species support themselves by means of their leaf tips, which are modified to twine and cling to any available support. The clematis 'flower' is not really a flower, but is composed of four or more sepals which are usually brightly colored and can vary in size from 1cm (½in) to 15cm (6in) across, depending upon the cultivar or species.

For container growing the *C. alpina* and *C. viticella* hybrids, with their dangling bell-like blooms, are ideal as they are very free-flowering and not too vigorous. *Clematis a.* **'Pamela Jackman'** has beautiful, large, blue and white flowers; *C.* **'Alba Luxurians'** bears mauve-tinted white blooms; *C.* **'Etoile Violette'** has violet-purple flowers; and *C.* **'Minuet'** has creamy-white flowers with a broad purple band on the tip of each sepal which it holds erect. Two unusual species are ***C. florida*** **'Sieboldii,'** which has passion-flower-like blooms produced on the current season's growth, and ***C. campaniflora***, which bears small, blue-tinted, white, bell-shaped flowers. Of the large-flowered hybrids, good performers include *C.* **'Vyvyan Pennell'**, with double flowers colored violet-blue with purple and carmine shades in the center; and *C.* **'Jackmanii Superba'**, which bears brilliant violet-purple blooms.

Cultivation Hardy; full sun/partial shade; water normally; mulch with bulky organic matter and top dress with slow-release fertilizer in spring; do not prune species that flower before midsummer unless you have to; prune by one third, in early to mid-spring, species that flower between mid- and late summer; cut back to 15cm (6in), in early spring, species that flower from late summer on; susceptible to powdery mildew and clematis wilt, and to attack by aphids, earwigs and slugs (see Diseases and Pests, pages 104–105); propagate climbing species from 10cm (4in) semi-ripe cuttings in mid-summer or 20cm (8in) hardwood cuttings in mid-autumn or early spring; propagate herbaceous species by division in early spring.

Special needs Prefer a well-drained alkaline growing medium. Keeps roots cool and moist.

Display ideas Very good for climbing through shrubs, which will provide the plant with natural support and winter protection. See page 19.

COBAEACEAE

COBAEA

Cobaea scandens

H Up to 3m (10ft) S 3m (10ft)

Cobaea is a genus of 18 species of evergreen and deciduous, woody-stemmed, half-hardy, perennial climbers. The evergreen species *C. scandens*, commonly known as **cathedral bells**, provides quick decorative cover for pergolas and trellises. This is a vigorous climber which supports itself by means of tendrils growing from the ends of the leaf stalks, just like the garden pea. At first the plant tends to grow vertically, but it spreads at the top and often hangs over to form a natural arch. The leaves are mid-to dark green and divided into four to six pairs of leaflets, very much like wisteria. The 7.5cm (3in) bell-shaped flowers which are produced from early summer to mid-autumn are creamy-green in color, turning purple as they age.

Cultivation Tender; full sun; water frequently when in full growth; feed with a base dressing of slow-release fertilizer when planting out; remove straggly or frosted growths in mid-spring; propagate from seed sown under protection in mid-spring.

Special needs Likes a sheltered position. Needs protection from frost. Prefers a well-drained growing medium with plenty of organic matter. Do not feed once established or the plant will put on large amounts of growth but will not flower. Dead-head regularly to ensure a succession of flowers.

Display ideas A vigorous-growing climber, ideal for growing up trellises.

CONVOLVULACEAE

CONVOLVULUS

H Up to 1m (3ft) S 1.2m (4ft)

Convolvulus is a large genus of hardy and half-hardy annuals, many of them trailing species, and deciduous and evergreen shrubby perennials of wide geographic distribution. All plants have wide-mouthed trumpet-shaped flowers, which are short-lived, carried either singly or in large clusters; these generally open in the early morning and fade and deteriorate as the day progresses. Useful annual species include ***C. tricolor***, a tough, free-flowering, bushy plant, with several named cultivars such as the trailing *C. t.* **'Royal Ensign'** which carries deep blue flowers with white and yellow centers. Among the perennial species worth consideration is ***C. cneorum***, a slightly tender, evergreen shrub of compact and low-growing bushy habit, with silvery, silky, narrow, pointed leaves on silver hairy stems. Its flowers, a soft pink in tight bud, opening to a pure white with a small golden-yellow eye in the center, are produced at the tips of the shoots from early summer to early autumn.

Cobaea scandens

Cultivation Tender; full sun/partial shade; water normally; liquid feed every two weeks from mid- to late summer; propagate annuals from seed sown in mid-spring; propagate perennials from 6cm (2½in) heel cuttings taken in mid-summer.

Special needs Prefer a free-draining growing medium. Perennial species must be protected from frost in winter to prevent their hairy leaves becoming wet and frozen. Dead-head immediately after flowering to provide a succession of flowers.

Display ideas Plant out in a mixed display among silver-leaved plants.

RUBIACEAE

COPROSMA

Coprosma × kirkii

H 1m (3ft) S 1.2–2m (4–6ft)

Coprosma is a genus of about 90 species of evergreen shrubs or small trees originating from Australia, New Zealand and South America. *Coprosma × kirkii*, commonly known as the **mirror plant**, is a natural hybrid which occurs in the wild. This plant is quite prostrate initially, but develops into a shrub with an erect habit as it matures. The leaves are a narrow oblong shape, glossy green in color with a thick leathery texture; they are arranged in opposite pairs on thin, erect, densely packed branches. The insignificant flowers are borne in large numbers in late spring. There are male and female plants, and the females often produce tiny, egg-shaped, translucent, red-speckled, white berries. Other cultivars include *C. × k.* **'Variegata'**, which has small mid-green leaves

Cordyline australis 'Purpurea'

with white margins, but is less hardy than the type; and *C.* × *k.* 'Prostrata' which forms a low, intricately branched mound.
Cultivation Tender; full sun; water normally; mulch with bulky organic matter in spring; remove straggly growths, if necessary, in mid-spring; propagate from 10cm (4in) semi-ripe cuttings in late summer.
Special needs Prefers a light, free-draining growing medium. You must have male and female plants to get the small fruits.
Display ideas Very good as a single specimen for dry, sunny, exposed sites; ideal for coastal districts. See page 74.

AGAVACEAE

CORDYLINE

Cordyline australis

H Up to 8m (26ft) S 3m (9ft)
Cordyline is a genus of 15 species and cultivars of evergreen palm-like shrubs and trees originating from Australasia, India and South America. They are tender or half-hardy plants grown for their handsome foliage, and may require protection in winter. *Cordyline australis*, commonly known as **cabbage palm** or **palm lily** forms a slender trunk with a large dense mass of sword-like, gray-green leaves at its tip. The small, fragrant, creamy-white flowers are produced after the plant is eight to ten years old in long spikes at the ends of the branches in early summer. There is a form with leaves flushed purple, *C. a.* 'Purpurea', and a species, *C. indivisa*,

Cotoneaster franchetii

with off-white flowers flushed purple, which appear in mid-summer.
Cultivation Hardy/slightly tender; full sun; water normally; mulch with organic matter in spring; prone to fungal leaf spot (see page 104); propagate from suckers in late spring or 7.5cm (3in) stem cuttings in early summer.
Special needs Prefers a free-draining growing medium.
Display ideas Good for a tropical effect. See page 41.

ROSACEAE

COTONEASTER

H 20cm–3m (8in–10ft)
S Up to 2.5m (8.5ft)
This is a genus of about 300 species of hardy evergreen and deciduous shrubs, with habits which range from low-growing types to large bushes and small trees. Many species and cultivars are grown for their rich autumn tints, and conspicuous berries in autumn. The berry colors range from yellow through coral-pink to deep scarlet. Many species also have attractive foliage, the leaves varying in color, but usually glossy and broad oval or oblong in shape. The prostrate cultivars are very popular for growing in containers, such as *C. horizontalis* with small red berries through winter; there is also a variegated form. Upright types are *C.* 'Coral Beauty', with glossy evergreen foliage and salmon-pink fruits in winter, and *C. lacteus*, with orange-red fruits.
Cultivation Hardy; full sun or partial shade; water normally; top dress with

Crocus chrysanthus 'Gipsy Girl'

slow-release fertilizer in spring; prune only to remove dead or damaged wood; propagate from semi-ripe heel cuttings, 7.5–10cm (3–4in) long, in late summer or early autumn; aphids (see page 105), which feed on the plants in the summer, may present a problem; fireblight may result in the death of the plant.
Special needs Will produce a better display of berries if kept slightly dry.
Display ideas Prostrate types are very good for providing a living mulch over the surface of the growing medium.

IRIDACEAE

CROCUS

H 10cm (4in) S 2.5–7.5cm (1–3in)
This genus of low-growing bulbous plants contains more than 70 species and countless varieties that flower outdoors from late summer through early spring. They produce long, erect tube-like flowers, opening out to form a funnel made up of six petals. Flower color varies from white through yellow to shades of pink and purple. The long, narrow, erect leaves are mid-green in color, with a white stripe running along the center. *Crocus chrysanthus* carries orange-yellow, scented flowers.
Cultivation Hardy; full sun; water normally; give plants a weak liquid feed at two week intervals after flowering until the foliage starts to yellow; propagate by removing small cormlets from the parent plant.
Special needs Prefer a well-drained growing-medium; flowers may need protection from birds.
Display ideas These plants are ideal for tubs and sink gardens, and are excellent for providing early spring color. The heavily scented cultivars are ideal for window boxes. See page 44.

CUPRESSACEAE

CUPRESSUS

H Up to 5m (15ft) s 2m (6ft)
The **cypresses**, as *Cupressus* are commonly known, are a genus of some 30 species of evergreen conifers, whose origins are widely distributed throughout the world. The majority form large trees of conical or columnar habit, often reaching well over 20m (65ft) in their native habitat, and are really not best suited to being grown in containers for any length of time. There are, however, numerous slower-growing clones and hybrids which have very attractive foliage or growth habits – these are highly desirable as container-grown plants. The most commonly grown cultivars are usually types of *C. glabra* (**smooth Arizona cypress**), an interesting plant for growing in tubs. The only true dwarf member of this species in cultivation is *C. g.* **'Compacta'**, which takes about 10 years to grow 45cm (18in) high and 40cm (16in) wide. Another useful species is *C. macrocarpa* (**Monterey cypress**), of which the very popular cultivar *C. m.* **'Goldcrest'** is fast-growing, with a conical habit and golden-yellow aromatic foliage held in plume-like sprays.
Cultivation Hardy; full sun/partial shade; water normally; mulch with bulky organic matter in spring; propagate from 5–7.5cm (2–3in) semi-ripe heel cuttings in early to mid-autumn.
Special needs Prefer a free-draining growing medium containing plenty of organic matter. Hardy once established, but give young plants winter protection in colder areas. Golden forms must be grown in full sun or they will lose their color and turn green.
Display ideas Very good for growing as individual specimens, or in matching pairs to frame a view. Useful support and protection for tender climbers.

PRIMULACEAE

CYCLAMEN

H Up to 30cm (1ft) s Up to 45cm (18in)
Cyclamen, commonly known as **sowbread**, is a small genus of hardy and half-hardy tuberous plants originating from the Mediterranean region, all grown for their unusual shuttlecock-shaped flowers. *Cyclamen hederifolium* (formerly *C. neapolitanum*), the most common hardy species, is grown for its 2.5cm (1in) long mauve-pink flowers produced on thin red stems 15cm (6in) long from late summer to late autumn. The persistent leaves, almost circular or ivy-leaf-shaped, dark green in color with silvery markings and reddish-purple underneath will often remain on the plant until early summer. There is also a white-flowered form, *C. h. album*. *Cyclamen coum* is another hardy species, but with heart-shaped mid-green leaves marked with silvery-white which are dark red beneath; the rounded flowers with broad petals are up to 7.5cm (3in) long, pink or occasionally white in color and produced from early winter to mid-spring.
Cultivation Hardy/half-hardy; full sun/partial shade; water normally; feed with bulky organic matter after flowering; susceptible to botrytis and viruses, and to attack by vine weevil and aphids (see page 105); propagate from seed sown in late summer.
Special needs Prefer a free-draining growing medium high in organic matter. Protect tubers from frost.
Display ideas Ideal for window boxes, planted up with heathers and trailing ivy. See page 59.

Dahlia 'Libretto'

Dahlia merckii

COMPOSITAE/ASTERACEAE

DAHLIA

H Up to 1.5m (5ft) s 1m (3ft)
Dahlia is a small genus of half-hardy herbaceous perennials with thick, fleshy, tuberous roots. All originate from a small number of Mexican species. The modern garden dahlias, considerably different from their ancestors, are grown mostly for the vast range of forms and colors which make them outstanding plants for container gardening. All dahlias have large mid- to

dark green leaves sub-divided into many smaller sections, very similar to the leaves of the ash tree (*Fraxinus*). The stems and branches of the plant are hollow, almost tube-like in appearance, but the tube is closed (chambered) at each leaf joint.

Dahlias are divided into two groups for ease of classification: bedding types, which are much more compact in height and habit, with 5–7.5cm (2–3in), single, semi-double or full-double flowers, and are usually grown as annuals; and border types, which are more vigorous and produce much larger flowers up to 30cm (1ft) across in a vast range of bright and softer pastel colors, and are usually grown as true perennials. Because of the large number of species, the border dahlias have been further subdivided into a classification based on flower type as follows: **Ball** types have small round flowers 7.5–10cm (3–4in) across; **Cactus** types have large spiky flowers with tube-like petals; **Decorative** types have fully-double flowers; and **Single** types have one row of petals and a central disc to the flower (like a daisy). There are also a number of cultivars with attractive foliage, such as **'Bishop of Llandaff'** with bronze-green leaves, and **'David Howard'** with purple-bronze leaves.

Cultivation Half-hardy; full sun; water normally; feed with a base dressing of slow-release fertilizer when planting out and liquid feed every two weeks during the flowering season; susceptible to viruses and sclerotinia, and to attack by aphids, slugs and earwigs (see Diseases and Pests, page 105); propagate bedding types from seed sown in mid-spring, and border types by softwood cuttings in early spring or division in mid-spring.

Special needs Prefer a free-draining growing medium. Must be protected from frost: lift tubers and store in a frost-free environment during the winter. Provide support for the taller cultivars.

Display ideas Ideal for a bold color display in a very hot, sunny position.

SCROPHULARIACEAE

DIASCIA

H Up to 30cm (12in) **S** 20–30cm (8–12in)

These half-hardy perennials, which originate from South Africa, make ideal summer-flowering container plants.

Diascia 'Ruby Field'

They have slender and low-growing habits, with dark green, broadly oval-shaped, glossy leaves, which have a row of teeth along the margin. The tube-shaped rosy-pink flowers, which are 7.5cm (3in) across and appear from mid-summer, have spotted throats. A popular cultivar is ***D.* 'Ruby Field'** with salmon-pink flowers.

Cultivation Half-hardy; full sun/partial shade; water sparingly, but keep slightly dry; liquid feed at half strength from late spring until early autumn; prune to encourage branching by pinching out the tips of shoots; propagate by sowing seeds indoors in early spring.

Special needs Must be cut back hard to promote a secondary flush of flowers. This process can be repeated after subsequent flushes to prolong flowering until late autumn.

Display ideas Superb for planting around the edges of large pots, as these plants will flop over the pot rim and partially trail down the sides.

AIZOACEAE

DOROTHEANTHUS

Dorotheanthus bellidiformis

H 7.5–15cm (3–6in) **S** 30–45cm (1ft–18in)

This is one of a large number of closely related species of succulent plants which until recently were classified under the genus *Mesembryanthemum*, but now have been assigned a new genus. Researchers have placed the plant described here in the genus *Dorotheanthus*. Originating from South

Dryopteris filix-mas 'Crispa Congesta' (back)

Africa and commonly known as the **Livingstone daisy**, it is well adapted to survive in dry, arid conditions. It has a low spreading habit, with narrow, light green, tube-like leaves which have a glistening appearance, making them look as if they are perspiring. These thick fleshy leaves can store large quantities of moisture and help the plant withstand drought. Given a good summer in a dry sunny position, the plant is unequalled for its sheer brilliant glowing color, producing masses of bright, 2.5cm (1in), daisy-like flowers whenever the sun comes out. Colors include white, pink, carmine, salmon-apricot and orange, and the display can last from mid-summer through to early autumn.

Cultivation Tender; full sun; water sparingly; feed with a base dressing of slow-release fertilizer when planting out; susceptible to basal rot caused by fungus, and to attack by slugs and ants (see page 105); propagate from seed sown indoors in mid-spring.

Special needs Prefers a hot site and a very free-draining growing medium. Keep slightly dry.

Display ideas Very good for hanging baskets, and ideal for those who forget to water now and again.

DRYOPTERIDACEAE

DRYOPTERIS

Dryopteris filix-mas

H 1m (3ft) **S** 1.2m (4ft)

Dryopteris is a large genus of hardy and tender perennial ferns whose name in Greek means 'oak fern', given because

Ensete ventricosum

of the shape of the scales protecting the spore capsules. The most commonly grown species is *D. filix-mas*, commonly known as the **male fern**, which occurs naturally in the northern hemisphere and is very hardy. The fronds (leaves) are deep glossy green, spear-shaped and finely divided into many smaller leaflets which arch gracefully, so that the tips almost touch the ground. This fern spores prolifically and soon colonizes whole areas. A similar species is *D. cristata*, the **crested buckler fern**, which is less vigorous but has a more spreading habit; its fronds are pale yellow-green with crinkled edges.

Cultivation Very hardy; partial shade/shade; water frequently; top dress lightly with bone meal in spring; leave the dead top growth to protect the crown over winter and cut back hard in mid-spring; propagate from spores sown in mid-spring or by division of rhizomes in late spring.

Special needs Prefers a peaty growing medium with a high fiber content. Avoid exposure to direct sunlight as the plant will quickly dry out and the leaves will shrivel. In hot weather spray the leaves with water each evening.

Display ideas Very good for giving life to a damp shaded corner.

ELAEAGNACEAE

ELAEAGNUS

H Up to 2m (6ft) S 2.4m (8ft)

Elaeagnus, commonly known as **oleaster**, is a genus of around 35 species of evergreen and deciduous shrubs and small trees originating from Asia, southern Europe and North America. These tough plants are suitable for exposed sites and are tolerant of both atmospheric pollution and coastal conditions. The small, white, highly scented flowers are produced in large numbers from early summer to late autumn and are followed by small, silvery-orange, egg-shaped fruits. The leaves of both the evergreen and deciduous species have a silver sheen to their oval-shaped leathery leaves, and thick golden stems. There are some very attractive variegated cultivars of ***E. pungens***: ***E. p.* 'Dicksonii'** is a slow-growing form with an irregular margin of pale golden-yellow around each leaf; and ***E. p.* 'Maculata'** has a vivid golden splash in the center of each leaf. Two forms of ***E. × ebbingei*** are also very popular: **'Gilt Edge'** has a deep green leaf with a broad margin of bright golden-yellow; and the glossy leaves of **'Limelight'** start green and develop a central blotch of golden-yellow as the plant matures.

Cultivation Hardy; full sun/partial shade; water normally; mulch with bulky organic matter in spring; cut out dead, diseased and straggly growths in late spring and remove green (reverted) shoots from variegated cultivars as they develop; propagate from 10cm (4in) semi-ripe cuttings in late summer.

Special needs These plants prefer a free-draining growing medium, which is rich in organic matter.

Display ideas Very good grown as a screen to protect more tender plants.

MUSACEAE

ENSETE

Ensete ventricosum

H Up to 5m (15ft) S 2m (6ft)

Commonly known as the **Abyssinian banana**, this species of evergreen perennial herbaceous plants originates from the east coast of Africa, southern China and South-east Asia. The large, paddle-shaped leaves, up to 4m (12ft) long and 1m (3ft) wide, are deep green in color with a bright red vein running through the center of each. The plant has a tree-like appearance but no trunk, the body of the plant being composed of a sheath of leaf-stalks. The flower heads emerge from within this 'trunk' in an erect globe-like cluster up to 30cm (1ft)

across. Each flower head may have up to 20 flowers, 5cm (2in) long, in a cluster of purple-brown bracts. These produce banana-like fruits up to 7.5cm (3in) long. The compact species *E. superbum* is very popular for containers.
Cultivation Half-hardy; full sun; water normally; mulch with bulky organic matter in spring; dead-head flowers; propagate by seed sown indoors in early spring.
Special needs Prefers a deep container with a well-drained growing medium high in organic matter.
Display ideas Very good for use in a display with a sub-tropical theme.

ERICACEAE

ERICA

H 30–45cm (1ft–18in)
S Up to 60cm (2ft)
The **heaths** or **heathers**, as they are commonly known, are a large genus of over 500 species ranging from dwarf shrubs to small trees. The largest number of species originates from South Africa but many of these are not frost-hardy; the Mediterranean species, however, are hardy. The cultivars are now very numerous, with many new ones swelling the ranks. Their popularity is based on the fact that it is possible to achieve an all-year-round effect by planting different cultivars.

Those cultivars with long flower spikes are ideal for container growing, and even when the red, pink or white blooms are dead and brown, they are still attractive. Other species are also grown for their attractive foliage, such as ***E. cinerea* 'Golden Drop'**, which has golden foliage whose new growths are copper-red; and ***E. carnea* 'Aurea'**, which has golden-yellow leaves and pink flowers.
Cultivation Hardy; full sun; water normally; liquid feed with a weak solution every two weeks from mid-summer to early autumn; dead-head immediately after flowering by clipping over with shears; propagate from 2.5cm (1in) hardwood cuttings in late summer.
Special needs Must have a lime-free growing medium, although certain species, such as *E. carnea*, ***E. erigena*** and ***E. terminalis*** and their cultivars will tolerate some lime.
Display ideas Grow around the base of wide-stemmed plants, such as *Betula*. See pages 44 and 53.

CELASTRACEAE

EUONYMUS

H 45cm–2m (18in–6ft) S 1–4m (3–12ft)
Euonymus, commonly known as **spindle**, is a genus of about 175 evergreen and deciduous species of low-growing shrubs, small trees, climbers and creepers. Most originate from Asia, although other species are distributed throughout Europe, North America and Australia. The flowers are insignificant, but the fruits, lobed and sometimes winged, can be brightly colored and very striking, while the deciduous species often produce very attractive autumn leaf color. Of the evergreen species, ***E. fortunei*** is very popular because it is so versatile and will grow almost anywhere. The following variegated forms are worth growing: ***E. f.* 'Silver Queen'** is a low-growing plant with silver-margined, dark green leaves which in winter are flushed pink and cream. Of the deciduous species, ***E. europaeus* 'Red Cascade'** has leaves which turn a rich scarlet in autumn and produces large quantities of rosy-red fruits that hang on the plant all winter.
Cultivation Very hardy; full sun/partial shade; water normally; top dress with slow-release fertilizer in mid-spring and mulch with bulky organic matter in mid-autumn; remove dead or damaged branches; young shoots susceptible to aphid attack (see page 105); propagate from 10cm (4in) semi-ripe heel cuttings in late summer.
Special needs Tend to be shallow-rooted, so do not allow to dry out.
Display ideas The evergreen species are very good for topiary.

EUPHORBIACEAE

EUPHORBIA

H Up to 1.2m (4ft) S Up to 1.2m (4ft)
Euphorbia, commonly known as **spurge**, is a large genus of about 2000 species of annuals, biennials, perennials and shrubs, widely distributed, particularly in sub-tropical regions. The perennial and shrubby species all have strap-like leaves with a prominent rib along the center and exude a milky-white sap when damaged. ***Euphorbia wulfenii*** is a popular evergreen sub-shrub with light blue-green, strap-like leaves and bright yellow, densely packed bracts on the tip of each shoot. ***Euphorbia griffithii*** makes a vivid display with its flame-red bracts in 10cm (4in) clusters produced from early to mid-summer. The upright ***E. schillingii***, with bright green leaves and buttercup-yellow blooms, creeps across the surface of the growing medium, rooting as it spreads.
Cultivation Hardy; full sun/partial shade; water normally; top dress with slow-release fertilizer in mid-spring; remove dead flowerheads and thin out overcrowded growth; propagate from 7.5cm (3in) cuttings in late spring.
Special needs Dislike a windy site. *The sap of these plants is a skin irritant and may cause severe blistering and, if swallowed, gastroenteritis.*
Display ideas The shorter species are very good for providing ground cover or for filling gaps.

Euonymus fortunei 'Silver Queen'

Euphorbia schillingii

Fatsia japonica 'Variegata'

ARALIACEAE

FATSIA

Fatsia japonica

H Up to 2.4m (8ft) S 3m (9ft)
This is a single-species genus, originating from the Far East – namely Japan and Taiwan, which forms an evergreen shrub of loose spreading habit. *Fatsia japonica*, or **Japanese figleaf**, produces leaves 15cm (6in) across, very much like those of ivy in shape; they are a glossy dark green color and held on long leaf stalks. The large clusters of white flowers, appearing in mid-autumn, are held on the tips of erect, thick, green stems. This plant is extremely tolerant of coastal conditions and copes well with the atmospheric pollution of a town or city garden. The variegated form, *F. j.* 'Variegata', with white tips to the leaves, is not as vigorous or as frost-hardy as the type.

Cultivation Hardy; full sun/partial shade; water normally; top dress with slow-release fertilizer in spring; remove any straggly, diseased or frost-damaged growths in late spring; propagate from stem cuttings (a single leaf and bud) in late spring.

Special needs Although hardy, this plant will benefit from the shelter of a south- or west-facing wall in colder districts. Prefers a free-draining growing medium containing plenty of coarse sand or gravel. Cover the base of the plant with straw or some form of matting in severe weather.

Display ideas Best grown as a single specimen, the large leaves bringing a subtropical atmosphere to a display.

Fuchsia 'Thalia'

MORACEAE

FICUS

H Up to 3m (10ft) S 3m (10ft)
This is a very large genus of well over 800 species of deciduous and evergreen trees, shrubs and woody vines originating from tropical and sub-tropical areas. Very few are hardy enough to be grown outdoors all year round, but some are used for outdoor displays and overwintered indoors, the main exception being *F. carica*, the edible fig, and its close relatives which, though not fully hardy, can survive outdoors with some winter protection. The thick stocky shoots of this species carry the large, dark green, deeply lobed leaves; and the fruits, borne on the shoot tips, may stay on the tree over winter if protection is given. ***Ficus c.* 'Brown Turkey'** is the most reliable cropping variety in a temperate climate. Less hardy is *F. benjamina*, commonly known as **weeping fig**, a beautiful evergreen plant with thin twiggy stems carrying narrow pointed leaves, which open as pale green and turn darker green with age. When left to develop, the plant makes a large ornamental shrub or small tree of 2m (6ft) or so, with gently curving, pendulous branches.

Cultivation Half-hardy/tender; full sun; water normally; liquid feed every two weeks from early summer to early autumn; no pruning required, apart from the removal of dead or damaged branches; propagate *F. carica* by layering in mid-summer, *F. benjamina* by air layering in early summer.

Fuchsia 'Leonora'

Fuchsia 'La Campanella'

Special needs Prefer a south-facing position. Avoid overwatering which can result in premature leaf fall. Provide winter protection for hardier species and move tender plants indoors.
Display ideas Very good for a subtropical theme.

ONAGRACEAE

FUCHSIA

H Up to 1.2m (4ft) S Up to 1.2m (4ft)
This is a genus of some 100 species of deciduous shrubs originating mainly from Central and South America. These plants, commonly known as **lady's eardrops**, are usually grown for their attractive, pendulous, bell-shaped flowers, which range from slender tubes through to double bicolored blooms with overlapping petals, and four spreading sepals forming a bell. The leaves are mostly a dull mid-green, with a prominent red leaf stalk continuing as a central rib running along the middle of each leaf. In coastal and milder inland areas, fuchsias appear fully hardy and in some places are even used as hedge plants. The less hardy varieties are popular as house plants. Many of the upright cultivars, such as *F.* **'Leonora'**, can be grown as bushes or standards, while the trailing and prostrate forms make attractive plants for hanging baskets and window boxes.

Of the hardy fuchsias *F. magellanica* is the hardiest, with its small tube-like flowers of dull red with a purple center, although the species has been superseded by hybrid plants or forms with more ornamental interest. *Fuchsia m.* **var.** *molinae* **'Sharpitor'**, for instance, has small white flowers tinted mauve, and gray-green leaves margined white. There are very attractive variegated forms, such as *F. m. gracilis* **'Variegata'**, which bears green leaves margined creamy-yellow flushed pink, or *F. m.* **'Versicolor'**, a low-growing shrub with young leaves which are gray-green tinted rose, and irregularly variegated creamy-white when mature. The half-hardy hybrids are increasing in number: two attractive examples are *F.* **'Thalia'**, with orange-red flowers and large, dark green leaves with a red underside; and *F.* **'Cascade'**, with pink and white flowers and a weeping habit.
Cultivation Half-hardy; full sun/partial shade; water normally; mulch with bulky organic matter and top dress lightly in spring with slow-release fertilizer; cut back to the base in spring after the worst frosts have passed; susceptible to rust, and to attack by aphids, capsids and whitefly (see page 105); propagate from 7.5cm (3in) softwood cuttings from mid-spring to late summer.
Special needs Plant out deeply and cover the roots with a deep mulch of bracken or straw. Standard specimens must be overwintered indoors.
Display ideas Excellent for displays in hanging baskets and window boxes. See page 16.

Galanthus nivalis

LILIACEAE/AMARYLLIDACEAE

GALANTHUS

Galanthus nivalis

H 7.5–20cm (3–8in) S Up to 15cm (6in)
Commonly known as the **snowdrop**, this popular plant is native to Britain and northern Europe, and one of a genus of small clump-forming bulbs grown for their white flowers produced in early spring. The pure white, or occasionally white spotted green (*G. n.* **'Viridapicis'**) pendulous flowers, produced from mid-winter onwards, are carried singly on a slender green stalk up to 20cm (8in) in length. The flat strap-shaped leaves are a dull mid-green with a slightly bluish sheen to them, arranged in pairs usually one either side of the flower stalk. Several lesser-known cultivars and species have started to become popular for container growing: *G. n.* **'Flore Pleno'**, which bears double white flowers; and *G. n.* **'Samuel Arnott'**, which is more vigorous than the type and produces larger flowers up to 4cm (1½in) in length. There is even an autumn-flowering species, *G. reginae-olgae*. The other common species is *G. elwesii*, which is larger and more vigorous with larger flowers flecked green in early to mid-spring.
Cultivation Very hardy; partial shade; keep moist in summer until the leaves die back; liquid feed from late spring to early summer; propagate by division of clumps immediately after flowering and before the leaves die back.
Special needs Never allow the bulbs to dry out when they are growing. Cut off the dead flower heads – **do not** allow the seeds to develop.
Display ideas Very early-flowering. Looks refreshing when planted under evergreen shrubs in large containers.

ERICACEAE

GAULTHERIA

H Up to 1m (3ft) **S** Up to 1.5m (5ft)
This genus of 20 species of hardy, ornamental, evergreen shrubs, formerly known as *Pernettya*, originates from South America and Australasia. The plants have a dense bushy habit when young, but may become open and leggy with age. The small leaves are a glossy, dark green color and oval in shape, ending in a sharp point; they are usually tightly packed on slender red stems. The flowers are usually white and carried in the leaf joints from early to mid-summer. These are followed by clusters of small round fruits, which range in color from white and pink to purple and red. A popular species is *G. mucronata*, which has interesting variations: *G. m.* 'Bell's Seedling' has red fruits, and *G. m.* 'Lilacina' bears pale lilac fruits.
Cultivation Hardy; full sun/partial shade; water normally; mulch with bulky organic matter in spring; cut back leggy growths hard in spring to encourage new growth; propagate from 5cm (2in) cuttings in late summer.
Special needs Water with rain water in hard-water areas. Plant in groups of three or five to assist pollination.
Display ideas The berries form a magnificent winter display. See page 53.

GERANIACEAE

GERANIUM

H 10–75cm (4in–2½ft) **S** Up to 1m (3ft)
Geranium is a large genus of hardy herbaceous perennials, many with a low spreading habit. These free-flowering plants, commonly known as **cranesbill**, produce large clusters of 2.5cm (1in) saucer-shaped blooms from early summer until the first frosts. Most geraniums have very finely divided leaves, which are carried above the main body of the plant on tough thin leaf stalks. In some varieties the foliage is the main attraction, and the few variegated types are sometimes also attractively blotched with purple in the flowers. A number are evergreen, spending the winter as tight clumps of foliage, and they even have autumn color, with leaves turning a deep orange-red as the days get shorter. *Geranium* × *magnificum* has deeply lobed leaves and lavender-blue flowers. For softly textured sage-green leaves, choose *G. renardii*, its white flowers are delicately veined purple, and it makes an ideal choice for a container. Among the most popular and easy-to-grow plants are *G.* 'Johnson's Blue', *G.* 'Wargrave Pink', and the white *G. macrorrhizum* 'Album'. There are also a few annual species worth growing and these often have brightly colored stems and leaves in autumn and winter: for example, *G. lucidum* and *G. robertianum*.
Cultivation Hardy; full sun/partial shade; water normally; mulch with organic matter in spring; cut back to soil level in spring; prone to rust, and to attack by slugs (see pages 104–105); propagate by division when dormant.
Special needs Prefer an open free-draining growing medium. Divide every three to four years to reduce disease and encourage vigorous growth.
Display ideas Excellent for a warm dry situation. Very good for trailing over the rims of tubs and pots.

GRISELINIACEAE

GRISELINIA

H Up to 4m (12ft) **S** Up to 3m (9ft)
A small genus of about six species of slightly tender trees and shrubs, *Griselinia*, commonly called **broadleaf**, is native to New Zealand, Chile and Brazil. The most common species is *G. littoralis*, a large, dense, evergreen shrub bearing thick, fleshy, leathery, pale apple-green leaves with a glossy sheen. It is slow-growing and fully hardy only in the mildest localities, but thrives in coastal regions and is very tolerant of sea-salt spray. For different leaved forms look for *G. l.* 'Variegata', with its creamy-white-edged, green foliage; and the more recent *G. l.* 'Dixon's Cream', whose leaves are splashed and marked creamy-white. Variegated-leaved forms are generally less hardy than green types.
Cultivation Half-hardy; full sun/partial shade; water normally; mulch with bulky organic matter in spring; cut out untidy or frost-damaged growths in late spring; propagate from 10cm (4in) semi-ripe heel cuttings in late summer.
Special needs Must be sheltered from cold winds.
Display ideas Very good for an evergreen screen, perfect as support for herbaceous climbers such as *Eccremocarpus*, *Tropaeolum*, and *Lathyrus* (the perennial sweet pea).

Geranium × magnificum

CARYOPHYLLACEAE

GYPSOPHILA

Gypsophila paniculata

H Up to 1m (3ft) **S** 1.2m (4ft)
This hardy herbaceous perennial, commonly known as the **chalk plant** or **baby's breath**, is widely distributed across Europe and as far east as Siberia. It has a large, thick, fleshy root system, which makes it very tolerant of dry conditions, and thin, strap-like, gray-green leaves, very similar to those of the carnation, which are useful for limiting moisture loss. Masses of 1cm (½in) flowers in large clusters are produced on flower stalks up to 1m (3ft) high from mid- to late summer. Dwarf and pink-flowered cultivars are available, such as *G. repens* 'Rosea' which is very low-growing, being just 10–15cm (4–6in) high, and spreading to form a mat-like covering over the surface of the pot. The 1cm (½in) rose-pink flowers are produced in abundance from mid-summer to early autumn.
Cultivation Hardy; full sun/partial shade; water normally; top dress with a slow-release fertilizer in spring; cut back to soil level in late winter/early spring; propagate from tip cuttings in late spring or mid-summer, or from root cuttings when dormant.
Special needs Prefers a free-draining, slightly alkaline growing medium; top dress with lime every second year. Use brushwood to provide support by the time the plants are 30cm (1ft) high.
Display ideas Very attractive when grown in a combination with sweet peas, which weave through the foliage.

Gypsophila repens 'Rosea'

Hebe albicans

Hedera helix 'Glacier'

SCROPHULARIACEAE

HEBE

H Up to 1m (3ft) **S** Up to 1.5m (5ft)
There are about 75 species and countless hybrids of *Hebe*, commonly known as **Veronica**, which originate from the southern tip of South America, Australia and New Zealand. All are evergreen; most are shrubs, a few are trees. Many have attractive or unusual foliage and most have long flowering periods from spring through to autumn. The small flowers are borne in large, densely packed clusters. Many species are suitable for growing in containers, and the whipcord types with their thin *Cupressus*-like growths are particularly good for exposed sites. These include *H. armstrongii*, with densely branched, olive-green stems tipped yellow, and small white flowers appearing in late summer; *H. cupressoides*, a small shrub of erect habit with slender gray-green branches and small, pale blue flowers produced in mid-summer; and *H. ochracea* **'James Stirling'**, a dense dwarf shrub with erect, glossy, cord-like golden stems. The broad-leaved types include some very attractive forms, such as *H. albicans*, an evergreen shrub that forms a dense mound of blue-gray foliage, and tight clusters of white flowers, and the variegated *H.* × *franciscana* **'Variegata'**, whose leaves are edged creamy-white (very good for window boxes); and *H. speciosa* **'Tricolor'**, with its young growths tipped rosy-pink in autumn. The most popular of the ground cover types is the evergreen *H. pinguifolia* **'Pagei'** which bears purplish stems, gray-blue leaves and masses of small, white flowers in late spring or early summer.

Cultivation Half-hardy; full sun; water normally; top dress with slow-release fertilizer in spring; remove frost-damaged stems in spring, dead-head throughout the flowering season; susceptible to downy mildew (see page 104) and leaf spot (particularly in wet years); propagate from 5–10cm (2–4in) semi-ripe cuttings taken in mid- to late summer.

Special needs Protect from freezing winds. These plants will grow well in coastal areas.

Display ideas Very good for growing in window boxes. Small species are useful for alpine displays.

ARALIACEAE

HEDERA

H Up to 3m (10ft) **S** Up to 3m (10ft)
Commonly known as **ivy**, *Hedera* is a small genus of plants distributed as far afield as Europe, North America and Japan. Most are evergreen climbers which support themselves by means of aerial roots, or by running horizontally over the ground.

There are a large number of forms with very interesting characteristics, suitable for a range of themes and situations. *Hedera helix*, a native of the British Isles, is found in several popular cultivars: *H. h.* **'Palmata'**, slow growing with palm-like leaves; *H. h.* **'Glacier'**, which has small silver-gray leaves with a narrow white margin; and, perhaps the most striking, *H. h.* **'Goldheart'**, with a large, conspicuous, golden splash in the center of each leaf. The larger-leaved *H. algeriensis*, a vigorous-growing species with stout reddish shoots, is rather more erect than *H. helix*, which makes it ideal for growing up a trellis. The leathery leaves are a bright glossy green in summer, turning bronze-green in winter. The top growth

Hibiscus rosa-sinensis cvs.

may be slightly tender and needs cutting back in severe winters. Often seen is the variegated cultivar *H. algeriensis* **'Gloire de Marengo'**, with leaves which are dark green in the center, merging through silver-gray to a white irregular margin. *Hedera colchica* **'Dentata'** is a rapidly climbing plant with thin, widely lobed, dark green, purple-tinted leaves, the largest of all the ivies. It is an excellent wall plant, as is *H. colchica* **'Dentata Variegata'** with leaves marked with cream-yellow and pale green. The latter is much hardier than the more common cultivar *H. algeriensis* 'Gloire de Marengo'.

Cultivation Hardy; full sun/partial shade; water normally; liquid feed every month from late spring to early autumn; prune only to remove old flower stalks and damaged shoots; young shoots susceptible to attack by scale insect; propagate from single leaf bud cuttings taken in mid- to late summer.

Special needs These plants may need some tying until the aerial roots take a hold. Plants propagated from the juvenile growth retain juvenile characteristics, whereas plants propagated from adult growth will retain adult characteristics; this is one of the reasons why so many different forms and varieties exist. Ivy can be cut back hard but juvenile growth usually develops as a result of this.

Display ideas Very good in tubs, hanging baskets and window boxes, to form ground cover or grow over the sides and break up the straight lines of the container. Ivies with green leaves are very shade tolerant, while variegated forms require more light. See pages 40–41, 53, 58–59 and 64–65.

CISTACEAE

HELIANTHEMUM

H 7.5–10cm (3–4in) **S** Up to 60cm (2ft)

This large genus of dwarf evergreen shrubs, commonly called **rock rose**, originates from Asia, North Africa, southern Europe and the Americas. Characteristically these plants have a range of growth habits which reflect their aptitude for survival in a dry harsh environment. They form low-growing clumps with thin tough stems and small, oval, pale green leaves covered in fine hairs, and are very resistant to

Helianthemum 'Fire Dragon'

drought. The 2cm (¾in) saucer-shaped flowers are produced in massed flushes, close to the ground on short stems, from mid- to late summer in a multitude of colors. There are numerous single-colored named cultivars which perform very well in tubs and window boxes, including *H.* **'Wisley Pink'** with pink flowers emerging through gray foliage; *H.* **'Wisley Primrose'**, which bears soft golden-yellow flowers; *H.* **'Wisley White'**, with pure white flowers and gray foliage; *H.* **'Fire Dragon'**, with green foliage and deep copper-orange flowers; and the interesting ***H. nummularium grandiflorum*** **'Variegatum'**.

Cultivation Hardy; full sun; water normally; trim after flowering to produce a second flush of blooms in autumn; propagate from 5–6cm (2–2½in) semi-ripe heel cuttings in late summer.

Special needs Require a free-draining growing medium. Do not allow to become waterlogged.

Display ideas Ideal for trailing over the edges of tubs and window boxes.

BORAGINACEAE

HELIOTROPIUM

H Up to 60cm (2ft) **S** Up to 40cm (15in)

Commonly known as **heliotrope**, this is a large genus of half-hardy and tender annuals and shrubs. The plant most commonly grown for display, *H.* × *hybridum*, is a tender evergreen shrub. Its thick-textured, wrinkled, oblong, dark green leaves are carried on dark green, almost black stems. The small flowers, which are formed in 10cm (4in) flat clusters, are very fragrant, with colors varying from a deep bluish-purple through to lavender and pure white. Named varieties include **'Princess Marina'**, with large clusters of

Hosta sieboldiana var. *elegans*

violet-blue flowers, and the pale mauve **'Florence Nightingale'**. Other types can be raised from seed, such as the newer **'Mini Marine'** with its compact, bushy habit, bronze-green leaves and violet-purple flowers.

Cultivation Half-hardy; full sun/partial shade; water normally; liquid feed every two weeks from early summer to early autumn; propagate from seed sown in mid-spring, named cultivars from 10cm (4in) cuttings in late summer.

Special needs Prefer a free-draining growing medium. Large plants which are wanted for the following year must be overwintered indoors.

Display ideas Grow as standards in pots and plunge into the display containers. These plants can be induced to flower all year round.

MALVACEAE

HIBISCUS

H 2–3m (6–9ft) **S** 2m (6ft)

This is a genus of some 200 species of tender and hardy herbs, trees and shrubs, widely distributed throughout the world. The leaves vary in size, but all are mid- to dark green in color and roughly oval-shaped with a toothed margin. The flowers vary in size from 7.5–12cm (3–5in) depending upon the species, but all are trumpet-shaped and usually brightly colored. The tender species are often overwintered indoors and grown outside in summer. Popular species include *H. rosa-sinensis*, which produces large crimson, orange-pink and yellow flowers up to 12.5cm (5in) across. Of the hardier species most are related to *H. syriacus*: **'Blue Bird'** is mid-blue with a red center; **'Snowdrift'** is pure white; and **'Woodbridge'** is a rich rose-pink.

Hosta fortunei var. *albopicta*

Cultivation Hardy to tender; full sun; water normally; mulch with bulky organic matter in spring; thin out spindly growths and reduce strong growths by one third in early spring; propagate tender species from 7.5–10cm (3–4in) heel cuttings in mid- to late summer, hardier species from 7.5–10cm (3–4in) semi-ripe heel cuttings in mid-summer.

Special needs If kept too dry, the flower buds will fall off.

Display ideas Large flowers make an instant impact in a sunny situation.

LILIACEAE/HOSTACEAE

HOSTA

H Up to 40cm (16in) **S** 60cm (2ft)

This hardy herbaceous perennial, commonly known as the **plantain lily**, and originating mainly from China and Japan, has small trumpet-like flowers carried on spikes of up to 60cm (2ft) in height between mid-summer and early autumn. In the main, however, plantain lilies are grown for their attractive foliage: leaf shapes range from long and narrow through to oval with a pointed tip, and colors can vary from a glaucous blue through to rich combinations of silver and green or green and golden variegations. They are renowned for their tolerance of shade, although the variegated forms produce the best color in light shade. Given the right conditions, they will grow for many years requiring little or no maintenance. ***Hosta sieboldiana* var. *elegans*** has broadly spear-shaped, glossy bluish-green leaves with prominent veins and soft lilac-blue flower spires. The very attractive, but slower-growing ***H. s.* 'Frances Williams'** bears broad, elongated oval-shaped, glossy, bluish-green leaves with a bold yellow margin; *H.*

Hosta sieboldiana 'Frances Williams'

fortunei var. *albopicta* is a vigorous hardy perennial with pale green leaves with cream centers.

Cultivation Hardy; partial shade/shade; water normally; remove all dead leaves in mid-spring and dead-head after flowering before seed pods form; prone to attack by slugs (see page 105); propagate by division in mid-spring.

Special needs Prefer a damp, peaty, free-draining growing medium. Mulch with gravel to keep the roots cool and prevent slugs getting to the plant.

Display ideas Display in old chimney pots, which will also keep the roots cool. See pages 2, 63 and 64.

CANNABACEAE

HUMULUS

Humulus lupulus

H Up to 10m (30ft) **S** 5m (15ft)

One of a small genus of self-supporting, herbaceous, perennial climbers and commonly called the **hop**, this plant is a native of the British Isles, much of Europe and Asia. The leaves are 10–15cm (4–6in) long, deeply lobed rather like those of the sycamore (*Acer pseudoplatanus*), bristly and toothed around the margins; they are carried in pairs on thin, bristly twining stems. The flowers are insignificant, but the fruit clusters are quite attractive in autumn. The most popular cultivar is ***H. l.* 'Aureus'**, a clone with soft golden-yellow leaves.

Cultivation Hardy; full sun/partial shade; water normally; mulch with bulky organic matter in spring; susceptible to hop mildew; propagate from semi-ripe cuttings in mid-summer.

Special needs Prefers a free-draining growing medium.

Display ideas Very good for growing over a tripod. See pages 6 and 64.

Hyacinthus orientalis 'L'Innocence'

Hydrangea macrophylla 'Hamburg'

Hydrangea macrophylla 'Europa'

LILIACEAE/HYACINTHACEAE

HYACINTHUS

H Up to 25cm (10in) S Up to 15cm (6in)
This very popular bulb originates from the Mediterranean region, where as many as 30 species are known to be growing wild, although only a few of these are commonly grown as ornamental plants. The latter are, in the main, selections from *H. orientalis*, and the typical species itself is now quite rare in cultivation. All hyacinths have strap-like mid-green leaves, finishing in a cupped pointed tip, with four to six leaves accompanying each flower. The flowers are carried in a compact, densely packed spike up to 15cm (6in) long. An ever-increasing number of varieties is available in a vast range of colors. One of the most popular cultivars is *H.* **'L' Innocence'**.

Cultivation Hardy; full sun/partial shade; water normally; liquid feed every two weeks from the end of flowering until the foliage turns yellow; susceptible to viruses and bulb rot, and to attack by aphids (see pages 104–105); propagate from seed sown outdoors in mid-summer.

Special needs Keep moist when flowering. Store with care to avoid distorted growth and damaged flowers.

Display ideas Ideal for forcing early indoors and plunging into display containers just before flowering. See page 69.

HYDRANGEACEAE

HYDRANGEA

H Up to 2m (6ft) S Up to 2.4m (8ft)
This is a genus of 23 species consisting of small trees, shrubs and climbers, grown mainly for their large flat or dome-shaped clusters of flowers.

The shape or structure of the flower is used as a simple basis for classifying the plants. The **Lacecap hydrangeas** are grown for their broad flat blooms, which are surrounded by one or two rows of colored sepals. These include cultivars such as *H. macrophylla* 'Tricolor' with attractive, variegated green, white and yellow leaves. The **Hortensia hydrangeas** have the familiar large mophead blooms, which range in color from soft pink through powder blue. Attractive cultivars of this type include *H. m.* **'Hamburg'**, which is rose-pink but turns blue if treated; *H. m.* **'Europa'**, with large, deep pink flowers; and *H. m.* **'Nigra'**, with rose-colored flowers on black stems.

Cultivation Hardy; partial shade; water normally; mulch with bulky organic matter in spring; dead-head in spring; propagate from 10–15cm (4–6in) cuttings in late summer.

Special needs Prefer an acid growing medium, especially if blue flowers are desired; water with rain water in hard-water areas.

Display ideas Very good for cool shaded corners. See page 71.

AQUIFOLIACEAE

ILEX

H Up to 3m (10ft) S Up to 2m (6ft)
Ilex, commonly known as **holly**, is a genus of 400 species of evergreen and, occasionally, deciduous large shrubs and small trees originating from almost every corner of the globe. All have inconspicuous, white, star-shaped flowers, with red, orange, yellow or white berries on the female plants in winter.

Ilex aquifolium 'J.C. van Tol'

Impatiens Novette Sevies

Ipomoea rubrocaerulea 'Heavenly Blue'

The characteristic leaves, which range in color, bear sharp spines around the margin. There are many cultivars, the two most popular being ***I. aquifolium* 'Silver Queen'**, which has purple young shoots and dark green leaves with a silver margin, and which, despite its name, is **male**; and ***I. a.* 'J. C. van Tol'** with its glossy green leaves and bright red berries.

Cultivation Hardy; full sun/partial shade; water normally; mulch with bulky organic matter in spring; remove green shoots from variegated cultivars and trim clipped specimens by late summer; susceptible to leaf miner; propagate from 10cm (4in) semi-ripe cuttings in late summer.

Special needs These plants like a free-draining growing medium. The North American hybrids, which include ***I. × meserveae* 'Blue Prince', 'Blue Princess', 'Blue Angel', 'Blue Boy'** and **'Blue Girl'**, require a neutral or slightly acid planting medium.

Display ideas Very good as clipped specimens for topiary.

BALSAMINACEAE

IMPATIENS

H Up to 60cm (2ft) **S** Up to 60cm (2ft)

For many years the hybrid *Impatiens*, commonly called **busy Lizzie**, has been popular as a half-hardy bedding plant. The glossy leaves are broadly spear-shaped and can vary in color from deep green through to bronzy-red. The five-petalled flowers open out flat to 4cm (1½in) wide and are available in a range of colors including pink, deep red, and even red and white stripes. The plant will bloom non-stop from mid-summer through until the first frosts of autumn, and at times it is impossible to see any leaves because they are covered by the flowers. The vigorous **Novette Series**, with their large flowers often 7.5cm (3in) across produced all summer long, are particularly useful.

Cultivation Tender; full sun/partial shade; water normally; liquid feed every two weeks from mid-summer to early autumn; dead-head regularly to prolong flowering; susceptible to attack by slugs (see page 105); propagate from seed sown in early to mid-spring.

Special needs Must have a well-drained growing medium. Water regularly but avoid overwatering, which can cause the leaves to turn yellow and result in premature leaf drop.

Display ideas Good for a bold display of summer color. Perfect for hanging baskets and window boxes. See page 38.

CONVOLVULACEAE

IPOMOEA

H Up to 2m (6ft) **S** 1.5m (5ft)

Ipomoea, commonly known as **morning glory**, is a large genus of shrubs and climbers, some annuals and others perennials. The species usually grown are climbers that support themselves with their twining stems. These plants are grown for their attractive trumpet-shaped flowers, and two very striking examples are ***I. alba***, a soft-stemmed plant with 20cm (8in), evergreen, oval-shaped leaves and fragrant, white, tube-shaped flowers that open in the evening; and ***I. rubrocaerulea* 'Heavenly Blue'**, a vigorous plant with thin twining stems and pale green heart-shaped leaves. The latter has a free-flowering habit, producing large, deep blue blooms up to 13cm (5in) in diameter from mid-summer to early autumn.

Cultivation Half-hardy; full sun; water normally; liquid feed every two weeks from mid- to late summer; susceptible to slugs and aphids (see page 105); propagate from seed sown in late spring.

Special needs Provide canes or other support for the plant to grow well.

Display ideas Grow through deciduous trees and shrubs.

IRIDACEAE

IRIS

H Up to 75cm (2½ft) **S** 60cm (2ft)

Iris is a genus of over 300 deciduous and evergreen herbaceous perennials, all originating from the northern hemisphere. The leaves vary in color from bright glossy green to dull grayish-green. The flowers range in size, but the basic shape varies very little, all petal arrangements being in multiples of three. The bearded irises are the most popular, mainly because they are so easy to grow and offer a vast color range. Many irises grown today are hybrids of ***I. germanica***. Other interesting species include moisture-loving types, such as ***I. pallida* 'Variegata'**, with its lavender-blue flowers and green-and-gold-striped leaves; and ***I. pseudacorus* 'Variegata'**, with buttercup-yellow flowers and gold-and-green-striped foliage, which will grow well in water-logged conditions.

Cultivation Hardy; full sun; water normally; mulch with bulky organic matter in spring; susceptible to rhizome rot, and to attack by stem and bulb eelworm (see page 105); propagate by division immediately after flowering.

Special needs These plants prefer a neutral growing medium.

Display ideas Very effective mixed with other tall plants, including grasses and ferns.

Jasminum nudiflorum

Juniperus scopulorum 'Skyrocket'

OLEACEAE

JASMINUM

H Up to 3m (10ft) **S** 2m (6ft)
Jasminum, or **jasmine** as it is commonly known, is a genus of more than 400 species of deciduous and evergreen climbers and shrubs, originating mostly from the Old World. These popular plants are grown primarily for their fragrant, small, tubular flowers, which open to form a star of five petals of yellow, white or pale pink. In most species the stems are green, turning brown with age, and have a square shape. The climbing species support themselves by means of twining stems. The small mid-green leaves are often divided into three leaflets. *Jasminum officinale* produces white flowers with a very heavy fragrance from mid-summer to early autumn. As a wall shrub *J. nudiflorum* is outstanding, the yellow flowers appearing from late autumn through late spring. There is also a golden-leaved cultivar, *J. n.* 'Aureum'.
Cultivation Hardy; full sun/partial shade; water frequently; top dress with slow-release fertilizer in spring; susceptible to botrytis in shoots damaged by frost, and to attack by aphids (see page 105); propagate from 7.5–10cm (3–4in) semi-ripe cuttings in late summer.
Special needs None.
Display ideas Grow near a door or window to benefit from aromatic scent.

CUPRESSACEAE

JUNIPERUS

H Up to 7m (23ft); slow-growing cultivars up to 1.5m (5ft) **S** 1m (3ft)
This important genus of conifers, commonly known as **juniper**, includes almost 50 species. Most are hardy, showing remarkable tolerance to heat and considerable resistance to drought. These plants undergo a change in foliage characteristics as they make the transition from juvenile to adult growth. As they mature the reddish-colored bark takes on a peeling habit, and many plants produce black, berry-like fruits.

A vast array of cultivars exists, selected for particular growth forms and habits or different foliage colors, the dwarf or slow-growing cultivars proving the most popular for containers. Some of the *J. chinensis* clones are very popular: for instance, *J. c.* 'Stricta', which displays foliage blue to slate-gray in color and a pyramidal habit; and *J. c.* 'Obelisk', with its erect upright habit and bluish-green foliage. One of the most commonly available junipers is *J. scopulorum* 'Skyrocket', which is probably the narrowest conifer in cultivation. It has silvery blue-green foliage, but is sometimes used as a clear stem with a prostrate cultivar grafted on to it to form a weeping standard. *Juniperus squamata* 'Pygmaea', a cultivar with yellow-green foliage, takes 10 years to reach 1m (3ft) high. An excellent architectural plant is *J. recurva* var. *coxii*, a slow-growing juniper taking 10 years to reach 3m (10ft) high, with green foliage and thread-like tips that weep downwards.
Cultivation Very hardy; full sun/partial shade; water normally, but do not allow root system to dry out; top dress with slow-release fertilizer in spring; propagate from 10–15cm (4–6in) semi-ripe heel cuttings in late summer.
Special needs Gold and variegated cultivars look their best displayed in bright sunlight.
Display ideas Make good individual specimens and clipped topiary bushes.

LABIATAE/LAMIACEAE

LAMIUM

H Up to 60cm (2ft) **S** Up to 3m (10 ft)
Lamium, commonly called **dead nettle** (USA), is a genus of 45 species of annuals and hardy herbaceous perennials originating from north Africa and Europe, and closely related to the nettle (but without the sting). The small tubular flowers, which may be purple, pink, yellow or white in color are carried in groups of three or four in the leaf joints from early to late summer. The large, mid-green, broadly oval leaves have a coarse texture and are covered with fine hairs, although some cultivars are grown for their silver or variegated leaves. Most plants have a lax habit and often trail along the ground, rooting at each leaf joint (node). The slower-growing *L. maculatum* has small green leaves with a central silver stripe, and small purple-pink flowers. *Lamium m.* 'Roseum' has pale-pink flowers, and *L. m. album* bears pure white flowers. There is a golden-leaved cultivar, *L. m.* 'Aureum'. The most commonly seen cultivar is *L. galeobdolon* 'Florentinum', with large yellow flowers and silver-flecked evergreen leaves, although this is rather vigorous for containers.
Cultivation Hardy; full sun/partial shade; water normally; top dress lightly with slow-release fertilizer in spring; propagate by division in winter.
Special needs Silver- and golden-leaved types will perform well in partial shade.
Display ideas Ideal for trailing over the sides of containers. See page 59.

VERBENACEAE

LANTANA

H Up to 1.2m (4ft) **S** 1m (3ft)
Commonly known as **shrub verbena**, *Lantana* is a genus of some 150 frost-tender, evergreen, perennial shrubs originating from South America and the

West Indies. The deep green leaves are an elliptical shape with a toothed margin, carried on square or angular green stems. The attractive dome-like flower heads, up to 5cm (2in) across, are made up of large numbers of small tubular flowers. They are seen from early summer to mid-autumn, with colors varying from red-pink, through to orange-yellow. *Lantana camara* 'Cloth of Gold' has yellow flowers and a loose, spreading habit, while *L. c.* 'Rose Queen' is more upright but produces soft-pink flowers from orange-salmon buds. The much lower-growing *L. montevidensis* has small heads of rose-lilac flowers.
Cultivation Half-hardy; full sun; water frequently; liquid feed every two weeks from mid-summer to early autumn; susceptible to attack by whitefly; cut back to 10cm (4in) in mid-spring; propagate from 7.5cm (3in) cuttings taken in late summer and overwintered indoors.
Special needs Must have a very free-draining growing medium. Should be overwintered indoors.
Display ideas Very good for growing in hot, dry situations.

LAURACEAE

LAURUS

Laurus nobilis

H Up to 6m (18ft) S 3–6m (10–18ft)
Laurus is a genus of two species of unisexual, hardy, evergreen, small trees or large shrubs commonly called **bay laurel** and originating from the Mediterranean. These plants thrive in coastal regions, though the leaves may be scorched by persistent cold winds. They are often grown as clipped and shaped specimens, or as standards or half-standards in containers. The dark green, glossy, sharply pointed, strap-like leaves of *L. nobilis* are aromatic and used in cooking. Clusters of small yellow-green flowers appear on both male and female plants in spring, but are so inconspicuous that they often go unnoticed. *Laurus n.* 'Aurea' has pale golden-yellow leaves, and when compared to the common green form actually looks rather unhealthy in appearance.
Cultivation Hardy; full sun/partial shade; water normally; mulch with bulky organic matter in spring; trim into shape two or three times during summer, and remove sucker shoots from the stems of standards as soon as they

Lantana camara 'Cloth of Gold'

develop; susceptible to attack by scale insect (see page 105); propagate from 10cm (4in) semi-ripe heel cuttings in late summer/early autumn.
Special needs Prefers a free-draining growing medium. Keep trimmed with secateurs/pruning shears to retain the desired shape.
Display ideas This makes an excellent specimen plant, or as a centerpiece in a herb tub.

LABIATAE/LAMIACEAE

LAVANDULA

H 45cm–1.2m (18in–4ft)
S Up to 1.2m (4ft)
Lavandula, or **lavender** as it is commonly known, is a genus of 20 species of evergreen shrubs, originating from the warmer regions of Europe and the Mediterranean. They are useful for containers as they prefer dry conditions, and are popular for their fragrant flowers and aromatic foliage. The leaves are long, narrow, silver-gray and are covered with fine felt-like hairs, which are very effective for reducing moisture loss. The small tube-like flowers are carried in narrow clusters (spikes) up to 5cm (2in) long, on tough square stems, from early summer to early autumn. The most commonly grown form is *L. angustifolia* 'Hidcote', which has strongly scented, deep purple-blue flowers and a compact bushy habit. There is also a white cultivar, *L. a.* 'Alba', and a pink cultivar, *L. a.* 'Rosea'. The French lavender, *L. stoechas pedunculata*, has increased greatly in

Lavandula angustifolia 'Hidcote'

Lavandula stoechas pedunculata

popularity over recent years, the most noticeable feature of this species being the tuft of bracts at the top of each flower spike.
Cultivation Hardy; full sun; water sparingly; top dress with slow-release fertilizer in spring; remove the dead flower stalks and lightly trim the plants in late summer to keep their shape; cut back straggly plants hard in mid- to late spring; susceptible to botrytis in frost-damaged tissues, and to leaf spot, and to attack by frog hoppers (in cuckoo spit on the stems).
Special needs Although hardy, may suffer frost damage in cold regions. Tend to grow leggy with age, so discard or replace after five or six years.
Display ideas Grow close to the house to benefit from the scent.

PORTULACACEAE

LEWISIA

H Up to 30cm (12in) S 25cm (10in)
This small genus of some 20 species and numerous varieties of herbaceous and evergreen semi-succulent perennials originates mainly from the Pacific coast of Canada and the United States. Most varieties are grown for their attractive flowers which appear in late spring and early summer. Evergreen plants produce a flat rosette of leaves, which radiate out from a central point, and bear large quantities of funnel-shaped flowers on thin, erect flower stalks up to 30cm (12in) high. The two most popular plants are *L. cotyledon alba*, which forms clumps of rosette-like, fleshy evergreen leaves with crinkled or toothed edges, and clusters of white flowers in early summer; and *L. tweedyi* with large, fleshy leaves and stout stems, and large flowers up to 5cm (2in) across that range in color from whitish-pink through pinky-orange.
Cultivation Hardy; full sun; water sparingly; apply a base dressing of organic matter at planting time; dead-head flowers; propagate by removing offsets in summer.
Special needs The thick fleshy necks are susceptible to rotting in wet winters. To combat rotting, place a thick layer of chippings around the plants to provide additional drainage.
Display ideas These plants are shown off to best effect in troughs and sinks.

Lewisia cotyledon alba

OLEACEAE

LIGUSTRUM

H 2m (6ft) S 1.5m (5ft)
Ligustrum, commonly called **privet**, is a genus of about 50 species of hardy, fast-growing, deciduous and evergreen shrubs, originating from Asia. Many species produce conspicuous white flowers in summer followed by small blue-black fruits in autumn. The leaves are glossy green, turning duller with age, and usually a broad oval shape tapering to a narrow point. They are carried in pairs on thin blackish-green stems, which become gray-brown with age. The variegated-leaved forms are the most interesting ones to grow, such as *L. lucidum* 'Tricolor', with deep green leaves marked with gray-green and edged with pale creamy-yellow; and *L. ovalifolium* 'Aureum' (**golden privet**), a vigorous evergreen shrub which bears bright golden leaves with a splash of yellow in the center.
Cultivation Hardy; full sun/partial shade; water frequently; mulch with organic matter and top dress with slow-release fertilizer in spring; trim over with shears every six weeks from late spring until early autumn to keep a tight habit; propagate from 15–20cm (6–8in) hardwood cuttings in winter.
Special needs None: very easy to grow.
Display ideas Ideal specimen plant. Very useful for topiary.

LILIACEAE

LILIUM

H Up to 2.4m (8ft) S 30cm (1ft)
Lilium, commonly known as **lily**, is a large genus of bulbous plants that are in the main hardy. They are widely distributed over the northern hemisphere. The shapely trumpet-like blooms with six petals curl back on themselves to produce a star shape which can vary in size from 2.5cm (1in) to 25cm (10in) across depending upon the variety. The flowers are arranged either in a loose cluster at the top of the stem or as a long spike. Many lilies have a heady fragrance, and their tall erect stems make them excellent for providing bold splashes of color. The leaves are pale to dark green and strap-shaped.

The most popular lily for container growing is *L. regale* and its hybrids, which are quite remarkable considering that the species was introduced to Europe from China only at the start of the 20th century. These strikingly beautiful plants are capable of reaching a height of 1.5m (5ft), but their tendency to form support roots at the base of the stem makes them unlikely to fall over. *Lilium regale* bears 12–15cm (5–6in) white flowers with a butter-yellow center and a rose-pink reverse; they are very heavily scented.
Cultivation Hardy; full sun/partial shade; water normally; susceptible to viruses, and to attack by aphids (see pages 104–105), and lily beetle (which damages both leaves and flowers); propagate by division of bulblets (see page 98) in early to mid-autumn.
Special needs None: most lilies are easy to grow, given a well-drained growing medium and reasonable care.
Display ideas Grow on in smaller pots and plunge in display containers when the flower buds emerge. See pages 13, 72 and 73.

CAMPANULACEAE

LOBELIA

H 10–75cm (4in–2½ft) S 30–60cm (1–2ft)
Lobelia is a genus of 200 or more species of hardy and half-hardy annuals, herbaceous perennials and sub-shrubs. The leaves are light green and a narrow oval shape. The flowers are basically tubular, opening out to a broad, three-to-five-lobed mouth; they range from white through to blue and red, and appear from early summer until mid-autumn. The annual species are much used in containers. *Lobelia*

Lupinus polyphyllus 'Inverewe Red'

Lysimachia nummularia 'Aurea'

erinus, for instance, is a dwarf spreading plant with 2.5cm (1in), pale blue, white or mauve flowers, which are produced in vast quantities from early summer until the first frosts. The bedding varieties are usually grouped according to habit, and classified as compact or trailing. Some of the herbaceous perennial species are suitable for container growing. ***Lobelia cardinalis***, the **cardinal flower**, for instance, which is not totally hardy, has an erect and branching habit and produces spikes of scarlet flowers in summer.
Cultivation Half-hardy/hardy; full sun/partial shade; water normally; liquid feed from mid- to late summer; propagate from seed sown under protection in early to mid-spring, or, for perennial species, by division in mid-spring.
Special needs The half-hardy perennial species need winter protection: cover the roots with leaves or straw, or lift the root ball in late autumn.
Display ideas Very good for trailing over the edges of hanging baskets, tubs and window boxes. See page 24.

CAPRIFOLIACEAE

LONICERA

H Up to 3m (10ft) S 2m (6ft)
Lonicera, commonly known as **honeysuckle**, is a genus of 200 species of deciduous and evergreen flowering shrubs and woody climbers. The outdoor species are hardy twining climbers, ideal for walls, fences, archways or for growing through shrubs and trees. The fragrant flowers are basically tubular, opening out to a broad mouth, and arranged in pairs either individually or in clusters; the colors range from white through pale-yellow to gold, pink and scarlet. The leaves vary in shape from broadly oval to almost circular, are pale to mid-green and borne in pairs on thin twiggy stems. The bushy types provide some of the best winter-flowering shrubs: ***L. × purpusii*** and ***L. standishii*** both produce fragrant creamy-white flowers from late autumn to mid-spring. Of the climbing plants, ***L. periclymenum*** cultivars are very popular, as is ***L. × tellmanniana*** with its orange flowers.
Cultivation Hardy; full sun/partial shade; water normally; mulch with organic matter in spring; cut back lateral shoots to 15cm (6in) in mid-spring and thin out old wood after flowering; susceptible to mildew, and to attack by aphids (see pages 104–105); propagate from 20cm (8in) hardwood cuttings in mid-autumn.
Special needs Plant climbing species with their roots in the shade and their tops in the sun.
Display ideas Very good for providing evening scent in the garden.

LEGUMINOSAE

LUPINUS

H 75cm–1.2m (2½–4ft) S 45–75cm (1½–2½ft)
A genus of 200 species of hardy and half-hardy annuals, herbaceous perennials and semi-evergreen sub-shrubs, these plants are closely related to the garden pea, *Pisum sativum*. Lupin flowers are carried in large numbers in tightly packed spikes, the flowers opening in rapid succession from the bottom upwards. The dark green leaves consist of six or eight narrow, strap-like leaflets, arranged in a circle which resembles the spokes of a wheel, held on narrow erect stalks. The herbaceous perennial cultivars are the most commonly grown, with named varieties in white, yellow, blue and the red ***L. polyphyllus* 'Inverewe Red'**. Of the shrubby species, the tree lupin ***L. arboreus*** is most commonly grown, its heavily fragrant clear, pale yellow flowers being carried on short spikes up to 15cm (6in) long in early summer.
Cultivation Hardy; full sun/partial shade; water normally; mulch with bulky organic matter in spring; dead-head flowers to prevent seeds developing; herbaceous perennials prone to viral infection (see page 104); propagate by cuttings 7.5–10cm (3–4in) long in early spring, placing them in a cold frame.
Special needs Heavy flowering stems may require staking using twiggy sticks.
Display ideas Very good for a bold display of summer color where some height is required. Woody species are ideal for coastal situations. See page 38.

PRIMULACEAE

LYSIMACHIA

H 5cm–1m (2in–3ft) S 60cm–1m (2–3ft)
This large genus of 150 species of summer-flowering, hardy annuals and herbaceous perennials, and a small number of sub-shrubs, is commonly known as **loosestrife**. The thin, strap-like leaves are arranged in opposite pairs and are mid-green in color, often turning orange-red in autumn. The star-shaped, cup-like flowers, which may be white or yellow, are arranged on the tips of the stems in spikes 15–20cm (6–8in) long. The most popular species include ***L. clethroides***, which is 30cm (1ft) high and produces multiple spikes of small, white, star-shaped flowers from mid-summer to early autumn; and ***L. nummularia***, a vigorous, evergreen, trailing plant that bears a profusion of small, bright yellow, cup-shaped flowers in mid-summer. There is also a yellow-leaved cultivar, ***L. n.* 'Aurea'**.
Cultivation Hardy; full sun/partial shade; water normally; liquid feed every

Magnolia stellata

Matteuccia struthiopteris

Milium effusum 'Aureum'

two weeks from early to late summer; cut back old growths in early spring; propagate by division while dormant.
Special needs *Lysimachia nummularia* requires careful watering as it is very shallow-rooted. Stake tall species with twiggy sticks. Leave the old stems to provide winter protection.
Display ideas Very good as a filler, and excellent for ground cover.

MAGNOLIACEAE

MAGNOLIA

H Up to 5m (15ft) S 3m (10ft)
Magnolia is a genus of some 125 species of hardy deciduous and evergreen trees and shrubs, originating from as far afield as the Himalayas and south-east Asia to North and Central America. It contains some of the most spectacular flowering trees and shrubs to be found anywhere. The flowers (often produced before the leaves) come in a variety of whites, pinks and purples, with exceptions such as the evergreen cultivars of *M. grandiflora* which have pale creamy-yellow blooms and large, glossy, green leaves similar to those of the laurel. The shapely tulip-like flowers vary in size from 7.5cm (3in) to 35cm (14in) across; they are produced from mid-spring to early autumn depending upon the variety. The leaves are usually a mid-green and spear-shaped, often with an orange or brown felt-like covering on the underside. *Magnolia stellata* is a slow-growing species and perfect for growing in a container. Its pale green leaves do not appear until after the highly scented, clear white, star-shaped flowers in mid- to late spring. There is a pink-flowered cultivar, *M. s.* 'Rosea'.
Cultivation Hardy; full sun/partial shade; water normally; mulch with bulky organic matter in spring; susceptible to coral spot (see page 104) on untended damaged branches; propagate by layering in late spring.
Special needs Most prefer a well-drained acid growing medium and will grow better if watered with rain water in hard-water areas (*M. delavayi* and *M. kobus* are lime-tolerant). A sheltered position, away from morning sun, is advisable to avoid damage to the flowers from spring frosts.
Display ideas Ideal for growing as a single specimen. *Magnolia grandiflora* makes a superb wall shrub.

DRYOPTERIDACEAE

MATTEUCCIA

Matteuccia struthiopteris

H Up to 1.5m (5ft) S 1m (3ft)
Matteuccia is a genus of three species of deciduous rhizomatous ferns, widely distributed throughout the northern hemisphere, but only *M. struthiopteris*, the **shuttlecock fern** (UK) or **ostrich feather fern** (USA), is in general cultivation. This elegant moisture-loving plant has an outer rim of pale yellow-green leaves (fronds), which are gently arching, up to 1m (3ft) long and sterile. The inner leaves are shorter, dark greenish-brown and capable of producing spores (seed). Both types of frond are extremely thin, very deeply cut and broadly spear-shaped with a blackish-brown middle rib holding the leaf together and supporting it. Each year the basal clump (crown) of the plant produces several knuckle-like black rhizomes (modified stems that grow horizontally through the growing medium) which form subsidiary crowns (offsets), and these are the plant's main means of reproduction.
Cultivation Hardy; partial shade/shade; water frequently; mulch with bulky organic matter in spring; propagate from offsets in spring.
Special needs Requires plenty of growing room for root development, and a strong container. Prefers shade but will tolerate sun.
Display ideas Very good for greening up a damp shaded corner.

GRAMINEAE

MILIUM

H Up to 1.5m (5ft) S 90cm (3ft)
This is a small genus of six species of annual or perennial grasses, originating from Eurasia and North America. The narrow light green leaves are usually about 30cm (1ft) long and 1.5cm (¾in) wide, and often give a translucent appearance in morning sunlight. The plant develops into a dense, tufted clump up to 60cm (2ft) wide, and quickly covers the surface of the container. Branched flowering stems, up to 1.5m (5ft) high, and usually pale green tinged with purple, are produced in summer. The most popular form is *M. effusum* (**wood millet**), which has several leaf variations providing different foliage effects: *M. e.* 'Aureum' has

golden-yellow leaves and flower spikes; and *M. e.* **'Variegatum'**, a rare slow-growing cultivar, has green-and-white striped leaves.
Cultivation Very hardy; full sun/partial shade; water normally; mulch with bulky organic matter in spring; trim flower stalks with hand shears to prevent plants becoming straggly; propagate by division in spring.
Special needs Prefers a free-draining growing medium, high in rich, bulky organic matter.
Display ideas Very good for an attractive evergreen display.

SCROPHULARIACEAE

MIMULUS

H Up to 75cm (2½ft) S Up to 1m (3ft)
Commonly called **monkey flower**, *Mimulus* is a genus of 100 species of hardy annuals and herbaceous perennials, originating from North, Central and South America. They have open-mouthed, snapdragon-like flowers from late summer to early autumn. The mid-green leaves are broadly oval-shaped, and carried in pairs on slender green, often slightly hairy stems. These plants like semi-waterlogged conditions, but will grow well in moist growing medium in a container. *Mimulus luteus*, with its yellow flowers marked inside with crimson spots, is the species most usually grown in containers. There are two interesting forms: *M. l.* **'Alpinus'**, which is very low-growing, forming a dense mat over the surface of the growing medium; and the strange-looking *M. l.* **'Duplex'**, which has the appearance of one flower growing inside another. An interesting color change is provided by *M.* **'Bees Scarlet'**, which is low-growing at 15cm (6in) and has deep scarlet flowers.
Cultivation Hardy; full sun/partial shade; water frequently; feed with a base dressing of slow-release fertilizer when planting out; delay cutting down old flower stems until the spring in colder districts; propagate by division or seed sown in mid-spring or from 5cm (2in) cuttings in late spring.
Special needs These are marginal plants and must be kept moist throughout the growing season or they will wilt very rapidly.
Display ideas These plants are ideal as part of a containerized water feature. See page 45.

Muscari armeniacum

LILIACEAE/HYACINTHACEAE

MUSCARI

H Up to 38cm (15in) S 5–7.5cm (2–3in)
Muscari, commonly known as **grape hyacinth**, is a genus of 60 species of hardy, dwarf, bulbous plants related to the hyacinth and originating from Asia, south-east Europe and the Mediterranean region. They have mid-green, thin, strap-like leaves with a grooved inner surface. The upper paler-colored flowers are sterile and do not open; the lower urn-shaped flowers are 5mm (¼in) long and colored white, yellow, blue or purple. *Muscari armeniacum* is the most popular species grown. This has compact and densely packed spikes of deep blue flowers with white rims from late spring to early summer. There are also some distinct named varieties including *M. a.* **'Cantab'**, which is paler blue; and *M. a.* **'Heavenly Blue'**, which is a much brighter color. Two sorts which are not so common but are very attractive are the yellow-flowered *M. macrocarpum*, which has blue-green leaves, and the white-flowered *M. botryoides* **'Album'**, which flowers from mid-spring to early summer.
Cultivation Very hardy; full sun; water normally; liquid feed every two weeks from early to mid-summer; propagate species from seed sown in late summer, cultivars by division in mid-summer when the leaves have died down.
Special needs None: very easy to grow.
Display ideas Force indoors, placing up to 12 bulbs in a 15cm (6in) container, and transplant outside just before flowering. See page 72.

MYRTACEAE

MYRTUS

Myrtus communis

H Up to 2m (6ft) S 1.2m (4ft)
Commonly known as **myrtle**, this genus of two species of hardy and half-hardy, aromatic, evergreen trees and shrubs originates from the Mediterranean and N. Africa. The five-petalled flowers are white, saucer-shaped and outstanding for their prominent display of thin golden stamens from early to late summer, followed by small, white, red, purple or black berries, produced singly from the upper leaf joints. The leaves are a glossy mid- to deep green or reddish-brown, oval- to spear-shaped and strongly aromatic. As they age some species develop a handsome stem with golden-brown bark, flaking to expose pale green patches. *Myrtus communis* is one of the hardiest species grown in temperate regions. This shrub flowers from mid- to late summer, and is perfect for growing in a pot. *Myrtus c. tarentina* is a dwarf variety with smaller leaves and flowers; *M. c.* **'Variegata'** has gray-green leaves with narrow creamy-white margins.
Cultivation Half-hardy; full sun; water normally; mulch with bulky organic matter in spring; remove frost-damaged shoots in late spring; propagate from 5–7.5cm (2–3in) heel cuttings taken in mid-summer.
Special needs Likes a sheltered position. Cover with straw and sacking in cold weather, or move indoors.
Display ideas Very good for coastal areas, or against a south-facing wall.

Narcissus 'April Tears'

BERBERIDACEAE

NANDINA

Nandina domestica

H Up to 1.2m (4ft) S 1m (3ft)
A single-species genus of hardy evergreen shrub, *Nandina*, commonly known as **Chinese sacred bamboo** or **heavenly bamboo** (USA), originates from China and Japan. It is closely related to *Berberis*, although it resembles a bamboo in habit and growth. *Nandina domestica* has leaves up to 45cm (18in) in length but divided into many small leaflets arranged in pairs like those of the ash (*Fraxinus*). These leaves are a bright reddish-purple when young, turning dark green in summer and taking on shades of purple in autumn. Wide-open spikes of white flowers up to 35cm (14in) across appear in the tips during mid-summer and may be followed by scarlet fruits in late summer. Two attractive cultivars are *N. d.* **'Firepower'**, a dwarf shrub with lime-green leaves in summer, turning a vivid orange-red in late autumn and winter; and *N. d.* **'Nana Purpurea'**, which forms a compact plant with reddish-purple leaves.

Cultivation Hardy; full sun; water normally; mulch with bulky organic matter in spring; remove dead and weak shoots after flowering; propagate from 10cm (4in) heel cuttings in late summer.

Special needs Prefers a well-drained growing medium. Provide a sheltered position as young growth may suffer frost damage in severe winters.

Display ideas Ideal for growing as an individual specimen.

Narcissus 'Paper White'

LILIACEAE AMARYLLIDACEAE

NARCISSUS

H Up to 60cm (2ft) S Up to 20cm (8in)
Narcissus, commonly called **daffodil**, is a genus of bulbs originating from Europe, Africa and Asia grown for their showy ornamental flowers. They vary in size and in flower form and color. Some are deliciously scented. *Narcissus* have narrow, strap-like, mid- to dark green leaves, usually appearing in groups of three or more with each bloom. Each flower has an inner frill-like trumpet or cup and an outer row of petals. Most are fully hardy. *Narcissus* are classified and grouped according to the arrangement of their floral parts: there are 12 divisions. The cultivars with smaller flowers appear to be more weather-resistant – *N.* **'April Tears'** has small yellow petals and a short, rounded, yellow cup; the late-flowering *N.* **'Cheerfulness'** produces up to three flowers per stem, with a creamy-white outer and an orange-rimmed center. Some sorts, such as *N.* **'Paper White'**, are heavily scented as well as being very attractive.

Cultivation Hardy/tender; full sun/partial shade; water normally; liquid feed every two weeks from the end of flowering until the foliage starts to turn yellow; dead-head to prevent seeds setting and remove dying foliage in mid- or late summer; susceptible to narcissus yellow stripe virus, and to attack by stem and bulb eelworm (see page 105) and large narcissus fly (which eats into the bulb, causing basal rot); propagate by division six weeks after flowering; species may be propagated by seed in autumn.

Nerium oleander

Special needs Lift and divide the dense clumps every three to five years, since excessive competition will lead to a gradual decline in the number of flowers produced.

Display ideas Superb for early spring displays in tubs and window boxes. Can be forced indoors and plunged into display containers. See page 40.

NEPHROLEPIDACEAE

NEPHROLEPIS

H Up to 75cm (2½ft) S 75cm (2½ft)
Nephrolepis is a genus of 30 species of frost-tender evergreen and semi-evergreen ferns, which increase rapidly by producing numerous stolons. Only three or four species, commonly known as the **sword** or **Boston fern**, are usually cultivated – generally as house plants, but they are also suitable for growing outdoors in summer. The most commonly grown is *N. exaltata*, which bears pale green, finely divided fronds (leaves), and its various cultivars such as *N. e.* **'Bostoniensis'**, which has broader fronds and is a more vigorous grower than the type, or *N. e.* **'Elegantissima'**, which produces unusual compact, bright green, finely cut leaves which are closely set and overlapping. ***Nephrolepis cordifolia*** has light green fronds narrowly divided on short leaf stalks.

Cultivation Tender; partial shade/shade; water frequently; liquid feed every two weeks from mid-summer to early autumn; propagate by division – separate young plantlets from the stolons as they develop.

Special needs Prefer moist conditions, but will tolerate an occasional drying-out.
Display ideas Use in hanging baskets.

APOCYNACEAE

NERIUM

Nerium oleander

H 2–3m (6–9ft) s 3–3.5m (9–10ft)
Commonly known as **oleander** or **rose bay**, this small genus of three species of tender evergreen flowering shrubs originates from the Mediterranean. *Nerium oleander* has long, leathery, narrow, strap-like leaves, deep green in color. The single white flowers are produced in clusters on the tips of the shoots from spring through autumn. Individual flowers, which are tubular with five petal-like segments that spread out flat, measure up to 3–4.5cm (1–1½in) across.
Cultivation Not fully hardy; full sun; water normally; mulch with bulky organic matter in spring, and give established plants a weak liquid feed at two week intervals from late spring to early autumn; pinch out growing points to form multi-branched plants; propagate by semi-ripe cuttings 10cm (4in) long taken in late summer.
Special needs This plant is frost tender and must be overwintered indoors in low temperatures; overwatering may cause the plant to shed all its leaves.
Display ideas Looks attractive combined with the perennial sweet pea, *Lathyrus latifolius*.

SOLANACEAE

NICOTIANA

H Up to 1.5m (5ft) s Up to 45cm (18in)
This tall bedding plant, commonly called the **tobacco plant**, is grown for its extravagant display of five-petalled, 3in (7.5cm) long, tube-like flowers, produced in large clusters from early summer. A multitude of flower colors exists, with white, cream, yellow and crimson being the most popular. All new flowers open in the evening, and most are heavily scented. The leaves are mid-green, oval-shaped with a pointed tip and slightly sticky to the touch. The tall *N. sylvestris*, which grows to 1.5m (5ft), is particularly attractive with its large spires of white, fragrant trumpet-shaped flowers up to 8cm (3½in) long.
Cultivation Half-hardy; partial shade; water normally; feed with a base dressing of fertilizer, and liquid feed every two weeks throughout summer; remove dead flower heads to encourage further displays; young plants are susceptible to slug damage (see page 105), and are also prone to gray mold (Botrytis) in very wet summers; propagate by seeds sown indoors in spring.
Special needs Prefer a well-drained growing medium; perform best in a warm, sunny position. Tall cultivars will need support on exposed sites.
Display ideas Position close to the door or window to benefit from the heady perfume. See pages 20 and 38.

Nicotiana sylvestris

NYMPHAEACEAE

NYMPHAEA

s Up to 2m (6ft)
Nymphaea, commonly known as **water lily** (UK) or **water nymph** (USA), is a genus of 50 species of hardy and tender perennial water plants, originating from Europe, North Africa and North America. They are generally grown for their rounded (usually floating) leaves and brightly colored flowers. The hardy water lilies, which flower over a long period in summer, have large blooms which are sometimes sweetly scented. The flowers and leaves rise from a thick rootstock anchored securely in the mud below the water by long tough roots. The round or heart-shaped leaves are reddish-green with a shiny leathery upper surface. The saucer-shaped flowers have several rows of petals; the outer layer opens out almost flat as the flower opens fully, to show conspicuous golden or orange stamens in the center. The dwarf species and cultivars are the most practical for container growing: *N.* **'Aurora'** grows well in shallow water (45–60cm/18in–2ft) deep – with flowers which change from pinkish-yellow to deep orange before becoming deep red; *N.* ***tetragona*** is the smallest white-flowered species, bearing 2.5cm (1in) flowers with yellow stamens.
Cultivation Hardy/tender; full sun/partial shade; feed with slow-release fertilizer in spring; susceptible to stem rot, and to attack by aphids (see page 105) and water lily beetle; propagate by division in late spring/early summer.
Special needs Must be grown in clear, well-oxygenated water at least 45cm (18in) deep.

Nymphaea tetragona

Origanum vulgare 'Gold Tip'

Display ideas The dwarf cultivars are ideal for growing in a half-tub, preferably in a warm, sunny position.

LABIATAE/LAMIACEAE

ORIGANUM

H Up to 45cm (18in) s 30cm (1ft)
Origanum, commonly known as **dittany**, **marjoram** or **oregano**, depending on the species, is a genus of 36 or so species of deciduous perennials and sub-shrubs. Some species are grown as

Osteospermum 'Whirligig'

Oxalis triangularis 'Cupido'

Oxalis tetraphylla 'Iron Cross'

culinary herbs, but many others for their clusters of pink tubular flowers and bracts (they look very similar to hop fruits) produced from mid-summer to early autumn. The small, aromatic, gray-white leaves are oval-shaped, ending in a pointed tip and carried in pairs on thin stems. Most species have arching or prostrate stems. *Origanum* **'Kent Beauty'** has large, dangling, pink bracts and white flowers. *O. vulgare* **'Gold Tip'** has yellow markings on some leaves, and the very low-growing *O. amanum* has small, deep rose flowers. This plant will survive an average winter outdoors if protected around the base with straw or a loose covering of well-rotted grass clippings.

Cultivation Half-hardy; full sun; water normally; feed with a base dressing of slow-release fertilizer when planting out; propagate from 5cm (2in) cuttings in late summer.

Special needs Prefer a sheltered position. These plants are usually overwintered as rooted cuttings: do not plant out in containers until early summer.

Display ideas Very good for trailing over the sides of pots or hanging baskets. See page 15.

OSMUNDACEAE

OSMUNDA

Osmunda regalis

H 1.2–1.5m (4–5ft) S 1.2–1.5m (4–5ft)

Osmunda regalis, commonly known as **royal fern** or **flowering fern**, is one of 10 species of hardy large ferns from a genus which grows all over the world with the exception of the Antipodes. The fronds (leaves), which arch gently from the base of the plant, are mid-green, broadly spear-shaped and very finely divided into many leaflets. The veins on the underside of the inner fronds are often covered with spore (seed) capsules which turn brown when ripe, usually in mid-summer. The outer fronds may also carry spore cases, but these are sterile and will not produce spores. For a slightly unusual effect try *O. r. purpurascens*, which has young fronds of deep coppery-pink changing to copper-green as they mature, with purple veins and midribs. The plant gradually builds up a mass of crowns and matted black roots which resemble a bird's nest.

Cultivation Hardy; partial shade/shade; water frequently; mulch with organic matter in spring; propagate from fresh spores sown in mid- to late summer, or by division in late spring.

Special needs A growing medium with a high fiber content is essential, which must be kept moist at all times. Avoid exposure to direct sunlight, which will dry the growing medium and cause the leaves to shrivel.

Display ideas Very good for giving life to a damp shaded corner, or for shading other plants.

COMPOSITAE/ASTERACEAE

OSTEOSPERMUM

Osteospermum jucundum

H 30–60cm (1–2ft) S 45–60cm (18in–2ft)

Commonly known as **African daisy**, and originally from South Africa, *O. jucundum* (also called *Dimorphotheca barberae*) is a deep-rooted plant with a lax sprawling habit, very drought-resistant and thus ideal for container culture. It is grown for its 5cm (2in) daisy-like flowers carried on long slender stems, which come in a myriad of colors, including pink, white, purple or creamy-yellow, sometimes with a contrasting violet disc or darker reverse petal colors. The plant flowers continuously from early summer through to early autumn, although the blooms will not open on dull days or if the plant is shaded. The narrow, strap-like, aromatic leaves are dull mid-green and covered with a layer of fine hairs, which help to reduce moisture loss in dry periods. *Osteospermum* **'Silver Sparkler'** is a cultivar with white flowers, and creamy-white-and-green-variegated leaves; the purple-flowered *O.* **'James Elliman'** is the most hardy cultivar available. The prostrate *O.* **'Cannington Roy'** with its white, purple-tipped flowers, and *O.* **'Whirligig'**, which has white propeller-like petals that create an almost mesmerizing effect, are both worth consideration.

Cultivation Tender; full sun; water normally; liquid feed once a month during the growing season; dead-head to ensure constant flowering; young plants susceptible to botrytis; propagate from semi-ripe cuttings in late summer or seed sown indoors in mid-spring.

Special needs Must have an open free-draining growing medium. Overwinter rooted cuttings indoors.

Display ideas Very good for dry, exposed conditions. Excellent for hanging baskets. See page 76.

OXALIDACEAE

OXALIS

Oxalis adenophylla

H 7.5cm (3in) **S** Up to 20cm (8in)

Commonly known as **pink oxalis**, this plant originates from Chile. *Oxalis adenophylla* is one of a genus of hardy and half-hardy annuals, tender shrubs, perennials and bulbous plants, which include some weeds. It has an unusual, fiber-covered, corm-like structure (it is in fact a rhizome). The foliage is arranged in rosettes, with crinkled leaves of bluish-green carried on thin pinkish leaf stalks. The small flowers, 2.5cm (1in) in diameter, which appear from early to mid-summer, are a soft apple-blossom pink with darker purple eyes and carried on stalks 5cm (2in) above the leaves. In winter the leaves die down so that only the tip of the rhizome remains visible.

Cultivation Hardy; full sun/partial shade; water sparingly; propagate by division in mid-spring.

Special needs Prefers a well-drained growing medium containing plenty of organic matter.

Display ideas These plants are ideal for planting around the edge of a container. See page 53.

CRUCIFERAE/BRASSICACEAE

PACHYPHRAGMA

Pachyphragma macrophyllum

H Up to 40cm (16in) **S** 25cm (10in)

Pachyphragma is a very small genus consisting of a single species of hardy herbaceous perennial, a member of the cabbage family. The bulk of this creeping, mat-forming plant is a rhizome, which is covered in old leaf scars, rather like that of an iris. The large round leaves are a glossy dark green color, have a crinkled margin, and are attached to the stem (rhizome) by 25 cm (10in) long leaf stalks. Masses of tiny, white, odorous flowers are produced in late winter and early spring on long slender stems, followed by small winged fruits in summer.

Cultivation Very hardy; full sun/partial shade; water normally; mulch with bulky organic matter in spring; propagate from seed sown in autumn, or by division in spring, or from 7.5cm (3in) cuttings in late spring.

Special Needs Prefers a free-draining growing medium. Grows better if divided every three to four years.

Display ideas Very good for covering the surface of a pot, forming a living mulch over the growing medium.

PASSIFLORACEAE

PASSIFLORA

H Up to 7m (21ft) **S** 3m (9ft)

Passiflora, commonly known as **passion flower**, is a large genus of mainly vigorous, self-supporting, flowering climbers which attach themselves by means of twining tendrils. The brilliantly colored and very intricate flowers start as a short green tube, before opening out to reveal the various delicate reproductive organs, laid out on a saucer of tepals (the area where the sepals and petals are joined together). The most attractive display comes from several rings of brightly colored, almost hair-like filaments. The leaves are mid-green in color and palm-like in shape, each with three to five sections.

Passiflora × caeruleoracemosa

Passiflora caerulea

Passiflora × caeruleoracemosa is a free-flowering hybrid with violet-blue tepals, rings of violet-purple filaments and a green and purple center. *Passiflora caerulea* is a hardy species that produces white flowers in late summer/ early autumn.

Cultivation Half-hardy; full sun/partial shade; water normally; mulch lightly with organic matter in spring; cut out one third of the main growths and cut back side shoots to 15cm (6in) in spring; susceptible to viruses (see page 104); propagate from 7.5–10cm (3–4in) heel cuttings in mid- to late summer.

Special needs Prefer a free-draining growing medium and a sheltered position. Tie in the young growths until the tendrils have taken hold. The top growth may be killed by frost, but new stems usually emerge from the container the following spring. Whole plants may be killed in very severe winters: this may be overcome by providing winter protection with cloches, brushwood or plastic sheeting.

Display ideas Perfect for growing up a trellis or through a small tree or large shrub, which will offer the plant winter protection.

Pelargonium 'Bird Dancer'

Pelargonium 'Frank Headley'

Pelargonium 'Royal Oak'

GERANIACEAE

PELARGONIUM

H 15–60cm (6in–2ft)
S Up to 75cm (2½ft)
This genus of some 280 species of tender sub-shrubs, commonly called **geranium**, though not to be confused with the herbaceous (or true) *Geranium*, originates mostly from the southern half of the African continent. They are possibly the most widely grown summer-flowering container plants. Some people dislike their smell, and will not grow them for this reason. However, they are extremely versatile, flourishing in almost any position. The cultivars are divided into groups.

Ivy-leaved geraniums, with their long, sprawling, trailing growths, are especially suitable for hanging baskets and window boxes. However, the stems are brittle, so damage easily. The flowers are produced in a range of shades from deep red to white, with lavender tints as well. Some varieties flower more readily than others, particularly the compact cascade varieties which become smothered in pink or red blooms, and of the stronger growers there are many cultivars to choose from. Possibly the most well-known cultivar is **'L'Elégante'**, with its creamy-edged green leaves which turn purple in autumn. The small delicate flowers are white, and all of these plants are descended from *P. peltatum*.

Zonal geraniums usually have an erect, often bushy habit. The circular leaves are sometimes marked with a dark C-shape, and there are other cultivars with golden or silver-variegated foliage, such as *P.* **'Frank Headley'**. Apart from the fine flowering cultivars, there are those with variegated, yellow and multi-colored leaves, such as *P.* **'Bird Dancer'**. Many of the old favorites have been overtaken in popularity by the seed-raised hybrids.

Regal pelargoniums have a vigorous, erect, branching habit, with palm-like leaves which often have a toothed puckered edge. The large, ruffled, often multi-colored flowers of the **'Grand Slam'** series make a very striking display. These flowers often shatter easily and should not be grown in exposed situations; they really do well outdoors only in a hot dry summer.

Scented-leaved geraniums vary greatly in size and habit according to their parentage. Many have leaves which are finely cut, toothed or finely lobed. Usually the flowers are much smaller than those of the hybrid plants, which are more commonly seen. The range of leaf aromas is truly amazing: peppermint, orange, lime and balm are among the most popular. One of the most reliable species is *P.* **'Graveolens'**,

Pelargonium 'Stellar Apricot'

Pelargonium 'Vancouver Centennial'

which has palm-shaped, rose-scented leaves covered in fine felt-like hairs. The habit is spreading and branching, and the large clusters of rose-pink flowers emerge from the leaf joints between mid-summer and mid-autumn. Alternatively *P. quercifolium* (the **oak-leaved geranium**) has an upright habit and triangular toothed leaves with a crinkled margin. The purple-veined pink flowers are carried in clusters from late spring to mid-summer. An interesting cultivar is *P.* **'Royal Oak'**, with slightly sticky spicy-scented leaves and pale mauve flowers.

Floribunda geraniums have been produced as a result of recent breeding programmes. This group bears such a profusion of flowers that it has been given a separate classification. Floribunda geraniums have strong growth and a compact habit, reaching a height of about 30cm (1ft).

Cultivation Tender; full sun/partial shade; water sparingly (these plants produce more flowers if kept slightly dry); liquid feed weekly from mid-summer to early autumn; dead-head regularly to ensure continued flowering; susceptible to rust and leafy gall

Penstemon 'Sour Grapes'

(which causes stunted distorted shoots at soil level), and to attack by whitefly and vine weevil (see page 105); propagate from seed sown from mid-winter to mid-spring or from 7.5–10cm (3–4in) cuttings in early to late summer.

Special needs None: very easy to grow.

Display ideas Plunge the plants in pots into the display containers so that they can be lifted and overwintered indoors. See pages 10, 20, 51, 53, 60 and 70.

SCROPHULARIACEAE

PENSTEMON

H Up to 1m (3ft) **S** 75cm (2½ft)

Penstemon, commonly known as **beard tongue**, is a genus of 250 species of evergreen and semi-evergreen hardy and half-hardy herbaceous perennials and sub-shrubs, originating from Mexico and the United States. The flowers are 2.5cm (1in) long, snapdragon-like and carried in spikes up to 15cm (6in) long on the tips of the shoots. The leaves are mid-green and vary in shape from broadly oval to narrow and strap-like, carried on thin, twiggy stems. The alpine species such as *P. davidsonii*, with its small ruby-red flowers, tend to form a low, spreading, trailing bush. The most commonly seen species is *P. barbatus*, which forms an erect plant with pink or red flowers produced from mid- to late summer; there is also a white-flowered form. Most of the garden hybrids, such as *P.* **'Sour Grapes'**, with its purple shot with blue flowers, are related to the now rare *P. hartwegii*. Many species are suitable for container growing, and the half-hardy ones will survive severe winters if provided with shelter and well-drained conditions.

Cultivation Half-hardy/hardy; full sun; water normally; liquid feed from mid-

Petroselinum crispum

summer to early autumn; cut back hard after flowering; propagate from 5–7.5cm (2–3in) cuttings in late summer.

Special needs Intolerant of wet conditions. Support the tall cultivars with brushwood or twigs.

Display ideas Ideal as fillers.

UMBELLIFERAE/APIACEAE

PETROSELINUM

Petroselinum crispum

H Up to 60cm (2ft) **S** 30cm (1ft)

The genus *Petroselinum*, which consists of three species of hardy biennial with a thick swollen base, commonly known as **parsley** or **corn parsley**, is related to carrot and celery. It is widely cultivated throughout the world, including tropical regions, and is tolerant of a fairly wide range of environmental conditions. The species *P. crispum* is a native of southern Europe and has become naturalized in many temperate countries. Its leaves, which are mid-green, densely curled and rather moss-like, are carried on green hollow stems. The greeny-yellow flowers are borne in clusters on hollow green stems in summer. The curly-leaved cultivars, such as **'Champion Moss Curled'**, provide an attractive decorative effect, as does the very hardy **'Darki'**, with its very dark green, tightly curled leaves.

Cultivation Hardy; full sun/partial shade; water normally; liquid feed every month from early summer to early autumn; susceptible to root rot and to attack by celery fly; propagate from seed sown in mid-spring.

Special needs Grows poorly in acid growing medium, and plants cannot tolerate poorly drained conditions.

Display ideas Good as a fringe around the edge of containers.

Petunia 'Express Ruby'

Petunia 'Surfinia'

SOLANACEAE

PETUNIA

H Up to 35cm (14in), but may trail or hang to 75cm (2½ft) **S** 60–75cm (2–2½ft)

This genus of some 40 species of annuals or herbaceous perennials originates from South America, although most of the plants now grown are hybrids and far removed from the original species. Petunias are colorful, with single or double trumpet-shaped flowers, 5–12cm (2–5in) across, produced from mid-summer until the first frosts. The flower colors include cream, white, pink, red, mauve, blue and a variety of striped forms. The leaves, carried on green hairy stems, are broadly oval, mid- to dark green and sticky to touch. The plants are treated as half-hardy annuals which flower in the first season from seed, and are one of the most popular plants for hanging baskets, tubs and window boxes. The varieties fall into four main groups: **Multifloras** are bushy plants, producing large numbers of comparatively small flowers; **Grandifloras** are plants of any size which produce fewer but much larger flowers; **Nana Compactas** are the dwarf varieties; and **Pendulas** are plants of long trailing habit, which include the very popular **'Surfinia'** types. Most of the seed available is F1 hybrid, which produces vigorous uniform plants.

Cultivation Half-hardy; full sun; water normally; feed with a base dressing of fertilizer when planting out and liquid feed every two weeks from mid-summer to early autumn; dead-head regularly to ensure continued flowering and prevent disease; susceptible to botrytis on dead flowers, and to attack by aphids (see page 105); propagate from cuttings or from seed sown in spring.

Special needs Prefer a free-draining growing medium.

Display ideas Use to trail over the edges of hanging baskets. See pages 1, 13, 49, 51, 56 and 68.

POLEMONIACEAE

PHLOX

H 10–60cm (4in–2ft) **S** 15–30cm (6–12in)

This genus of hardy herbaceous perennials, half-hardy and hardy sub-shrubs and annuals mostly originates from the western and mid-western United States. The plants are grown for their attractive, bright pink, red, purple or white, star-shaped flowers carried in dense clusters on the tips of the shoots. They are suitable for growing as alpines, herbaceous perennials and bedding annuals. The leaves are usually narrow and strap-like, mid- to light green in color and carried in pairs on slender green stems. The popular annual varieties are related to *P. drummondii*, which is a half-hardy annual, with an erect upright habit and white, pink, purple, lavender and red flowers produced in dense heads up to 7.5cm (3in) across from mid-summer to early autumn. *Phlox drumondii* is available in mixed colors or named cultivars. The herbaceous perennial species *P. paniculata* has led to numerous garden varieties, including *P. p.* **'Fairy's Petticoat'**, with pale pink flowers with dark pink centers; and the vigorous *P. p.* **'Harlequin'**, with variegated foliage and bright purple flowers. The low-growing *P.* **'Chattahoochee'** has saucer-shaped lavender flowers with a red eye.

Cultivation Half-hardy/hardy; full sun/partial shade; water normally; feed annuals with a base dressing of fertilizer when planting out; top dress perennials with slow-release fertilizer in spring; dead-head regularly for continued flowering; susceptible to mildew and to attack by eelworm (see pages 104–105); propagate alpines and other perennials from 5–7.5cm (2–3in) stem cuttings in mid-spring to mid-summer, annuals from seed sown in mid-spring.

Special needs Prefer a free-draining growing medium.

Display ideas Very good for providing bright drifts of color.

PITTOSPORACEAE

PITTOSPORUM

Pittosporum tenuifolium

H 5m (15ft) **S** 1.5–2.1m (5–7ft)

Pittosporum is a genus of around 200 species of evergreen flowering shrubs and small trees originating from Australia. Although chiefly grown for their attractive, ornamental foliage, some have colorful, small, fragrant, bell-shaped flowers with spreading petals. *Pittosporum tenuifolium*, commonly called **tawhiwhi** or **kohuhu**, and its various leaf forms is the most commonly grown species, mainly because it

Phlox 'Chattahoochee'

Pittosporum tenuifolium 'Silver Queen'

is one of the hardier types. It has pale green, oblong leaves, with very prominent wavy margins, arranged on thin bluish-black branches. The small flowers, which appear in spring, are chocolate-purple in color and have the gentle aroma of warm honey. There are several cultivars which provide interesting variations in leaf color, including *P. t.* **'Abbotsbury Gold'**, which has leaves with a blotch of bright greenish-yellow in the center when young, which fades as they age; *P. t.* **'Purpureum'**, which is very attractive as the pale green leaves slowly turn bronze-purple as they age; and *P. t.* **'Silver Queen'** with silver-gray variegated leaves.

Cultivation Hardy; full sun/partial shade; water normally; mulch with bulky organic matter in spring; cut out long straggling shoots as they develop; propagate from 10cm (4in) semi-ripe heel cuttings in mid-summer.

Special needs Provide shelter from north and east winds.

Display ideas Excellent for growing as a single specimen.

PLUMBAGINACEAE

PLUMBAGO

H Up to 2m (6ft) S Up to 2m (6ft)

This South African native has long been grown by gardeners in other parts of the world as a pot-grown ornamental climber because of its long flowering season. *Plumbago auriculata* (formerly *P. capensis*, and commonly known as **South African leadwort**) has 25cm (10in) bunches of pale blue primrose-like flowers 2.5cm (1in) in diameter, each one on the end of a small tube, which are produced continuously from early summer through until mid-autumn. The leaves are 7.5–10cm (3–4in) long, a dull mid-green color and elliptical in shape, often borne sparsely on the vigorous stems. Its thin twiggy growths are too weak to support the plant and some form of trellis or pole has to be provided for training. There is also a white-flowered variety, *P. a.* var. *alba*, which is not quite so vigorous.

Cultivation Half-hardy; full sun/partial shade; water normally; liquid feed every two weeks from mid-summer to early autumn; cut back all growth by at least two thirds immediately after flowering; susceptible to whitefly (see page 105) and red spider mite when overwintering indoors; propagate by 10cm (4in) semi-ripe heel cuttings in mid-summer.

Special needs Some support must be provided. Overwinter indoors.

Display ideas Grow in pots and plunge these into the display containers. Try growing the plants through other shrubs and climbers so that only the flowers are visible.

DRYOPTERIDACEAE

POLYSTICHUM

H Up to 1m (3ft) S 60cm (2ft)

This genus of over 130 species of hardy evergreen ferns, originating mainly from Asia, Europe and North America, is grown for its attractive glossy foliage. The most widely grown plant is *P. acrostichoides*, commonly referred to as the **Christmas fern**, which has very attractive deep green fronds, narrowly divided into many small leaflets that hold their color throughout winter. As the new fronds develop they are covered with glistening white scales, giving the plant a beautiful frosted appearance. *Polystichum munitum*, commonly known as **sword fern**, is more unusual, with long, strap-like leaves, which are notched at the edges.

Cultivation Hardy; partial shade; water normally; mulch with bulky organic matter in spring, and liquid feed at monthly intervals from late spring to early autumn; propagate by division of crowns in spring.

Special needs Grow in partial shade in a non-acidic growing medium rich in organic matter, preferably containing lime or chalk. Some winter protection may be needed.

Display ideas Very good for brightening a cool, shaded corner.

ROSACEAE

POTENTILLA

H 30cm–1.2m (1–4ft) S 1–1.2m (3–4ft)

Potentilla is a genus of about 500 species of mainly herbs, hardy annuals and perennials, deciduous flowering shrubs and sub-shrubs. The shrubby potentillas, or **cinquefoils** as they are commonly known, have single or double flowers of white, pink, yellow, orange and, more recently, red, which are roughly saucer-shaped and usually carried in loose clusters on the tips of shoots and side shoots. Most of the herbaceous perennials are hardy hybrids from the rarely cultivated species *P. atrosanguinea* or *P. nepalensis*, and some excellent named cultivars have been produced, such as *P.* **'Gibson's Scarlet'**, a low-growing plant with bright red, single flowers; and the taller *P. nepalensis* **'Miss Willmott'**, which has cherry-pink flowers with a deep pink eye. Most gardeners are more familiar with the shrubby species, such as *P. fruticosa* **'Elizabeth'**, which is covered in rich canary-yellow flowers from late spring until early autumn; and *P. f.* **'Red Ace'** with its deep orange-red flowers which fade as they age.

Cultivation Hardy; full sun/partial shade; water normally; top dress with slow-release fertilizer in spring; cut back perennials immediately after flowering; remove thin spindly growths,

Primula × *polyantha*

Primula 'Blue Sapphire'

diseased stems and old non-flowering stems from shrubs in spring; propagate shrubs from 7.5cm (3in) semi-ripe cuttings in late summer, and perennials by division in mid-autumn.

Special needs The red-flowered cultivars will hold their color better if grown in partial shade.

Display ideas Excellent flowering shrubs for growing in a mixed display.

PRIMULACEAE

PRIMULA

H Up to 1m (3ft) S Up to 60cm (2ft)

This genus consists of about 400 species of deciduous and evergreen, hardy and half-hardy annuals, biennials and perennials, widely distributed throughout the northern temperate zone. The common names are **polyanthus**, **primula** and **primrose**. All species have rosettes of basal leaves and tubular, bell- or primrose-shaped flowers. The flower stems, leaves and sepals are, in some species and their hybrids, covered with a waxy powder. Primulas are very colorful and therefore popular for container gardening. They are grouped according to various botanical sections, as outlined below.

Candelabra primulas are those with tubular, flat-faced flowers borne in tiered whorls up the stem.

Polyanthus primulas are derived from *P. vulgaris* crossed with *P. veris*, *P. juliae* and other species, and they are usually grown as biennials. Polyanthus flowers are produced in large umbels on stout stems. Some cultivars are grown for their attractive, double flowers, such as *P.* **'Blue Sapphire'**.

Auriculas may be considered as three sub-groups: Alpine, Border and Show. The flowers of all Auricula types are carried in an umbel on a stem above the foliage and are individually flat and smooth in appearance.

Primula vulgaris (primrose) has bright green, corrugated leaves arranged in a flat rosette; the flowers are 2.5cm (1in) across and pale yellow with a golden center. The Polyanthus types are very popular plants for container growing because of their showy flowers in scarlet, orange, rose, pink, blue, yellow and white at a time of year when few plants are flowering. Among the Candelabra types are some very interesting species, such as *P. bulleyana*, which has a rosette of strap-like, dark green leaves, and deep orange flowers produced in tiers on stems 75cm (2½ft) high in mid-summer.

Cultivation Hardy/tender; full sun/partial shade; water normally; mulch with bulky organic matter in spring; susceptible to attack by slugs (see page 105) and root aphid; propagate species by seed sown when fresh or in spring, hybrids by division or root cuttings when the plants are dormant, auriculas by offsets in early spring or autumn.

Special needs Prefer a free-draining growing medium with a high content of organic matter.

Display ideas These plants are ideal for growing in a raised water feature, or for seasonal displays in tubs, hanging baskets and window boxes. See pages 44 and 52.

ROSACEAE

PYRACANTHA

H Up to 4m (12ft) S 2.5m (8ft)

Commonly called **firethorn**, this small genus of about ten species of hardy evergreen shrubs is grown for its masses of small white flowers, produced in early summer, and its bright berries which persist throughout autumn and winter. The spiny branches are clad with broadly oval-shaped leaves which have toothed margins from early autumn to early spring. Popular cultivars include *P.* **'Mohave'**, a vigorous shrub which produces large clusters of orange-red fruits throughout winter; *P.* **'Shawnee'**, a small-leaved, densely branched shrub, which bears deep yellow berries from late summer onwards; and *P.* **'Sparkler'** a slow-growing variegated form, which has green leaves with white margins, often with a slight pink tinge when young.

Cultivation Hardy; full sun/partial shade; water normally; top dress with a slow-release fertilizer in spring; no regular pruning required; remove and burn any plants which show symptoms of fireblight; propagate from 10–15cm (4–6in) hardwood cuttings in autumn, placing them in a cold frame.

Special needs Pots may need some protection in extreme cold to protect the plants roots.

Display ideas These plants are ideal for a north- or east-facing site. Keep them tightly clipped for the first two years to form an upright column of flowers and berries, or train them in a fan shape against a framework.

RANUNCULACEAE

RANUNCULUS

H Up to 45cm (18in) s 60cm (2ft)
Commonly known as **buttercup** or **crowfoot**, this large genus of some 400 species of annuals, aquatics and herbaceous perennials originates from Europe, Asia and North Africa. Most are hardy plants, some are evergreen or semi-evergreen. The leaves are mid- to deep green, often deeply notched and held above the base of the plant on thick erect leaf stalks. Among the most popular plants are *R. montanus* 'Molten Gold', a clump-forming and compact hardy-perennial with large cup-shaped flowers of bright golden-yellow, produced from late spring to early summer; *R. lyallii* is an evergreen plant of stout, upright habit with large leaves and large clusters of white, shallow, cup-shaped flowers, borne in summer. *Ranunculus aconitifolius* 'Flore Pleno', commonly known as **fair maids of France**, has mid-green, deeply toothed, palm-shaped leaves and shining white flowers about 1cm (½in) across, produced in early summer.

Cultivation Hardy to half-hardy; full sun/partial shade; keep well-watered; give established plants a weak liquid feed every two weeks from late spring to early autumn; prune back dead leaves in spring; generally trouble-free of pests and diseases; propagate by division from mid-autumn to late-spring.

Spring needs Thrive in a loam-based growing medium.

Display ideas Ideal for a cool wet area or a raised water feature.

POLYGONACEAE

RHEUM

H Flower spikes up to 2m (6ft) s 1m (3ft)
Rheum, commonly called **ornamental rhubarb**, is a genus of 50 species of hardy herbaceous perennials, many originating from China and the Himalayan region, which are closely related to the *Polygonum*, usually with a thick basal clump (crown). The large leaves, which extend to 45–60cm (18in–2ft) across, are generally glossy mid-green and held clear of the crown on thick fleshy stalks. The large paper-thin bracts covering the flowers are carried above the leaves on tall spikes, which are often pink or red. Two species commonly seen are: *R. alexandrae*, grown for its 1m (3ft) high flower spikes which carry large papery discs resembling drooping tongues which cover the small inconspicuous flowers; *R. palmatum*, which has deeply cut, hand-shaped leaves and greenish-yellow flowers; and *R. p.* **'Atrosanguineum'**, with reddish-purple leaves which fade to a glossy bronze tinted green after the deep red, bead-like flowers have appeared in mid-summer.

Rhododendron simsii 'Perle de Noisy'

Rhododendron leptanthum

Cultivation Hardy; full sun; water normally; apply a liquid feed every three weeks from mid-summer to early autumn; cut the flower spikes back to ground level immediately after flowering; propagate by division from mid-autumn to early spring.

Special needs Leave plants undisturbed for as long as possible. Use a strong container as the roots can exert considerable pressure.

Display ideas Make superb architectural plants: grow as single specimens.

ERICACEAE

RHODODENDRON

H Up to 3m (10ft) s Up to 5m (15ft)
Rhododendron is one of the largest and most diverse genera of ornamental plants in cultivation, with some 800 species. They are among the most spectacular of the spring- and early summer-flowering shrubs. The genus includes not only evergreen large-flowered hybrids but also dwarf small-leaved shrubs, and the deciduous species (formerly called *Azalea*). The flowers are usually single, but may be semi-double or even double, including hose-in-hose (one flower tube inside another); there is a wide range of colors. Although there are many hybrids in cultivation, some of the species are very attractive: *R. luteum* is a deciduous suckering shrub, with matt green leaves which turn scarlet in autumn, and forsythia-yellow fragrant flowers produced in early to mid-summer; *R. atlanticum* has oval gray-green leaves and funnel-shaped white flowers tinged with pink in late spring/early summer; the evergreen *R.* × *obtusum* forms a low mound of glossy green leaves, with scarlet flowers in early summer; *R. indicum*, with its glossy leaves and brown stems covered in fine hairy bristles, produces bright-red or pink flowers in mid-summer; and *R. i.* 'Balsaminiflorum' is a double-flowered salmon-pink cultivar.

Cultivation Hardy to tender; full sun/partial shade; water normally; mulch with bulky organic matter in autumn, and apply iron sequestrene (flowers of sulphur) at six-week intervals from early spring to mid-autumn; dead-head after flowering; susceptible to azalea gall (which causes distortion of the young leaves and flower buds) and bud-blast fungus (which causes the buds to become brown, covered in black fungal bristles and die), and to attack by vine weevil (see page 105) and leafhopper.

Special needs Water with rain water in hard-water areas. These plants are shallow-rooted, so it is advisable to grow them in a broad shallow container.

Display ideas Very good for town gardens as tolerant of pollution.

Rosa **'Climbing Pompon de Paris'**

Rosa **'Boule de Neige'**

ROSACEAE

ROSA

H Climbers up to 3m (10ft); bushes up to 1m (3ft) **S** Climbers up to 2m (6ft); bushes up to 75cm (2½ft)

This is a genus of some 250 species and numerous naturally occurring varieties of deciduous or semi-evergreen shrubs and scrambling climbers. Roses are grown for their profusion of blooms, many of which are fragrant and which occur in a variety of forms: single (4–7 petals), semi-double (8–14 petals), double (15–30 petals), and fully double (over 30 petals). The leaves are usually divided into as many as seven oval, sometimes toothed leaflets, and the stems are often, but not always, clad with curved spine-like thorns. Climbing

Rosmarinus officinalis **'Sissinghurst Blue'**

roses in particular lend themselves to pot culture: for instance, *R.* **'Zéphirine Drouhin'**, which has semi-double, cerise-pink blooms and is thornless; and the beautiful semi-evergreen *R.* **'Mermaid'**, which has large, bright yellow, single blooms with prominent amber stamens. The delicate climbing **'Pompon de Paris'** is shorter and more compact than cluster-flowered roses (floribundas). The vigorous, upright old Bourbon rose **'Boule de Neige'** is heavily scented and more vigorous than the above.

Cultivation Hardy; full sun/partial shade; water normally; mulch with organic matter and top dress with slow-release fertilizer in late winter/early spring, and liquid feed every three weeks in spring and summer; remove dead and damaged wood from bush roses in late autumn/early spring; when pruning bush roses always cut to an outward-facing bud in order to direct the growth away from the center of the plant (this process reduces the risk of attack by pests and diseases, as it opens up the center of the bush, thus providing a good air flow); prune climbing and rambling roses similarly but more lightly; susceptible to blackspot, mildew and rust, and to attack by aphids and sawfly (see pages 104–105); propagate from 25cm (10in) hardwood cuttings in autumn or 10cm (4in) semi-ripe cuttings in mid-summer.

Special needs None.

Display ideas Climbers are excellent for providing color on walls and trellis; patio roses make good pot-grown specimens. See pages 9 and 70.

LABIATAE/LAMIACEAE

ROSMARINUS

H Up to 2m (6ft) **S** Up to 1.5m (5ft)

This small genus of hardy and half-hardy shrubs, commonly known as **rosemary**, has long been cultivated in western Europe. The narrow, aromatic, evergreen leaves are used, either fresh or dried, for culinary purposes. The small tubular flowers range in color from white through blues, pinks and mauves. The most commonly grown species is *R. officinalis*, which has an erect open habit and tough, narrow, mid- to dark green leaves with pale green undersides. *Rosmarinus o.* **'Sissinghurst Blue'** has particularly brilliant blue flowers from spring to autumn. A newer slow-growing cultivar is *R. o.* **'Severn Sea'**; it is compact and develops into a dwarf shrub with masses of brilliant blue flowers borne on arching branches. This particular cultivar is less hardy than the parent species and may need protection in severe winters.

Cultivation Hardy; full sun; water normally; mulch with bulky organic matter in spring; trim back straggling shoots but avoid hard pruning; propagate from 7.5–10cm (3–4in) semi-ripe cuttings in late summer.

Special needs Must be grown in an open free-draining growing medium. May suffer from root rock on exposed sites. To guard against root rock, stake plant and keep growing medium firm around base of plant.

Display ideas Perfect for providing shelter or grown as a screen.

RUTACEAE

RUTA

H 1m (3ft) **S** 60cm (2ft)

Ruta, commonly known as **rue**, is a genus of seven species of hardy, aromatic, evergreen perennials and subshrubs, originating from Europe and south-west Asia. The blue-green leaves are roughly oval in shape, deeply divided to give a fern-like appearance, and give off a pungent aroma. The small mustard-yellow flowers are carried on thick blue-green stems in large flat clusters on the tip of each shoot, rather like the elder (*Sambucus*) flower, from early summer to early autumn. *Ruta graveolens* **'Jackman's Blue'** is a compact cultivar, with bright blue-gray foliage.

Salvia officinalis 'Tricolor'

Cultivation Hardy; full sun; water normally; mulch with bulky organic matter in spring; cut back to old wood in late spring to retain the bushy shape or to prevent legginess, and remove dead flowers in autumn; propagate from 7.5cm (3in) cuttings in late summer.
Special needs *Handle with care: these plants have sap which is a skin irritant and can cause severe skin blistering.* They are fairly drought-tolerant.
Display ideas Grow in a terracotta pot as an individual specimen, or with *Tropaeolum speciosum* (flame creeper), which will use it for support.

LABIATAE/LAMIACEAE

SALVIA

H Up to 1.2m (4ft) **S** 60cm (2ft)
This genus of some 900 species of deciduous and evergreen hardy, half-hardy and tender annuals, perennials and sub-shrubs, originating from Asia, Europe, the Mediterranean, Central and South America, is commonly known as **sage**. The (sometimes aromatic) dull green leaves have a roughly textured surface; they are arranged in pairs on erect square stems, which often have a reddish tinge to them. The flowers, which are tubular, opening into a funnel shape at the mouth, are produced in clusters from mid-summer to early autumn. These plants are divided into three groups depending upon their habit and hardiness.

The half-hardy annuals include ***S. splendens*** and its cultivars, with red-, purple-, pink- and white-flowered

Saxifraga sempervivum

strains available, the bright red **'Rambo'** being an outstanding cultivar.

The tender perennials include ***S. microphylla*** (formerly ***S. grahamii***), which has a shrubby habit and dark crimson-red flowers, ageing to purple, produced from mid-summer until the first frosts of winter.

The hardy perennials include the true sage, ***S. officinalis***, which has both purple and variegated-leaved cultivars, such as ***S. o.* 'Tricolor'**. An interesting plant is ***S. sclarea* var. *turkestanica***, which produces 60cm (2ft) spikes of pink stems and large white or lavender-purple flowers.
Cultivation Hardy to tender; full sun/partial shade; water sparingly; liquid feed every 10 days from early summer to early autumn; dead-head regularly to ensure continued flowering, and cut back perennials in spring after the frosts have finished; propagate perennials from 7.5cm (3in) cuttings in early or late summer, annuals from seed sown in early to mid-spring.
Special needs If the plants are kept slightly dry, they produce more flowers.
Display ideas Very good for window boxes and hanging baskets. See page 15.

COMPOSITAE/ASTERACEAE

SANTOLINA

H 75cm ($2^1/_2$ft) **S** 1m (3ft)
Santolina, commonly called **cotton lavender**, is a genus of 18 species of hardy evergreen shrubs, originating from the Mediterranean region. The leaves are silvery-gray, hairy, aromatic and finely cut. Small, yellow, button-shaped flowers are borne in mid-summer on slender twiggy stems. ***Santolina chamaecyparissus*** has silver foliage and lemon-yellow flowers, while ***S. c.* var. *nana*** is a more compact form with dense growth. ***Santolina rosmarinifolia*** is an attractive species with green thread-like leaves and lemon-yellow flowers. ***Santolina pinnata* subsp. *neapolitana* 'Edward Bowles'** has silver-gray foliage and creamy-white flowers.
Cultivation Hardy; full sun; water normally; mulch with organic matter in spring; remove dead flowers and trim the plants to prevent them becoming straggly; propagate from 5–7.5cm (2–3in) semi-ripe heel cuttings in late summer.
Special needs Prefer a well-drained growing medium and dry conditions.
Display ideas Form very attractive, low-growing shrubs.

SAXIFRAGACEAE

SAXIFRAGA

H Rosettes up to 5cm (2in); flowers up to 60cm (2ft) **S** 2–15cm (3/4–6in)
This genus of 370 species of evergreen low-growing hardy and half-hardy annuals and perennials, commonly known as **saxifrage** (UK) or **rockfoil** (USA), originates from mountainous regions such as the Alpine, North American and Himalayan ranges. The leaves vary greatly from mid- to gray-green; some are finely divided to give the plant a moss-like appearance, and others have small, oval-shaped, silvery leaves often encrusted with lime. All saxifrages have their leaves arranged in broad, flat or slightly raised rosettes to give the whole plant a mound-like appearance, but the rosette which bears the flower stalk dies off after flowering. An outstanding alpine is ***S. sempervivum***, with arching spikes of pink flowers. ***Saxifraga* × *urbium*** has thick fleshy leaves and pink star-shaped flowers produced on thin red stalks.
Cultivation Very hardy; full sun/partial shade; water normally; apply a base dressing of slow-release fertilizer when potting on; susceptible to rust, and to attack by vine weevil and root aphid (see pages 104–105); propagate by division.
Special needs Prefer a well-drained loam-based growing medium.
Display ideas Very good for dry exposed sites. See page 57.

Scabiosa 'Pink Mist'

Sedum acre 'Aureum'

DIPSACEAE

SCABIOSA

H 45–60cm (18in–2ft) **S** 60cm (2ft)
Originating from central, eastern and southern Europe, **scabious** or **pincushion flower**, as it is commonly known, is a clump-forming herbaceous perennial with 5–7.5cm (2–3in) chrysanthemum-like flowers borne on long, slender, leafless stems. The leaves form a low matted cluster reaching no more than 20–30cm (8in–1ft) high. The most popular cultivars are ***S. caucasica***, the rich blue ***S. c.* 'Clive Greaves'** being a reliable grower and a long-established favorite, with **'Miss Willmott'** being the most popular white cultivar. Two more recent introductions are the more compact **'Butterfly Blue'** and **'Pink Mist'**.
Cultivation Hardy; full sun; water normally; mulch with bulky organic matter in spring; dead-head regularly to maintain continued flowering and good flower size; susceptible to powdery mildew and to attack by slugs (see pages 104–105); propagate annuals by seed in spring, and perennials from 5cm (2in) softwood cuttings in spring, rooted indoors or under protection outdoors.
Special needs Do not allow to become waterlogged. Will stand some exposure. Staking with brushwood (or similar) may be needed for the large-flowered forms, particularly in wet weather.
Display ideas Very good for covering the edges of containers. See page 8.

GOODENIACEAE

SCAEVOLA

H 1–2.5m (3–8ft) **S** 2m (6ft)
Commonly known as the **beach berry**, this genus comprises almost 100 species of herbaceous perennials and shrubs, originating mostly from Australasia. The clump-forming plants have broad, bright green, strap-like leaves, rising from the base, arranged alternately to form a spreading succulent shrub. Some variations produce a downy felt covering on younger leaves. The fragrant flowers are borne in clusters of six to eight in the leaf joint, each bloom measuring about 2cm (¾in) long, and creamy-white streaked with purple. White fleshy berries about 1cm (½in) long are produced after flowering. ***Scaevola plumieri*** has short, spatula-shaped leaves, with 2.5cm (1in) long, pinkish-white flowers, followed by black berries.
Cultivation Half-hardy; full sun; water normally; mulch with leaf mold in spring; remove flower stalks after seeds have fallen; propagate by semi-ripe cuttings 15cm (6in) long, placed in a closed case, in spring.
Special needs Prefer a sand-based, free-draining growing medium. These plants must be provided with a frost-free environment.
Display ideas Grow as a single specimen in a very hot situation. Alternatively, combine with cordylines and yuccas for an exotic effect.

CRASSULACEAE

SEDUM

H 60cm (2ft) **S** 1m (3ft)
Sedum, commonly known as **stonecrop** or **ice plant**, is a genus of 300 species of hardy, half-hardy and tender annual and perennial, deciduous and evergreen succulents, widely distributed throughout Asia, Europe and North Africa. These plants are very drought-tolerant. They are grown for their attractive star-shaped flowers, in various shades, which are borne in broad flat-topped clusters on the tips of the shoots. They also produce some very interesting leaf shapes, forms and colors. The most commonly seen herbaceous species is ***S. spectabile***, and its various cultivars have broadly oval leaves, white-green in color, carried in pairs at each leaf joint. The pink flower heads, tinged mauve, are 10cm (4in) or more wide and appear from early to mid-autumn. ***Sedum acre* 'Aureum'** is one of the best ground cover sedums; it has small flower heads which are yellow on first opening, later fading to cream.
Cultivation Hardy; full sun; water sparingly; top dress with slow-release fertilizer in spring; cut the stems back to soil level in early spring; susceptible to root rot in wet conditions, and to attack by slugs (see page 105); propagate by division in spring.
Special needs Do not allow to become waterlogged.
Display ideas *Sedum spectabile* is good for attracting butterflies. See page 57.

CRASSULACEAE

SEMPERVIVUM

Sempervivum tectorum

H Rosettes 5cm (2in); flowers up to 20cm (8in) **S** Up to 30cm (1ft) across
This is a genus of 42 species of hardy and half-hardy, rosette-forming, evergreen succulents commonly known as **houseleeks**. Many are alpines, originating from mountainous regions around the world, such as the Alpine, Carpathian and Pyrenean ranges. Many species are fully hardy. The fleshy green leaves, often tipped with red or purple, are oval to elliptical in shape, ending in a sharp point, and

Sempervivum tectorum

curve around each other to form tight rosettes. The star-like flowers are carried in sprays on thick fleshy flower stalks. The rosette which bears the flower stalk dies off after flowering.

Cultivation Hardy; full sun; water sparingly; feed with a base dressing of slow-release fertilizer when planting out; susceptible to rust (see page 104); propagate by division – remove and pot on offsets in early autumn or late spring.

Special needs Prefers a loam-based growing medium containing plenty of coarse sand. Protect newly planted rosettes against uprooting by birds.

Display ideas Ideal for covering the surface of the growing medium to form a living mulch. Will tolerate drought. See pages 55 and 57.

COMPOSITAE/ASTERACEAE

SENECIO

H 60cm–1.2m (2–4ft) **S** Up to 1.2m (4ft)

Senecio, commonly known as **daisy bush** (UK) or **groundsel** (USA), was until recently a very large genus of some 1500 species of perennial herbs, shrubs and climbers, widely distributed throughout the world. Many of the shrubby members from New Zealand are now called *Brachyglottis* by some authorities; they are mostly evergreen and produce daisy-like flowers in summer. The shrubby species are hardy in most areas, and many have broadly oval-shaped leaves which appear silver-gray as a result of a hairy felt-like covering. *Senecio* (*Brachyglottis*) **'Sunshine'** has attractive gray foliage and forms a broad mound-like shrub with broad heads of golden-yellow flowers. *Senecio cineraria bicolor* is an excellent annual, the leaves deeply notched and a vivid silver-gray; the best colored cultivar is **'Silver Dust'**.

The Cineraria hybrids are ideal for providing a bold splash of color from later summer through until mid-autumn. They are available in a wide range of sizes and colors – from light and dark blue to white, pink, crimson and scarlet; many have a white base to each petal making the flower look like an eye.

There are both large-flowered (Grandiflora) and small-flowered (Stellata) varieties, and, more recently, shorter growing types (Nana multiflora) with a much more compact growth habit. All types are treated as annuals and grown from seed every year for flowering in the second half of the year. Although they are very easy to grow, these plants have very specific watering requirements and will start to wilt in a matter of hours if they are not watered on a regular basis. If overwatered, these plants will rot very quickly at the base of the stem and collapse. However, Cineraria hybrids are ideal for the beginner as they do not like to be pampered.

Cultivation Hardy/half-hardy; full sun/partial shade; water normally; mulch shrubs with bulky organic matter in spring; liquid feed half-hardy plants every 14 days; cut out straggly growths in spring and dead-head regularly as the flowers fade; propagate half-hardy plants from seed sown under protection in mid-spring, shrubs from 10cm (4in) cuttings in late summer.

Special needs Protect half-hardy plants from frost.

Display ideas Annuals make an excellent display in hanging baskets and window boxes.

RUTACEAE

SKIMMIA

H Up to 1m (3ft) **S** 1.2m (4ft)

This small genus of four species of hardy, evergreen, slow-growing, aromatic shrubs is native to east Asia and the Himalayas. The thick leathery foliage is oval to strap-shaped, ending in a pointed tip. The small, star-like flowers are carried in spikes on the tips of the shoots in mid- to late spring; these are followed by orange-red berries. It is advisable to grow both sexes – ideally one male to four female plants. ***Skimmia japonica* subsp. *reevesiana*** is female. It forms a compact shrub with glossy mid- to deep green leaves and creamy-white flowers, followed by crimson fruits in late summer. ***Skimmia japonica* 'Rubella'** is a male clone with deep red flower buds and green foliage with a red margin. The male ***S. j.* 'Fragrans'** is a free-flowering plant, which carries the scent of lily-of-the-valley.

Cultivation Hardy; full sun/partial shade; water normally; mulch with bulky organic matter in spring; propagate from 10cm (4in) semi-ripe heel cuttings in mid- to late summer.

Special needs Protect the young leaves against frost damage in severe weather. *Skimmia reevesiana* does best in a lime-free growing medium.

Display ideas Ideal for inner-city use as very tolerant of atmospheric pollution.

COMPOSITAE/ASTERACEAE

TAGETES

H Up to 1m (3ft) **S** 60cm (2ft)

This genus includes 50 species of annual and perennial herbaceous plants, commonly known as **marigolds**. Both single- and double-flowered forms are available. The single flowers are daisy-shaped; the doubles resemble small carnations. When bruised the foliage gives off a pungent smell. *Tagetes erecta* (commonly called the **African marigold**) is a vigorous, erect, well-branched, half-hardy annual with deeply cut, glossy, dark green leaves and lemon-yellow, broad-petalled, daisy-like flowers, up to 5cm (2in) across, produced from mid-summer until the first frosts. The ***T. patula*** (**French marigold**) hybrids are fast-growing annuals, but only about half the size of the African marigolds, and come in a range of colors including yellow, orange and red. The **Mischief** hybrids are very popular with their single bicolored flowers.

Cultivation Half-hardy; full sun; water normally; feed with a base dressing of slow-release fertilizer when planting out; dead-head regularly; propagate from seed sown under protection in mid-spring.

Special needs None: easy to grow.

Display ideas Very good for edging containers, and ideal for hanging baskets and window boxes.

LABIATAE/LAMIACEAE

THYMUS

H Up to 30cm (1ft) **S** Up to 1m (3ft)
This genus, commonly known as **thyme**, includes around 350 species of hardy, evergreen, herbaceous perennials and sub-shrubs, with aromatic leaves, originating from Europe, Asia and the Mediterranean. Some are mat-forming and prostrate, with small, strap-like, gray-green leaves held just above soil level. Others make small shrubby plants with tough, thin, twiggy stems; their narrow, oval-shaped leaves are usually gray-green, sometimes hairy and arranged close together. The plants most commonly grown for their ornamental qualities are closely related to *T.* × *citriodorus*, whose mid-green lemon-scented leaves are an elongated oval shape and up to 1cm (½in) long. Its minute flowers are a pale lilac and appear in small clusters on the tips of the branches from mid- to late summer. There are a number of cultivars with attractive leaf forms, including *T.* × *c.* **'Aureus'**, which has golden foliage tinged with green; and *T. vulgaris* **'Silver Posie'**, with silver-and-green variegated leaves.
Cultivation Hardy; full sun/partial shade; water normally; clip over after flowering to remove dead blooms and keep the plants bushy; propagate from 5cm (2in) heel cuttings in mid-summer.
Special needs Prefer a free-draining growing medium.
Display ideas Grow in low pots close to pathways, so that the aroma is released as people brush past the plants.

SAXIFRAGACEAE

TOLMIEA

Tolmiea menziesii

H 15cm (6in) **S** 45cm (18in)
Commonly known as **mother of thousands**, **pickaback** or **piggy-back**, *T. menziesii* originates from the United States. It is, unusually, just as much at home as a house plant as it is in a container growing outdoors, even though it is a very hardy evergreen and will overwinter outside. *Tolmiea m.* **'Taff's Gold'** has small, hairy, almost nettle-like leaves (which do not sting), which are mottled pale green and yellow, and carried on equally hairy leaf stalks. The particular interest with this plant is that it produces small plantlets on the upper sides of the leaves, often with 30–40 young plants actually growing on the back of a single leaf. The 1cm (½in) tubular flowers are greeny-white flushed with red, have hair-like growths at the tips of the petals and are carried high above the leaves on flower stalks 45–60cm (18in–2ft) high.
Cultivation Hardy; full sun/partial shade; water normally; liquid feed monthly from early summer to early autumn; propagate by layering.
Special needs None.
Display ideas Excellent for covering the surface of a container and forming a living mulch, and for trailing over the sides of tubs and window boxes.

Thymus vulgaris 'Silver Posie'

PALMAE/ARECACEAE

TRACHYCARPUS

Trachycarpus fortunei

H 2–3m (6–10ft) **S** 2m (6ft) or more
A member of a small genus of evergreen palms originating from the Himalayas and East Asia, *T. fortunei*, commonly known as **windmill palm** or **chusan palm**, is hardy in temperate regions. This slow-growing plant gradually develops into a single furry trunk made up of the fibrous remains of old leaf stalks, which helps to give it a palm-like appearance. The fan-shaped mid-green leaves, about 1m (3ft) wide, are pleated and split at the ends to give a jagged tooth-like appearance, and are held on sharply toothed leaf stalks up to 1m (3ft) long. The small yellow flowers appear in early to mid-summer in large dense bunches up to 60cm (2ft) long. In a very hot summer these flowers may produce 1cm (½in) diameter, black, date-like fruits.
Cultivation Hardy; full sun/partial shade; water normally; mulch with bulky organic matter in spring; propagate by division in early summer.
Special needs Shelter from strong north and east winds.
Display ideas Brilliant choice for a sheltered sunny spot; very good for a Mediterranean theme.

Tropaeolum majus Jewel Series

COMMELINACEAE

TRADESCANTIA

H 60cm (2ft) **S** 75cm (2½ft)
Tradescantia, commonly called **trinity flower** or **spiderwort**, is a genus of perennial plants, some hardy and others very tender. The hardy perennial species *T. virginiana* is popular for container growing as it requires very little care and attention. This attractive plant, with dull green, strap-shaped leaves which taper to a narrow point, bears unusual triangular-shaped flowers with three petals 2–4cm (1–1½in) across, giving the blooms a very symmetrical appearance. The flowers are produced in small clusters from mid-summer to early autumn and vary in color. There are numerous named cultivars, including *T.* × *andersoniana* 'Blue Stone', with deep blue flowers; the very popular *T.* × *a.* **'Isis'**, rich royal-purple; and *T.* × *a.* **'Osprey'**, clear white with a powder-blue center.
Cultivation Hardy; full sun/partial shade; water normally; top dress with slow-release fertilizer in spring or liquid feed every two weeks from late spring to early summer; susceptible to attack by slugs (see page 105); propagate by division in mid- to late spring.
Special needs Keep the roots cool applying a layer of mulch.
Display ideas Very good for use in low containers, or as a filler between shrubs. See page 49.

TROPAEOLACEAE

TROPAEOLUM

H 60cm–5m (2–15ft) S Up to 1.2m (4ft)
Tropaeolum is a genus of 90 species of hardy annual and herbaceous perennial plants, mostly of climbing habit, originating from South America. All species have short, broadly trumpet-shaped flowers, some with petals of different sizes and a prominent spur at the base of each bloom, opening from midsummer to mid-autumn. The smooth leaves are almost circular with a crinkled edge, and mid- to light green in color. *Tropaeolum majus* (**nasturtium**) is an annual with a sprawling habit and flower colors which range from pale yellow through to scarlet-orange. The mixture **Alaska** has creamy-white variegated leaves and multi-colored flowers. *Tropaeolum speciosum* (**flame creeper**) is a deciduous perennial with narrow twining stems, mid-green leaves which have notches in the margins, and brilliant scarlet flowers made up of five rounded wavy petals which open out flat.
Cultivation Hardy/half-hardy; full sun/partial shade; water sparingly; for annuals, feed with a base dressing of slow-release fertilizer when planting out; for perennials, top dress with a slow-release fertilizer in spring; propagate annuals from seed sown outdoors in late spring, perennials by division in mid-spring.
Special needs To encourage flowering, keep the plants slightly dry and hungry. Some support must be provided for the climbing species.
Display ideas Climbing perennials are excellent for growing through other plants; annuals are ideal for trailing over the sides of hanging baskets.

LILIACEAE

TULIPA

H 10–50cm (4–20in)
S Up to 25cm (10in)
Tulips were first introduced into Europe from Turkey in the 16th century. Within 100 years these bulbous perennials had become so popular in the Netherlands that the term tulipomania was coined for the passion with which collectors fought over the new cultivars. Tulips are still extremely popular today, both as species plants and as cultivated forms.

Tulipa clusiana

Tulipa 'Van der Neer'

Tulipa 'Fringed Elegance'

Tulipa 'Flaming Parrot'

Some years ago, the Royal General Dutch Bulbgrower's Association in conjunction with the Royal Horticultural Society took on the Herculean task of classifying tulips into different divisions according to type. There are currently 15 divisions, of which the most well-known are the **Darwin Hybrids** (both large-flowered tall tulips); the **Lily-flowered tulips**, with their pointed petals; **Rembrandt tulips**, a broken color form of Darwin types; **Parrot tulips**, with their crinkly-edged petals, usually streaked with other colors; and the **Species tulips**, which usually have more open, cup-shaped flowers and tend to be much smaller than the cultivated forms. Of the species tulips *T. kaufmanniana*, the dwarf water lily tulip with bowl-shaped flowers, and *T. clusiana*, with long white flowers flushed deep pink on the outside, are easy to grow and very attractive.

Tulips all grow from a rounded or ovoid bulb, which has a thin skin. Most bulbs produce one flower on an erect stem, but a few bear two, three or more. The flowers are goblet-shaped, with six petals, and vary from slender to pointed to broadly rounded. The leaves are strap-like. Nearly every form of tulip flowers profusely in the first year after planting.
Cultivation Very hardy; full sun; water at planting time, thereafter keep bulbs moist until leaves start to yellow after flowering; feed with liquid fertilizer once a week after flowering until watering stops; susceptible to attack by slugs which feed on the bulbs, eelworm, tulip fire, and aphids (see page 105); propagate by offset.
Special needs Prefer a free-draining growing medium. Protect from strong winds. In wet weather, lift bulbs once leaves have died and store in a dry, frost-free environment until autumn.
Display ideas Tulips can make a stunning display if you plant more than one container of the same kind and color.

Viburnum carlesii 'Diana'

Viola 'Little Liz'

VERBENACEAE

VERBENA

Verbena × hybrida

H 15–45cm (6–18in) **S** 30cm (1ft)
This genus includes 250 species of annual and perennial herbaceous plants, originating from South America. These bushy plants produce mid- to dark green oval-shaped serrated leaves. The clusters of small, fragrant flowers, 7.5cm (3in) or more across, are carried on the top of the stems from mid-summer until the first autumn frosts. The colors range from white, pink and red to blue and lilac. *Verbena × hybrida* **'Silver Anne'** is just one of numerous outstanding varieties.

Cultivation Half-hardy; full sun; water normally; liquid feed every two weeks from late spring to early summer; susceptible to attack by aphids (see page 105); propagate annuals from seed sown indoors in spring, perennials from 7.5cm (3in) cuttings in late summer.

Special needs *Verbena rigida* and *V. bonariensis* can be overwintered outdoors if some protection is provided.

Display ideas Useful for window boxes and tubs. See pages 53 and 56.

CAPRIFOLIACEAE

VIBURNUM

H Up to 3m (9ft) **S** 2m (6ft)
Viburnum is a large genus of both deciduous and evergreen shrubs and small trees. Many have fragrant white flowers, some produce brightly colored fruits during the autumn. Many of the deciduous species bear their flowers in winter on naked wood, and some produce very attractive autumn leaf color. Among the most popular varieties are *V. carlesii*, a hardy deciduous species which has dull green, broadly oval leaves and heavily scented, waxy white flowers in late spring and early summer. Cultivars include ***V. c.* 'Aurora'** and the more compact ***V. c.* 'Diana'**. Also worth looking out for is ***V. tinus***, which has a bushy habit and smooth, dark green, oval-shaped, evergreen leaves. The flowers are pink in the bud, turning white as they open from late autumn to early summer.

Cultivation Hardy; full sun/partial shade; water normally; mulch with bulky organic matter in spring; susceptible to gray mold and fungal leaf spot, and to attack by aphids and white fly (see pages 104–105); propagate from 7.5–10cm (3–4in) semi-ripe heel cuttings in mid- to late summer or by layering in mid-summer.

Special needs Shelter from cold sites where early-morning sun after spring frosts may damage young growths.

Display ideas Position these plants close to the house in order to benefit from their heady scent.

VIOLACEAE

VIOLA

H 15–20cm (6–8in) **S** Up to 30cm (1ft)
One of a large genus of hardy annual and herbaceous plants, the **pansy** or **violet**, as it is commonly known, is one of the most popular container-grown plants. Most of the cultivars are hybrids of *V. × wittrockiana*; they have larger flowers and are stronger growers than the smaller species violas. The flowers range in size from 2.5–7.5cm (1–3in) across, with colors varying from golden-yellow, through to reds, whites, blues, blacks and deep violets. The narrow oval-shaped leaves have toothed edges, are a glossy mid-green and are carried on lax spreading stems.

Viola septentrionalis f. *alba*

Cultivation Hardy; full sun/partial shade; water normally; feed with a base dressing of slow-release fertilizer when planting out; susceptible to mosaic virus and rust (see page 104); propagate annuals from seed sown in mid-summer and perennials from softwood cuttings in spring.

Special needs Prefer a well-drained growing medium.

Display ideas Perfect for early spring displays of bright colors. Trailing varieties are ideal for hanging baskets. See pages 11, 29, 40 and 42.

VITACEAE

VITIS

H Up to 20m (60ft) **S** 4m (12ft)
The ornamental vines are part of a genus of mainly vigorous climbing plants which support themselves by twining tendrils. The leaf type and size vary considerably with each species, and several species are grown for brilliant autumn color, while others have attractive peeling bark. One of the most popular vines is *V. coignetiae* (**crimson glory vine**), the finest of all ornamental vines. It is a hardy vigorous species with thick mid-green leaves. Each up to 30cm (1ft) across and heart-shaped, they turn yellow, orange-red and eventually purple-crimson in autumn. *Vitis vinifera* (**common grape vine**), is also popular; there is also a cultivar with purple leaves, *V. v.* 'Purpurea'.
Cultivation Very hardy; full sun/partial shade; water normally; top dress with bulky organic matter in spring; thin out old growths and shorten young growths in late summer; susceptible to attack by scale insect and vine weevil (see page 105); propagate from hardwood cuttings in mid-autumn to early winter.
Special needs Prefer a deep container of moist loam-based growing medium enriched with plenty of manure. Provide a hot dry situation.
Display ideas Grow up a pole and allow to cascade back down over the pot.

LEGUMINOSAE/PAPILIONACEAE

WISTERIA

H Up to 10m (30ft) **S** 15m (50ft)
This genus of hardy deciduous climbers, originating from East Asia and North America, includes some of the most beautiful of all climbing plants when draped in their cascades of white, pink, blue or mauve, pea-like flowers 20–30cm (8–12in) long. The normal flowering season is early to mid-summer. The most commonly grown species is *W. sinensis* (**Chinese wisteria**) with dark to mid-green leaves divided into small leaflets. There are other forms, such as the white-flowered *W. s.* 'Alba'; *W. s.* 'Plena' has double mauve flowers.
Cultivation Hardy; full sun; water normally; mulch with bulky organic matter in spring; cut back side shoots to two or three buds in early spring, and check vigor with mid-summer pruning if necessary; susceptible to fungal leaf spot (see page 104), and to attack by red spider mite; propagate from 7.5–10cm (3–4in) heel or nodal cuttings in late summer or by layering in early summer.
Special needs Prefer a south- or west-facing position. Water with rain water in hard-water areas to avoid lime-induced chlorosis. Do not allow to dry out or the flower buds will drop.
Display ideas Can be trained to form a small standard.

Yucca filamentosa 'Variegata'

AGAVACEAE

YUCCA

Yucca filamentosa

H 60cm–1m (2–3ft) **S** 1–1.2m (3–4ft)
This plant, commonly known as **silk grass**, **Adam's needle** (UK) or **needle palm** (USA), is one of a genus of long-lived hardy and tender evergreen shrubs and small trees. They thrive in poor sandy conditions and are particularly at home in coastal gardens, as well as being tolerant to exposed windy sites. The strap-shaped leaves are up to 1m (3ft) in length, bluish-green, and usually dried and brown at the tip which forms a sharp spine-like point. The reddish-brown flower spikes are often 1.5–2m (5–6ft) high, their top two thirds covered with a display of white, drooping, bell-shaped, lily-like blooms each 6–8cm (2½–3½in) across. There are two variegated cultivars *Y. f.* **'Bright Edge'**, which has a narrow golden margin to the leaf edge; and *Y. f.* **'Variegata'**, with a creamy-white margin to the leaves. It is worth noting that the colored-leaf forms are generally not as hardy as the type.
Cultivation Half-hardy; full sun; water sparingly; top dress with slow-release fertilizer in spring; susceptible to leaf spot (see page 104); propagate by division in mid- to late spring – remove and pot on rooted suckers.
Special needs Cover the crown of the plant with straw to protect from frost in severe winters.
Display ideas Very good for creating an architectural feature; can be grown as a single specimen; a 'must' if a Mediterranean effect is desired. See page 20.

Zantedeschia aethiopica

ARACEAE

ZANTEDESCHIA

Zantedeschia aethiopica

H 45cm–1m (18in–3ft) **S** 60cm (2ft)
Commonly called the **arum lily** (UK), **florist's calla** or **garden calla** (USA), this is one of a genus of half-hardy herbaceous perennials originating from South Africa. The large, glossy, dark green leaves are an arrow-head shape. The white flowers with a central yellow spike are borne from mid-spring to mid-summer. There is a hardier cultivar, *Z. a.* 'Crowborough'.
Cultivation Hardy; partial shade; water normally; mulch with bulky organic matter in spring; prone to viruses (see page 104); propagate by division in autumn.
Special needs Prefers a peaty growing medium. Keep the roots cool.
Display ideas Grow in pots to achieve early flowering before plunging into the display containers.

INDEX

Page numbers in *italics* refer to illustrations and captions

A
Abutilon, 108
A. 'Apricot', *41*
A. megapotamicum, 108
A. pictum 'Thomsonii', 108
Abyssinian banana, 122–3
Acer palmatum, 21, 108
A.p. 'Beni-tsukasa', 108, *108*
A.p. var. *dissectum*, *38*, *66*
A.p. var. *dissectum* 'Atropurpureum', 108, *108*
A.p. 'Senkaki', 108
Acorus calamus 'Variegatus', *54*
A. gramineus 'Variegatus', *54*
Actinidia, 108
A. deliciosa, 17, 108
A. kolomikta, *17*, 62, 108
Adam's needle, 155
Adiantum capillus-veneris, 109
Aegopodium podagaria 'Variegatum', *63*
aeration, 102
African lily, *20*, 109
African marigolds, *151*
Agapanthus, *50*
A. africanus, *20*
A. campanulatus, 109
A. Headbourne hybrids, 109
Agave, 66
A. americanum, 109
A.a. 'Marginata', 109, *109*
A. attenuata, *6*
Akebia quinata, 70
Alchemilla mollis, 15, *26*, 66, 109, *109*
Allium, *27*, 110
A. albopilosum, 109
A. moly, 110
alpines, 42–3, 57–8
alyssum, 43
Amelanchier, 93
Anaphalis margaritacea, *12*
Anchusa azurea, 110
A. a. 'Blue Angel', *43*
A.a. 'Dropmore', 110
A.a. 'Loddon Royalist', 110, *110*
A.a. 'Morning Glory', 110
A.a. 'Opal', 110
Anemone, 52, 58, 110
A. blanda, *58*
A.b. 'White Splendour', *52*
A. coronaria, 110
A. hupehensis, 110
A. × *hybrida*, 110, *110*
Anemone contd.
A. × *hybrida* 'Honorine Jobert', *26*
angel's trumpet, 40, 72
Anthriscus sylvestris 'Moonlight Night', *62*
Antirrhinum, *42*
ants, 105
aphids, 105
Aponogeton distachyos, *54*
apple trees, *83*
applemint, *63*
Arabis alpina, *48*
Argyranthemum, *10*, 19, *42*, 116
A. foeniculaceum, *6*
A. frutescens, *75*
A. 'Mary Wooton', *6*
Aristolochia, 63
A. durior, 26, 62
A. macrophylla, 110–11
Artemisia, 10, 15, *38*, 62, 93
A. absinthium, *15*
A. pontica, 111, *111*
Arundinaria nitida, 111
Astelia chathamika, *61*
Aster, 40
Astrantia major, *68*
Athyrium filix-femina 'Victoriae', *26*
Atriplix hortensis var. *rubra*, *62*
Aucuba, 93
A. japonica, 71
auriculas, 146
Australian tree fern, 11, 38, 64
autumn crocus, 43
azalea, 21, 147

B
baby's breath, 126
Bacopa, 42, 111
B. caroliniana, 111
B. monnieri, 111
bacterial leaf spot, 104
balconies, 16–19
Ballota acetabulosa, *51*
bamboo, 21, 63, 111
banana plant, 40
barberry, 112–13
basil, 11, 15
bay, *8*, 133
beach berry, 150
beard tongue, 143
Begonia, *46*, 48, *68*, 112
B. 'Harlequin', 112
B. 'Jamboree', 112
B. masoniana, 112
B. 'Olympia', 112
B. rex, 112
B. semperflorens, 112
B.s. New Generation, 112
B.s. Thousand Wonders, 112
B.s. 'Volcano', 112
B. × *tuberhybrida*, *25*, 112, *112*
bellflowers, 73, 115
Bellis perennis, 43, *45*
Berberis, 14, 93, 112–13
B. *prattii* var. *laxipendula*, 113, *113*
B. thungergii, *22*
Betula 'Trost's Dwarf', 38
black spot, 104
blackfly, 105
bog gardens, 56–7
Boston fern, 138
Bougainvillea, 93
box, *8*, 10, *13*, 18, 21, *36*, 42, 44, 62, 63, 64, *66*, 67, 114
Brachycome, 11, 44, *45*, 72
B. iberidifolia, *43*, 113, *113*
Brachyglottis, *41*, 151
Brassica olearacea, *44*, *113*
B.o. 'Sekito', 113
B.o. 'White Peacock', *41*
broadleaf, 126
Browallia speciosa, 113
B.s. 'Heavenly Blue' 113
Brugmansia, 12, 18, 40, 72
B. × *candida* 'Knightii', 114, *114*
Buddleja, *12*, *70*, 72, 94
B. globosa, 93
bulbs, division, *98*
nutrients, 87
planting, *82*
bush basil, *15*
bush violet, 113
busy Lizzies, 10, 14, *16*, 26, *38*, 40, *42*, *43*, *131*
Butomus umbellatus, *54*
buttercup, 147
Buxus sempervirens, *8*, 10, *13*, 21, *22*, *36*, 42, 44, 62, 67, 93, 114
B.s. 'Aureovariegata', 23
B.s. 'Elegantissima', 114
B.s. 'Latifolia Maculata', 114
buying plants, 78–9

C
cabbage palm, 119
Calendula officinalis, 114
C.o. 'Lemon Beauty', 114, *114*
C.o. 'Pacific Beauty', 114
Calla palustris, *54*
Calluna, 93
Caltha palustris, *54*
Camellia, 65, 70, 93, *97*
C. japonica, 114–15
C. × *williamsii*, 115
C. × *williamsii* 'Anticipation', *115*
Campanula, *51*, 73, 115
C. alpestris, 115
C. carpatica, 115
C.c. 'Isabel', 115
C.c. 'Jewel', 115
C.c. 'White Star', 115
C. isophylla, 115
C. medium, *9*
C. portenschlagiana, *59*
C. rotundifolia, 68
Campsis, 115
C. radicans, 70
C. × *tagliabuana*, 115
C. × *tagliabuana* 'Madame Galen', *115*
candelabra primulas, 146
canker, 104
Canna, *22*, 44, 115
C. × *generalis* 'Wyoming', 115
C. 'Lucifer', 115
capsid bugs, 105
Caragana, 93
Cardamine trifolia, *45*
cardinal flower, 135
Carpinus betulus, 17
carrots, 52
Caryopteris, 93
caterpillars, 105
cathedral bells, 118
catmint, *9*, *12*, *17*
century plant, 109
Cercis, 93
Chaenomeles, 93
chafers, 105
chalk plant, 126
Chamaecyparis, *20*, 116
C. lawsoniana, 116
C.l. 'Grayswood Pillar', 116
C.l. 'Pygmaea Argentea', 116
C. nootkatensis 'Pendula', 116
C. obtusa, 116
C.o. 'Nana Gracilis', 116
Chimonanthus, 93
Chinese sacred bamboo, 138
Choisya, 65, 93
C. ternata, 14, 19, 64, 93, 116
C.t. 'Sundance', 116
Christmas box, *53*
Christmas fern, 145
Chrysanthemum see *Argyranthemum*, *Leucanthemum*, *Rhodanthemum*, *Tanacetum*
chrysanthemums, 40, 43, 58, 73, 116
chusan palm, 152
cineraria (*Senecio*), *52*
Cineraria pericallis, *52*
cinquefoils, 145
Cistus, 93, 117
C. × *aguilari* 'Maculatus', *116*, 117
C. 'Silver Pink', 117
× *Citrofortunella microcarpa*, 117
Citrus, 117, *117*
C. ichangensis, 117
clary sage, *68*
cleaning, 102
Clematis, 21, 71, 73, 74–5, 94, 117
C. 'Alba Luxurians', 118
Clematis contd.
C. alpina, 118
C.a. 'Pamela Jackman', 118
C. armandii, 21
C. campaniflora, 118
C. 'Caroline', *19*
C. 'Etoile Violette', 118
C. florida 'Sieboldii', *117*, 118
C. 'Hagley Hybrid', 71
C. 'Jackmanii', *19*
C. 'Jackmanii 'Superba', 118
C. 'Madame Julia Correvon', *19*
C. 'Margot Koster', *19*
C. 'Marie Boisselot', *19*
C. 'Minuet', 118
C. montana, 17, *19*, 75
C, 'Princess of Wales', *19*
C. 'Victoria', *19*
C. viticella, 10, 75, 118
C.v. 'Purpurea Plena Elegans', *19*
C. 'Vyvyan Pennel', 118
climbers,
balconies, 16–17, *17*, 19
pruning, 91–2
roof gardens, 21
Clivia, 64
C. miniata, 66
Cobaea scandens, 118, *118*
Coleus, *50*
Columnea, *51*
composts, 85
conifers 18
Consolida ambigua, 14
containers, 81
aeration, 102
choosing, 30–7
cleaning, 102
decorating, 37
designing with, 29–59
drainage, 102
large, 38–41
materials, 30
moving, *102*, 103
positioning, 36–7
shape, 36
size, 35–6, 81
small, 42–5
unusual, 54–9
weathering, *37*
winter protection, *103*
Convolvulus, 118
C. cneorum, *12*, *118*
C. tricolor, 118
C.t. 'Blue Flash', *43*
C.t. 'Royal Ensign', 118
Coprosma × *kirkii* 'Kirkii', 44, 118–19
C. × *kirkii* 'Prostrata', 118–19
C. × *kirkii* 'Variegata', *74*, 118–19
Cordyline, 11, 19, 64, 66
C. australis, *41*, 119
C.a. 'Purpurea', 119
C. indivisa, 119

Coreopsis tinctoria, *43*
corn parsley, 143
Cornus alba, 93
C.a. 'Elegantissima', 7
Corylus avellana, 93
Cosmos bipinnatus, 41
Cotinus coggygria, 93
Cotoneaster, 26, 119
C. atropurpurea 'Variegata', *63*
C. 'Coral Beauty', 119
C. franchettii, *119*
C. horizontalis, 119
C. lacteus, 119
cottage garden schemes, *18*, *22*
cotton lavender, 149
cranesbills, 126
crested buckler fern, 122
crimson glory vine, 155
Crinum × *powellii*, 66
Crocus, *44*, *45*, *82*, 119
C. chrysanthus 'Gipsy Girl', 119
crowfoot, 147
culinary herbs, *15*
Cuphea hyssopifolia, *38*
Cupressus, 120
C. glabra, 120
C.g. 'Compacta', 120
C. macrocarpa, 120
C.m. 'Goldcrest', 120
curbing growth, *84*
cuttings, 96–8
Cyclamen, *41*, 43, *44*, 50, 58, 120
C. coum, *59*, 120
C. hederifolium, 119
C.h. album, 119
cypress, 116, 120

D

daffodil, 43, 58, *82*, 138
Dahlia, 120–1
D. 'Bishop of Llandaff', 121
D. 'David Howard', 121
D. 'Libretto', *120*
D. merkii, *120*
daisy bush, 151
daisy-gone-crazy, 11
damping off, 104
Daphne, 93
Datura, 12, 18, 72
dead nettle, 132
dead-heading, 94
design, with containers, 29–59
with plants, 61–75
Deutzia, 93
Dianthus, *12*, *42*, *70*
D. barbatus 'Nigra', *42*
D. 'Little Jock', *57*
Diascia, 44, *48*, *49*, 73, 121
D. 'Ruby Field', *24*, 121, *121*
D. vigilis, *51*
Dicentra, 64
Dicksonia antarctica, 11, 38, 64
Dimorphetica pluvalis, *43*
diseases, 104
dittany, 139–40
division, 98–9
Dorotheanthus bellidiformis, 121

downy mildew, 104
Dracaena, *66*
drainage, 102
Dryopteris cristata, 122
D. filix-mas, 121–2
Dutchman's pipe, 26, 62, 63, 110–11

E

earwigs, 105, *105*
Eccremocarpus scaber, *17*
eelworms, 105
Eleagnus, 122
E. × *ebbingei*, 14, *23*, 122
E. × *ebbingei* 'Gilt Edge', 122
E. × *ebbingei* 'Limelight', 122
E. pungens, 122
E.p. 'Dicksonii', 122
E.p. 'Maculata', 122
Elodea canadensis, *54*
Enkianthus, 93
Ensete, 40
E. ventricosum, 122–3, *122*
entrances, 8–11
equipment, choosing, *78*
Erica, 93, 123
E. carnea, 123
E.c. 'Aurea', 123
E.c. 'Gracilis', *41*, 44, *44*, *53*
E.c. 'Golden Drop', 123
E. erigena, 123
E. terminalis, 123
Erigeron aureus 'Canary Bird', *59*
E. karvinskianus, 11, *45*, *59*
Escallonia, 93
Eucalyptus, 93
Eucryphia, 93
Euonymus, 71, 93, 123
E. europaeus 'Red Cascade', 123
E. fortunei, 93, 123
E.f. 'Emerald 'n' Gold', *22*
E.f. 'Silver Queen', 123, *123*
Euphorbia, 11, 19, 72, 123
E. dulcis 'Chameleon', *62*
E. griffithii, 123
E. robbiae, *27*
E. schillingii, 123, *123*
E. wulfenii, 123
Euryops chrysanthemoides, *41*

F

fair maids of France, 147
false acacia, 63
Fargesia nitida, 111
Fatsia japonica, *27*, 124
F.j. 'Variegata', 124, *124*
feeding plants, 86–7
Felicia, *16*, *49*
F. bergeriana, *43*
F. capensis, *24*
ferns, 21, *46*, 64
fertilizers, 86–7
feverfew, 15
Ficus, 124
F. benjamina, 124

Ficus contd.
F. carica, 124
F.c. 'Brown Turkey', 124
fig, 124
fireblight, 104
firethorn, 146
flame creeper, 71, 153
floribunda geraniums, 143
flowering fern, 140
flowering maple, 108
focal points, 72
foliage, 62–7, *80*
as backgrounds, 63–4
color, 63, 64–5
seasonal, 66–7
shape, 63, 65–6
texture, 65–6
variegated, *63*
foliar feeds, 86, 87
Forsythia, 93
French marigolds, 151
Fuchsia, *16*, 19, *38*, 48, *49*, 68, 72, 75, 93, 125
F. 'Annabel', 75
F. 'Dollar Princess', 75
F. 'Estelle Marie', 75
F. 'La Campanella', *125*
F. 'Leonora', 75, 125, *125*
F. magellanica, 125
F.m. var. *gracilis* 'Aurea', *6*
F.m. gracilis 'Variegata', 125
F.m. var. *molinae* 'Sharpitor', 125
F.m. 'Versicolor', 125
F. 'Pink Galore', 75
F. 'Thalia', 62, 75, *124*, 125
fungal leaf spot, 104
fungal diseases, 104

G

Galanthus, *125*
G. elwesii, 125
G. nivalis, *45*, 125, *125*
G.n. 'Flore Pleno', 125
G.n. 'Samuel Arnott', 125
G.n. 'Viridiapicis, 125
G. reginae-olgae, 125
Gaultheria, 125
G. mucronata, *53*, 126
G.m. 'Bell's Seedling', 126
G.m. 'Lilacina', 126
Gazania, *38*
Gentiana, *45*
Geranium, 126
G. 'Johnson's Blue', 126
G. lucidum, 126
G. macrorrhizum 'Album', 126
G. × *magnificum*, 126, *126*
G. renardii, 126
G. robertianum, 126
G. 'Wargrave Pink', 126
geraniums (pelargoniums), 10, *10*, 11, *16*, 19, *20*, 22, 26, *38*, 40, 42, 48, *49*, 50, *50*, *51*, *53*, 68, *70*, 73, *96*, 142–3
Geum rivale, *54*
Glechoma hederacea 'Variegata', *24*, *51*, 58

golden-leaved hop, *15*, 19, 26, 63, *64*
golden-leaved marjoram, *15*
grape hyacinths, 10, 14, 18, *24*, *45*, *72*, 137
grape vine, 155
grapefruit, 117
greenfly, 105
Griselina, 16, 23, *23*, *126*
G. littoralis, 126
G. l. 'Dixon's Cream', 126
G.l. 'Variegata', 126
ground elder, *63*
ground ivy, 58
groundsel, 151
growing medium, 85
growth rates, 84
Gunnera, 63
Gypsophila, 126
G. elegans, *43*
G. paniculata, 126
G. repens 'Rosea', 126, *127*

H

Hamamelis, 93
hanging baskets, 46–9
watering, 89
heathers, *44*, *53*, 123
heaths, 123
heavenly bamboo, 138
Hebe, 50, 127
H. albicans, 127, *127*
H. armstrongii, 127
H. cupressoides, 127
H. × *franciscana* 'Variegata', 127
H. ochracea 'James Stirling', 127
H. pinguifolia 'Pagei', 127
H. speciosa 'Tricolor', 127
Hedera, *17*, 48, *58*, 127–8
H. algeriensis, *65*, 127
H.a. 'Gloire de Marengo', 128
H. colchica 'Dentata', 128
H.c. 'Dentata Variegata', *63*, *65*, 128
H.c. 'Sulphur Heart', 26, 62
H. helix, *40*, *41*, *48*, 53, *53*, *59*, 127
H.h. 'Kolibri', *65*
H.h. 'Adam', *65*
H.h. 'Glacier', 127, *127*
H.h. 'Goldheart', *17*, 127
H.h. 'Palmata', 127
H.h. 'Sagittifolia', *17*, *24*, 62
H.h. 'White Knight', *65*
H. hibernica 'Deltoidea', *65*
hedges, roof gardens, 23
Helianthemum, *41*, 93, 128
H. 'Fire Dragon', 128, *128*
H. nummularium grandiflorum 'Variegatum', 128

Helianthemum contd.
H. 'Wisley Pink', 128
H. 'Wisley White', 128
H. 'Wisley Yellow', 128
Helianthus, 98
Helichrysum, 11, 14, *46*, *48*, *49*, 93
H. bracteatum 'Dargan Hill Monarch', *41*
H. italicum, *12*
H. petiolare, 48
H. 'Sulphur Light', 26
Heliotropium, 128–9
H. 'Florence Nightingale', 129
H. × *hybridum*, 128
H. 'Mini Marine', 129
H. 'Princess Marina', 128
Helleborus, *101*
H. argutifolius, 11
herbs, patios, 14–15
Hibiscus, 129
H. rosa-sinesis, *128*, 129
H. syriacus 'Bluebird', 129
H.s. 'Snowdrift', 129
H.s. 'Woodbridge', 129
holly, 18, 67, 130–1
honeysuckle, 8, 19, 21, 26, 72, 135
hop, 129
hornbeam, 17
hortensia hydrangeas, 130
Hosta, 11, 21, *27*, 64, 66, 129
H. fortunei var. *albopicta*, *64*, 129, *129*
H.f. var. *aureomarginata*, *63*
H. 'Halcyon', *63*
H. sieboldiana var. *elegans*, 65, 67, 72, 92, 129, *129*
H.s. 'Frances Williams', 129, *129*
hot colors, *38*
houseleeks, 55, 150
Humulus lupulus, 129
H.l. 'Aureus', *6*, 19, 26, *64*, *129*
Hyacinthus, 40, *82*, 130
H. orientalis, 130
H.o. 'L'Innocence', 130, *130*
Hydrangea, 14, 16, 70, *70*, 72, 93, 130–1
H. macrophylla 'Europa', 130, *130*
H.m. 'Hamburg', 130, *130*
H.m. 'Nigra', 130
H.m. 'Tricolor', 130
Hydrocharis morsus-ranae, *54*
Hypericum, 93
Hypoestes phyllostachya, *53*
Hyssopus officinalis, *15*

I

Iberis umbellatus, *43*
ice plants, 150
Ichang lemon, 117
Ilex, 67, 130–1
I. aquifolium, 93
I.a. 'J.C. van Tol', 131, *131*

Ilex contd.
I.a. 'Silver Queen', 131
I. × *meserveae* Blue Series, 131
Impatiens, 10, 14, 131
I. New Guinea hybrids, *38*, *43*
I. Novette series, 131, *131*
I. walleriana, *42*
Ipomoea, 131
I. alba, 131
I. rubrocaerulea 'Heavenly Blue', 131, *131*
Iris, 66, 131
I. ensata, *54*, 56
I. germanica, 131
I. gracilis, *45*
I. laevigata, 56
I. pallida, *54*
I. pseudacorus, *54*
I. verna, *43*
ivy, 17, 18, 26, 27, *40*, *44*, 48, *48*, *49*, 50, *53*, *58*, 63, *64*, *65*, 67, *77*, 127–8
ivy-leaved geraniums, 142

J
Japanese anemones, *26*, 27
Japanese cut-leaved maple, 38
Japanese figleaf, 124
Japanese maples, 21, 64, 66, 108
Japanese style, *55*
jasmine, 21, 132
Jasminum, 132
J. nudiflorum, 132, *132*
J.n. 'Aureum', 132
J. officinale, 132
Juniperus, 132
J. chinensis, 132
J.c. 'Obelisk', 132
J.c. 'Stricta', 132
J. recurva var. *coxii*, 132
J. scopulorum 'Skyrocket', 132, *132*
J. squamata 'Pygmaea', 132

K
Kalanchoe, *49*
Kerria, 93
kiwi fruit, 108
kohuhu, 144
Kolwitzia, 93

L
lacecap hydrangeas, 130
lady's eardrops, 125
lady's mantle, 15, 109, *109*
lamb's ears, 65
Lamium, *63*, 132
L. galeobdolon 'Florentinum', 132
L. maculatum, *63*, 132
L.m. album, 132
L.m. 'Aureum', 132
L.m. 'Cannon's Gold', *59*
L.m. 'Roseum', 132
Lantana, 132–3
L. camara 'Cloth of Gold', 133, *133*

Lantana contd.
L.c. 'Rose Queen', 133
L. montevidensis, 133
larkspur, 14, 73
Lathyrus odoratus, 8, *101*
L.o. 'Knee Hi', *43*
Laurus nobilis, *8*, 133
L.n. 'Aurea', 133
Lavandula, 11, 12, 14, 93, 133
L. angustifolia 'Alba', 133
L.a. 'Hidcote', *22*, 133, *133*
L.a. 'Rosea', 133
L. stoechas pedunculata, 133, *133*
Lavatera trimestris 'Mont Blanc', *43*
layering, 99
leaves *see* foliage
Lemna minor, *54*
lemon, 117
lemon thyme, *15*
lettuce, 52
Leucanthemum maximum, 116, *116*
L.m. 'Esther Read', 116
L.m. 'Horace Read', 116
L. vulgare, 8
Lewisia, *45*, 134
L. cotyledon alba, 134, *134*
L. tweedyi, 134
light, 80–1
Ligularia, 57
Ligustrum, 67, 134
L. delavayanum, *77*
L. lucidum 'Tricolor', 134
L. ovalifolium, *23*
L.o. 'Aureum', *134*
Lilium, 19, 72, 134
L. 'Casa Blanca', *73*
L. longiflorum, *13*
L. 'Pink Perfection', *72*
L. regale, 10, 72, 134
Limnanthes douglasii, *43*
liquid feeds, 86
Livingstone daisy, 121
Lobelia, 11, 17, *46*, *49*, 72, *101*, 134
L. cardinalis, 135
L. erinus, 135
L.e. 'Crystal Palace', *43*
L. 'Sapphire', *25*
Lonicera, 8, *17*, 135
L. hildebrandtianum, *17*
L. periclymenum, *17*, 135
L. × *purpusii*, 135
L. serotina 'Honeybush', 26
L. × *standishii*, 135
L. × *tellmanniana*, 135
loosestrife, 135
Lotus berthelotti, *12*
Lupinus, *38*, 40, 94, 135
L. arboreus, 135
L. polyphyllus 'Inverewe Red', 135, *135*
Lychnis coronaria, *12*, 65
Lysimachia, 135–6
L. clethroides, 135
L. nummularia, *51*, 135, 136
L.n. 'Aurea', 135, *135*

M
Macleaya cordata, 64
Magnolia, 65, 93, 136
M. delavayi, 136
M. grandiflora, 136
M. kobus, 136
M. stellata, 136, *136*
M.s. 'Rosea', 136
Mahonia 'Charity', *23*
maidenhair fern, 109
maintenance, 77–105
male fern, 122
Malva sylvestris 'Primley Blue', *38*
marguerites, *10*, *19*, *42*, *75*, 116
marigolds, *101*, 151
marjoram, 11, 52, 139–40
Matteuccia struthiopteris, 136, *137*
Matthiola Brompton series, *43*
M. incana, *12*
medicinal herbs, *15*
Melianthus, *38*
Mentha suaveolens 'Variegata', *63*
metal containers, 32
Mexican orange blossom, 62, 116
Milium, 136–7
M. effusum, 136
M.e. 'Aureum', 136–7, *136*
M.e. 'Variegatum', 137
Mimulus, 137
M. 'Bees Scarlet', 137
M. guttatus, *54*
M. luteus, 137
M.l. 'Alpinus', 137
M.l. 'Duplex', 137
M. Malibu series, *45*
mint, 15, 52
mirror plant, 118
monkey flower, 137
montbretia, 66
Monterey cypress, 120
morning glory, 131
mother of thousands, 152
Muscari, 10, 14, 24, *45*, *72*, 137
M. armeniacum, 137, *137*
M.a. 'Cantab', 137
M.a. 'Heavenly Blue', 137
M. botryoides 'Album', 137
M. macrocarpum, 137
myrtle, 67, 137
Myrtus communis, 67, 137
M.c. tarentina, 137
M.c. 'Variegatus', 137

N
Nandina domestica, 138
N.d. 'Firepower', 138
N.d. 'Nana Purpurea', 43, 45, 138
Narcissus 'April Tears', 138, *138*
N. 'Bridal Veil', *10*, *45*
N. 'Cheerfulness', *40*, 138
N. 'Paper White', 138
N. 'Tête-à-Tête', *45*
nasturtiums, 17, 32, *49*, *51*, 58, 68, *101*, *153*

needle palm, 155
Nemesia, *101*
Nepeta × *faassenii*, *9*, *12*, 17
Nephrolepis, 138–9
N. cordifolia, 138
N. exaltata, 138
N.e. 'Bostoniensis', 138
N.e. 'Elegantissima', 138
Nerium oleander, *138*, 139
Nicotiana, 10, *27*, 73, 139
N. × *sanderae*, *38*
N. sylvestris, 14, 20, 139, *139*
night-scented stocks, 14
nutrients, 86–7
deficiencies, 87
Nymphaea, 139
N. 'Aurora', 139
N. pygmaea, *54*
N. tetragona, 139, *139*

O
oak fern, 121–2
oak-leaved geraniums, 143
objets trouvé, 58
Ocimum basilicum, *15*
O.b. var. *minimum*, *15*
O.b. purpurascens, *15*
Oenothera cheiranthifolia, *16*
oleander, 139
oleaster, 122
orange, 117
oregano, 15, 139–40
Origanum, 139–40
O. amanum, 140
O. 'Kent Beauty', 140
O. vulgare, 139–40
O.v. 'Aureum', *15*
O.v. 'Gold Tip', 140
ornamental cabbage, 113
ornamental kale, *41*, *44*
ornamental rhubarb, 147
Osmunda regalis, 140
O.r. purpurascens, 140
Osteospermum, *50*
O. 'Cannington Roy', 141
O. 'James Elliman', 141
O. jucundum, *76*, 140–1
O. 'Silver Sparkler', 141
O. 'Whirligig', *140*, 141
ostrich feather fern, 136
ox-eye daisies, 8
Oxalis adenophylla, 141
O. articulata, *53*
O. tetraphylla 'Iron Cross', *140*
O. triangularis 'Cupido', *62*, *140*

P
Pachyphragma macrophyllum, 141
palm lily, 119
pansies, *9*, *11*, 14, 15, 18, *29*, *38*, *40*, 42, 44, 48, *48*, 50, *68*, *154*
parsley, 15, 44, *64*, 143
Parthenocissus, *21*, 63
P. henryana, *28*
P. tricuspidata, 17
Passiflora, 141
P. caerulea, 141, *141*
P. × *caeruleoracemosa*, 141, *141*

passion flower, 141
paths, 24–7
patios, 12–15
peat, 85
Pelargonium, 10, *10*, 11, *16*, 19, *20*, 22, 26, *38*, 40, 42, 48, *49*, 50, *50*, *51*, *53*, 68, *70*, 73, *96*, 142–3
P. 'Angel', *53*
P. 'Bird Dancer', 142
P. 'Frank Headley', 142, *142*
P. 'Grand Slam', 142
P. 'Graveolens', 142
P. peltatum, 141
P. quercifolium, 143
P. 'Royal Oak', *142*, 143
P. 'Stellar Apricot', *143*
P. 'Vancouver Centennial', *61*, *143*
Pellaea falcata, *59*
Penstemon, 73, 143
P. barbatus, 143
P. davidsonii, 143
P. hartwegii, 143
P. 'Sour Grapes', 143, *143*
pernettyas, *41*, *44*, *53*, 126
perpetual stocks, *12*
pests, 105
Petroselinum crispum, 64, 143, *143*
Petunia, 11, *49*, 50, *50*, 68, *68*, *77*, 144
P. 'Express Ruby', *114*
P. 'Surfinia', *68*, 144, *144*
pH tests, 85
Philadelphus, 93
Phlox, 144
P. 'Chattahoochee', 144, *145*
P. drummondii, 144
P. paniculata, 144
P.p. 'Fairy's Petticoat', 144
P.p. 'Harlequin', 144
Photinia, 93
pickaback, 152
pickerel weed, *54*
Pieris, 93
piggy-back, 152
pincushion flower, 150
pinks, *12*, *70*
Pittosporum tenuifolium, 144–5
P.t. 'Abbotsbury Gold', 145
P.t. 'Purpureum', 145
P.t. 'Silver Queen', 145, *145*
plantain lily, *27*, 129
planting, 82
Plumbago, 145
P. auriculata, *145*
P. a. var. *alba*, 145
polyanthus, *44*, *52*, 146
Polypodium vulgare 'Cornubiense', *46*
Polystichum, *59*, 145
P. acrostichoides, 145
P. munitum, 145
P. setiferum, 27
Pontederia cordata, *54*, 56
pot marigold, *114*
Potamogeton crispus, *54*

Potentilla, 145–6
P. atrosanguinea, 145
P. fruticosa 'Alba', *42*
P.f. 'Elizabeth', 145
P.f. 'Red Ace', 145
P. 'Gibson's Scarlet', 145
P. nepalensis, 145
P.n. 'Miss Willmott', 145
potting, 82–4
powdery mildew, 104
pricking out, *100*
primroses, *44*, 48, *52*, 146
Primula, *40*, *48*, *52*, *96*, 146
P. 'Blue Sapphire', 146, *146*
P. bulleyana, 146
P. malacoides, *52*
P. obconica, *52*
P. polyantha, *146*
P. vulgaris, 146
privet, 18, 21, 62, 64, 66, 67, 154
propagation, 96–101
pruning, 90–4
Prunus laurocerasus, 14
P. trilobus, 93
Pyracantha, 26, 146
P. 'Mohave', 146
P. 'Shawnee', 146
P. 'Sparkler', 146

Q
Queensland arrowroot, 115

R
radishes, 52
Ranunculus, 147
R. aconitifolius 'Flore Pleno', 147
R. lyallii, *147*
R. montanus 'Molten Gold', 147
regal lilies, 10
regal pelargoniums, 142
repotting, 82–4
Rheum, 98, 147
R. alexandrae, 147
R. palmatum, 64, 147
R.p. 'Atrosanguineum', 147
Rhodanthemum gayanum, 72
R. hosmariense, *48*, 116
Rhodochiton atrosanguineus, 70
Rhododendron, 21, 70, 93, 147
R. atlanticum, 147
R. indicum, 147
R.i. 'Balsaminiflorum', 147
R. leptanthum, *147*
R. luteum, 147
R. × *obtusum*, 147
R. ponticum, 93
R. simsisi 'Perle de Noisy', *147*
Ribes, *93*
Robinia pseudoacacia, 63
rock rose, 128
rockfoil, 149
Rodgersia, 57
roof gardens, 20–3
root rot, 104
roots, *80*, *81*
Rosa, 73–4, 93, 94, 148
R. 'Albertine', 71
R. 'Boule de Neige', 148, *148*
R. glauca, *62*
R. 'Mermaid', 148
R. 'Pompon de Paris', 148, *148*
R. 'Roseraie de l'Haÿ', *70*
R. 'White Pet', 74
R. 'Zéphirine Drouhin', 8, 71, 74, 148
rose bay, 139
rosemary, 11
Rosmarinus, 148
R. officinalis, *22*, 148
R.o. 'Severn Sea', 148
R.o. 'Sissinghurst Blue', 148, *148*
royal fern, 140
rue, 10
runner beans, 15
rust, 104
Ruta, 148–9
R. graveolens 'Jackman's Blue', 10, 148

S
sage, 11, 14, *15*, 149
Sagittaria japonica, *54*
Salvia, 50, 149
S. microphylla, 149
S. officinalis, *15*, 149
S.o. 'Icterina', *15*
S.o. 'Purpurascens Group' *62*
S.o. 'Tricolor', 149, *149*
S. patens, 14
S. sclarea var. *turkestanica*, 149
S. splendens, 149
S.s. 'Rambo', 149
S. viridis, 68
Sambucus, 93
Santolina, 10, 11, 44, 93, 149
S. chamaecyparissus, 149
S.c. var. *nana*, 149
S. pinnata subsp. *neapolitana* 'Edward Bowles', 149
S. rosmarinifolius, 149
sapphire flower, 113
Sarcococca, 93
S. hookeriana, *53*
Saxifraga, 42, *45*, 149
S. oppositifolia subsp. *latinia*, *57*
S. sempervivum, 149, *149*
S. × *urbium*, 149
Scabiosa, 72, 150
S. 'Butterfly Blue', *9*, 150
S. caucasica, 14, 150
S.c. 'Clive Greaves', 150
S.c. 'Miss Willmott', 150
S. 'Pink Mist', 150, *150*
Scaevola, 150
S. aemula, *68*
S. plumieri, 150
scented-leaved geraniums, 142
Schizanthus 'Hit Parade', *43*
Scilla, *45*, *48*, 50
seasonal displays, 40–1, 44, 66
Sedum, 42, *45*, 150
S. acre 'Aureum', 150, *150*
S. spathulifolium 'Capo Blanco', *57*
S.s. 'Purpureum', *57*
S. spectabile, 150
seed sowing, 99–101
Sempervivum, 44, *55*
S. 'Alpha', *57*
S. tectorum, 150–1, *151*
Senecio, *12*, 19, *41*, 151
S. cineraria bicolor, 151
S.c. 'Silver Dust', 151
S. 'Sunshine', *12*, 65, 151
shade-loving plants, *26*, *27*
shrub verbena, 132–3
shrubs, planting, *83*
pruning, 92–4
shuttlecock fern, 136
Silene pendula, *24*
silk grass, 155
silver-leaved plants, *12*
Skimmia, *41*, 70, 93, 151
S. japonica 'Fragrans', 151
S.j. subsp. *reevesiana*, 151
S.j. 'Rubella', 70, 151
slugs, 105, *105*
smooth Arizona cypress, 120
smuts, 104
snails, 105
snapdragon, *42*
snowdrops, *45*, 82, 125
soil tests, 85
Solanum crispum 'Glasnevin', 70
South African leadwort, 145
sowbread, 120
sowing seed, 99–101
Sphaeralcea, *51*
spiderwort, 152
spindle, 123
spurge, 123
Stachys byzantina, 65
S. macrantha 'Superba', 14
staking, *79*, 95, *95*
standards, *74*, *75*
stems, *80*
steps, 11
stocks, *43*
stomata, *80*
stone containers, 33–4
stonecrops, 150
strawberries, 15
Streptocarpus, *96*
Sutera diffusa, 10, 43, *51*
Swan River daisy, 113
sweet basil, *15*
sweet peas, 8, *43*, *101*
sweet William, *42*
sword fern, 138, 145
Syringa, 93

T
Tagetes, *101*, 151
T. erecta, 151
T. 'Gold Coin', *43*
T. patula, 151
Tanacetum parthenium 'Aureum', *15*
T. vulgare, *15*
T. parthenium, 116
tansy, *15*
tarragon, 52
tawhiwhi, 144
Taxus baccata, *23*, 67
Tellima grandiflora, *26*, *63*, 66
temperature, 81
terracotta containers, 30–2, *33*, 34
Teucrium fruticans, *22*
Thuja plicata, *23*
Thymus, 11, 14, 52, 152
T. × *citriodorus*, *15*, 152
T. × *citriodorus* 'Aureus', 152
T. vulgaris 'Silver Posie', 152, *152*
tobacco plants, *10*, *14*, 19, *20*, *27*, 73, 139
Tolmiea menziesii, 26, 48, *49*, 99, 152
T.m. 'Taff's Gold', 152
tomatoes, 15
topiary, 67, *67*, 93
Trachycarpus fortunei, 19, 38, 64, 152
Tradescantia, *49*, 152
T. × *andersoniana* 'Isis', 152
T. × *andersoniana* 'Osprey', 152
T. virginiana, 152
T. zebrina, *51*
trees, planting, *83*
trellis, *21*, 22–3
Trifolium repens 'Purpurascens Quadrifolium', *62*
trinity flower, 152
Tropaeolum, *51*
T. majus, *101*, 153
T.m. 'Alaska', 153
T.m. Jewels series, *152*
T. speciosum, 71, 153
trumpet creeper, 115
trumpet vine, 115
Tulipa, 18, *82*, 153
T. clusiana, 43, 153, *153*
T. 'Flaming Parrot', *153*
T. 'Fringed Elegance', *153*
T. kaufmanniana, 153
T. 'Van der Neer', *153*
T. 'White Triumphator', *40*
Typha latifolia, *54*

U
urns, *5*, *29*, *31*, *61*

V
Verbascum phoenicum, 70
Verbena, 14, 44, *48*, *49*
V. bonariensis, *68*, 154
V. × *hybrida*, *53*, 154
V. rigida, 154
Verbena contd.
V. 'Rose du Barry', *24*
V. 'Silver Anne', 154
V. 'Sissinghurst', *56*
verdigris effect, *37*
veronicas, 127
Versailles tubs, 32, *32*
Viburnum, 70, 93, 154
V. × *bodnantense*, 70
V. carlesii, 154
V.c. 'Aurora', 154
V.c. 'Diana', 154, *154*
V. tinus, 93, 154
vine weevils, 105
Viola, *9*, *29*, *38*, 40, *48*, *68*, 154
V. 'Little Liz', *154*
V. 'Molly Sanderson', *24*, *42*, *59*
V. riviniana 'Purpurea Group', *62*
V. septentrionalis alba, 154
V. ×*wittrockiana*, 154
violets, *68*, 154
viral diseases, 104
Vitis, 155
V. coignetiae, *17*, 62, 155
V. vinifera, 155
V.v. 'Brant', *17*
V.v. 'Purpurea', 155

W
walkways, 24–7
wall pots, *24*
wallflowers, 71
water, 81
water avens, *54*
water gardens, *54*, 56–7
water hyssop, 111
water lily, 56, 139
water nymph, 139
watering, 88–9
weathering a pot, *37*
weeping birch, 38
weeping fig, 124
Weigela, 93
whitefly, 105
windflower, 110
windmill palm, 152
window boxes, 50–3
winter protection, *103*
Wisteria, 19, *50*, *92*, 155
W. sinensis, 155
W.s. 'Alba', 155
W.s. 'Plena', 155
wood millet, 136
wooden containers, 32, *32*
wormwood, *15*, 62, 111

Y
yew, 62, 63, 66, 67
Yucca, 19, *20*, 64, 66
Y. filamentosa, 155
Y.f. 'Bright Edge', 155
Y.f. 'Variegata', 155, *155*

Z
Zantedeschia, 64
Z. aethiopica, 155, *155*
Z.a. 'Crowborough', 155
zonal geraniums, 142

ACKNOWLEDGMENTS

The photographer, publishers and authors
would like to thank the following people and organizations who kindly
allowed us to photograph their gardens (page numbers in brackets):
Randal Anderson (107); Barnsley House (44, 136/b, 138/c, 153/bl);
Barter's Farm (133/l); Blenheim Palace (115/b); Bosvigo House (61, 121/r); Bourton House
(47, 140/tr); Penelope Bray (137, 143/r); Frank Cabot (7); Robert Cooper (29);
Nicholas & Pam Coote (22, 108/l, 124/l, 130/tr, 130/br, 131/c); The Garden House (108/t);
Gothic House (72, 119/r, 125/r, 131/r, 144/l); Hadspen Gardens (63); Pam Hummer (1, 68);
Anne Huntington (117/r, 153/tr); Kiftsgate Court (39); Nell Maydew (154/tr);
Eve Meares (129/l); Carolyn McNab (147/l); The National Trust (Powis Castle 4, 16/r);
Anthony Noel (2, 13, 36, 75, 119/l, 153/br, 154/b);
Fiona and John Owen (108/b, 153/r); Jenny & Richard Raworth (14, 43, 65, 66,
70/r, 111, 129/r, 134, 145/l, 148/tl); RHS Rosemoor (124/r, 138/l); RHS Wisley (130/l);
Rodmarton Manor (125/b); Sticky Wicket (121/l, 142/l, 143/b, 150/l, 151/l, 152/l);
Beth Straus (20/r, 51, 112/r); Sudeley Castle (109/r); Rosemary Taffinder (56);
Louise Trepanier (9, 123/t, 127/l, 135/l, 138/r, 148/b, 150/l);
Whichford Pottery (64, 69, 129/c).

The publishers are also grateful to the following for their help and support in
the production of this book: Merrist Wood College, Worplesdon, Surrey;
The Garden Picture Library, Ransome's Dock, London SW11; Jenny Raworth for
designing the seasonal plantings; Ruth Baldwin for copy-editing the text;
Peter Green at the Royal Botanic Gardens in Kew for identifying the plants;
Tony Lord for checking the nomenclature of plants;
Richard Bird for compiling the index; Erddig for identifying the ivies;
and Caddick's Clematis Nursery for identifying the clematis.

Suppliers of containers, plants, tools and equipment

Clifton Nurseries, Clifton Villas, Little Venice, London W9 2PH;
Granville Garden Centre, London NW2; Spear and Jackson, Handsworth Road,
Sheffield S13 9BR; Rayment, The Forge, Durlock, Thanet, Kent CJ12 4HE; JC Castings,
38 Derwent Road, London SW20 9NH; Geo Goostock, Llangwyryfon, Dyfed SY23 4HD;
Jonathan Garratt, Hare Lane Pottery, Cranborne, Dorset; Annie Sloan, Relics,
35 Bridge Street, Witney, Oxon OX8 6DA.

Picture Credits

All location photographs by Andrew Lawson, except the following:
Lynne Brotchie, GPL, 50; Linda Burgess, GPL, 18; John Glover, GPL, 25; Jerry Harpur, 20/l;
Jacqui Hurst, 10/r, 70/l, 71; Ron Sutherland, GPL, 16/l, 46/t; Harold Taylor,
Oxford Scientific Films, 80/boxed; Brigitte Thomas, GPL, 10/l, 27.